*Dedicated
to the
Women of Sierra Leone
and to
Maduka Jr.*

Women and Collective Action in Africa

Development, Democratization, and Empowerment, with Special Focus on Sierra Leone

By

Filomina Chioma Steady

WOMEN AND COLLECTIVE ACTION IN AFRICA
© Filomina Chioma Steady, 2006.

All rights reserved. No part of this book may be used or reproduced in any manner whatsoever without written permission except in the case of brief quotations embodied in critical articles or reviews.

First published in 2006 by
PALGRAVE MACMILLAN™
175 Fifth Avenue, New York, N.Y. 10010 and
Houndmills, Basingstoke, Hampshire, England RG21 6XS
Companies and representatives throughout the world.

PALGRAVE MACMILLAN is the global academic imprint of the Palgrave Macmillan division of St. Martin's Press, LLC and of Palgrave Macmillan Ltd. Macmillan® is a registered trademark in the United States, United Kingdom and other countries. Palgrave is a registered trademark in the European Union and other countries.

ISBN 1–4039–7082–3
ISBN 1–4039–7083–1 (pbk.)

Library of Congress Cataloging-in-Publication Data

Steady, Filomina Chioma.
   Women and collective action in Africa : development, democratization, and empowerment / by Filomina Steady.
     p. cm.
   Includes bibliographical references and index.
   ISBN 1–4039–7082–3—ISBN 1–4039–7083–1 (pbk.)
    1. Women—Africa—Societies and clubs. 2. Non-governmental organizations—Africa. 3. Women in development—Africa. 4. Feminism—Africa. I. Title.

HQ2017.S74 2005
305.42′096—dc22
                                                      2005045946

A catalogue record for this book is available from the British Library.

Design by Newgen Imaging Systems (P) Ltd., Chennai, India.

First edition: January 2006

10 9 8 7 6 5 4 3 2 1

Printed in the United States of America.

Contents

List of Illustrations	ix
Preface	xi
Acknowledgment	xii
Acronyms	xiii

Chapter One   The Challenge and Conceptual Framework — 1

Introduction	1
The Challenge	2
Review of Relevant Concepts and Theories	3
Definition, Nature, and Origin	4
Collective Identities	5
Institutional Manifestations	6
Women's Collective Action and Social Movement Theory	8
The Nexus of Collective Action and Reaction: A Conceptual Model	9
The Development–Underdevelopment Nexus	10
The Democratization–Authoritarianism Nexus	15
Empowerment	18

Chapter Two   The Context and Background — 21

Introduction	21
The People	24
Social Organization	25
Women's Associations in Freetown	29
Trends in Women's Associations	31
Tensions Between International and Local NGOs	33
Federations and Associations of National Unity	33
Sierra Leone Association of Non-Governmental Organizations	34
National Policy for Women	34
Results of a PostWar Survey on Focus Groups	35

Chapter Three Collective Action for Political Empowerment — 37

Introduction — 37
Authoritarianism, War, and Economic Decline — 39
Women, Democratization, and the Quest for Peace — 42
    The Role of Women's Associations, Especially the Women's Forum — 43
    The National Commission for Democracy — 49
    National Conferences on Democratization — 49
    Female Activism in the 1996 Elections—A Test of Unity — 50
Gender Bias in the Body Politic — 53
1997 Coup d'Etat and Escalation of the War — 54
The Peace Process Renewed — 54
Militarism, Peacekeepers, and Gender Vulnerabilities — 55
End of the War, the 50/50 Group and the 2002 Election — 56
The Impact of Women on Democratization and the Peace Process — 57

Chapter Four Collective Action for Economic Empowerment — 59

Economic Profile of Women in Freetown — 59
Management and Decision-Making Positions — 60
Women's Economic Associations — 61
    The Women's Movement — 61
    The Gara Thrift and Credit Society and Similar Associations — 65
    Rotating Credit Associations—Indigenous Mechanisms for Micro-Credit — 66
    Cooperatives — 69

Chapter Five Collective Action for Educational and Occupational Empowerment — 75

Introduction — 75
Gendered Education — 75
Contributions of Individual Women to Female Education — 78
Education and Female Professional Advancement — 79
Educational Challenges for Women — 80
The Educational Role of Women's Associations — 83
Vocational Education — 84
    The Young Women's Christian Association (YWCA) — 85
    Kankalay — 88
    Forum for African Women Educationalists (FAWE)—Sierra Leone Chapter — 88
Guilds, Professional Associations, and Alumnae Associations — 91
Female Education and Socioeconomic Development — 91
    Human Resource Development — 92

CONTENTS / VII

Chapter Six "Traditional" Associations — 95

Secret Societies — 95
  Poro and Sande — 96
  Sande/Bondo Female Society — 96
Sande and Social Change — 100
  The Urbanized Secret — 101
  Sande and the Health Care System — 103
Female Circumcision: A Global Practice — 103
Social Change, Female Circumcision, and Controversies — 105
The Enduring Features of Sande — 108

Chapter Seven Mutual-Aid Associations — 111

Introduction — 111
Secret Society Affiliation — 111
Tensions Between Statutory and Customary Marriages — 113
Imposition of Moral Sanctions — 114
Multiple Mothering — 114
Recreational Activities — 115
Promoting Solidarity and Political Expression — 115

Chapter Eight Islam and Women's Associations — 119

Introduction — 119
Women and Islam — 121
Limitations of Islamic Influence on Social Institutions — 123
Examples of Muslim Women's Associations — 125
Political Links Through Hajas — 127

Chapter Nine Christianity and Women's Associations — 129

Introduction — 129
Christian Women's Associations — 133
  Mobilizing Resources — 133
  Maintaining the Ideology of Womanhood and Christian Marriage — 134
  Promoting Healing — 137
  Facilitating the Power of Prayer — 138
  Female Charisma and Religious Leadership — 138
  Promoting Unity — 139

Chapter Ten Comparative Insights at the National Level — 141

The Development–Underdevelopment Nexus — 141
The Democratization–Authoritarianism Nexus — 145
  The Role of the State — 148
  Nationalism — 150

Religious Fundamentalism and Democratization ... 153
Summary ... 155

**Chapter Eleven Comparative Insights at the Regional/
Pan-African and International Levels** ... **157**

Introduction ... 157
The Development–Underdevelopment Nexus ... 158
The Democratization–Authoritarianism Nexus ... 162
Conflict Prevention and Peace Building ... 164
The Contribution of African Women to the International Women's
  Movement ... 167
Summary ... 171

Chapter Twelve Conclusion ... **173**

Engendering and Enhancing Development and Democratization ... 173

Notes ... 179

Bibliography ... 189

Index ... 201

Illustrations

Map of Africa.	xv
Map of Sierra Leone.	xvi
1.1 Women and collective action model.	10
1.2 Economic growth in Africa UN-NADAF projection and actual.	14
1.3 Adults and children newly infected with HIV (millions), 2002.	14
1.4 Africa's foreign debt (US$ bn).	15
1.5 Aid to Africa (net official development assistance, $ bn).	16
1.6 Foreign direct investment to Africa as percent of total world flows.	16
Aerial view of Freetown. Courtesy: Christopher Greene.	23
Freetown Cotton Tree—A historical landmark in Sierra Leone. Courtesy: Azania Steady.	26
Sunset at Sussex Beach—Sierra Leone peninsula. Courtesy: Azania steady.	28
Colonel Yvette Gordon, Republic of Sierra Leone Military Forces—Freetown. Courtesy: Sierra Leone Information Services.	41
Madam Ella Koblo Gulama, Paramount Chief of Kaiyamba Chiefdom. Courtesy: Obai Kabia.	45
Dr. June Holst-Roness, former Mayor of Freetown. Courtesy: Sierra Leone Information Services.	51
Law Courts Building—Freetown. Courtesy: Azania Steady.	52
An association's school marching in a Freetown parade. Courtesy: Sierra Leone Information Services.	77
An old girls (alumnae) association Thanksgiving Service. Courtesy: Sierra Leone Information Services.	82
Children playing at Lumley Beach. Courtesy: Azania Steady.	116
Federation of Muslim Women marching for peace on International Women's Day in Freetown. Courtesy: Christopher Greene.	120
Mrs. Blanche Benjamin in *print en enkincha* traditional dress. Courtesy: Azania Steady.	132
Maroon Church—A historic landmark in Freetown. Courtesy: Christopher Greene.	136

Dr. Wangari Maathai of Kenya, winner of the 2005 Nobel Peace Prize, being congratulated by Kofi Annan, Secretary-General of the United Nations, a fellow Nobel Peace Prize Laureate. Courtesy: United Nations.   159

Sarah Daraba Kaba of the Mano River Women's Peace Network of West Africa (second from left) with other members of the Mano River Women's Peace Network. Courtesy: UN Photo/Stephenie Hollyman, New York.   165

Nkosazana Dlamini Zuma, Minister for Foreign Affairs of the Republic of South Africa, addressing participants at the Forty-Ninth Session of the Commission on the Status of Women, today at UN Headquarters. Courtesy: UN Photo/Mark Garten, New York.   168

Preface

This book is aimed at a wide audience and will be particularly useful for students and academics in Women's studies, African studies, Diaspora Studies, International Relations, and the social sciences in general. It is roughly divided into four main sections. The first section establishes the problem and conceptual framework. The next section is an in-depth study of women's associations in Freetown, Sierra Leone, based primarily on fieldwork. The third section draws comparative insights from four other countries, representing the four subregions of Africa. These are Kenya, Nigeria, South Africa, and Algeria. A fourth section shows how these themes are amplified and replicated by women's collective action at the regional or pan-African level and also at the international level.

The research is based on extensive fieldwork in Freetown, Sierra Leone, that was conducted at various intervals for several years. Participant observation, in-depth interviews and life histories were the main methods used to study 80 women's associations. The last fieldwork in 2002 included administering open-ended questionnaires to a sample of focus groups of associations. The objective was to assess the impact of the rebel war on these selected associations. While many were not functioning fully during the war, they have been very active in the postwar period. Fieldwork was supplemented by library and archival research, which included study of government documents, United Nations documents, and publications by Non-governmental Organizations. Interviews were also conducted outside of Sierra Leone in South Africa and during United Nations meetings in New York.

For national and pan-African comparisons, reliance was placed on studies conducted primarily, but not exclusively, by women from these countries and sub regions. They were supplemented by library research and by telephone or direct interviews with members of these associations between 1995 and 2002. Research was also conducted within the United Nations system as well as in the archives and libraries of national and pan-African associations.

The book represents the first comprehensive study of women's associations in Sierra Leone. Despite the ten-year rebel war, much of the history of this country has been peaceful. It is one of the most picturesque countries in Africa and has some of the most beautiful beaches in the world. Sierra Leonean women have played and continue to play an important role in their country's development, and many have achieved prominence both at home and abroad. We shall meet some of them in this book.

Acknowledgment

It has taken many years to bring this book to fruition. I wish to thank the Wenner-Gren Foundation for Anthropological Research for their sponsorship of the preliminary research that made this book possible. I also wish to thank the Association of African Women for Research and Development (AAWORD), the Social Science Research Council, and Wellesley College for awards that allowed me to conduct additional field research. I am grateful to my students for their insights and enthusiastic contributions to class discussions. My gratitude also goes to the women's associations studied both directly and indirectly, over a number of years, for their support and hospitality and for generously sharing their knowledge, wisdom, and advice.

As an African woman, I am honored for having had the opportunity to work at the international level, notably at the United Nations. I was able to make a contribution to the development of four international Plans of Action and to the implementation of the *Convention on the Elimination of All Forms of Discrimination Against Women*. These have been accomplished in my positions as a Director of the United Nations Division for the Advancement of Women and as Special Advisor on Women, Environment and Development to the United Nations. I have also been an active member of non-governmental organizations (NGOs), as a founding member of the Association of African Women for Research and Development (AAWORD) and as President of the Women's World Summit Foundation, an international NGO based in Geneva, Switzerland. Finally, I would like to thank my husband, Henry, and my children, Maduka, Azania, and also Chinaka for their devotion and support.

I would like to thank Christopher Geene, Azania Steady, Obai Kabia, the United Nations and the Sierra Leone information Services for use of the photographs reproduced in this book.

Acronyms

AAWORD	Association of African Women for Research and Development
ADB	African Development Bank
AFWE	African Federation of Women Entrepreneurs
ANC	African National Congress (South Africa)
AU	African Union
AWLI	African Women Leadership Institute
CEDAW	Convention on the Elimination of all forms of Discrimination Against Women
CMS	Church Missionary Society
CREDIF	Center for Research Documentation and Information on Women
CSO	Civil Society Organization
DAWN	Development Alternatives for Women in a New Era
ECA	Economic Commission for Africa
FAS	*Femmes Africa Solidarité*
FAWE	Forum for African Women Educationalists
FDI	Foreign Direct Investment
FEMNET	African Women's Development and Communication Network
FEMSA	Female Education in Mathematics and Science in Africa Association
FERFAP	*Fédération de Reseau des Femmes Africaines pour la Paix*
FOMWAN	Federation of Muslim Women's Associations in Nigeria
FOWODE	Forum for Women in Democracy
GATT	General Agreement on Tariffs and Trade
GERDDES	The Study and Research Group on Democracy and Economic and Social Development in Africa
HIPC	Highly Indebted Poor Countries Initiative
IMF	International Monetary Fund
IPAM	Institute for Public Administration and Management
LAWYERS	Legal Access through Women Yearning for Equality, Rights and Social Justice
MARWOPNET	Mano River Women's Peace Network
MYWO	*Maendeleo Ya Wanawake* Organization
NCW	National Commission for Women (Nigeria)

NCWK	National Council of Women of Kenya
NCWS	National Council of Women's Societies (Nigeria)
NEPAD	New Partnership for African Development
NGO	Non-Governmental Organization
NPRC	National Provisional Ruling Council (Sierra Leone)
PAC	Pan-African Congress of Azania
PAWLO	Pan-African Women's Liberation Organization
PRP	People's Redemption Party (Nigeria)
RUF	Revolutionary United Front (Sierra Leone)
SAP	Structural Adjustment Program
SLADEA	Sierra Leone Adult Development Educational Association
SLANGO	Sierra Leone Association of Non-Governmental Organizations
SLAUW	Sierra Leone Association of University Women
SLPP	Sierra Leone People's Party
UNAMSIL	United Nations Peacekeeping Mission in Sierra Leone
UNDAF	United Nations New Agenda for Development in the 1990s
UNDP	United Nations Development Programme
UNESCO	United Nations Educational, Scientific and Cultural Organization
UNIFEM	United Nations Development Fund for Women
UNTW	National Union of Tunisian Women
WBI	Women's Budget Initiative (South Africa)
WDF	Women's Development Foundation
WILDAF	Women in Law and Development in Africa
WOMEN	Women Organized for a Morally Enlightened Nation

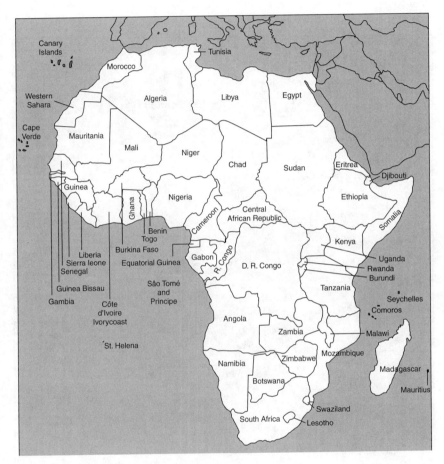

Map of Africa.

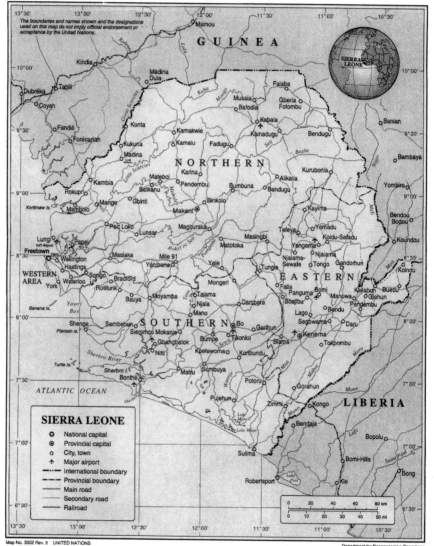

Map of Sierra Leone.

CHAPTER ONE

THE CHALLENGE AND CONCEPTUAL
FRAMEWORK

Introduction

The main purpose of this study is to understand the dynamic tension between women's collective attempts to promote development and democratization and their resistance to underdevelopment and authoritarianism. It is argued that the outcome of this tension, *the nexus of action and reaction*, is what characterizes women's associations or movements and shapes their agenda, strategies, and quest for empowerment.

As this study will show, women's collective action is rooted in three main factors. The first is the indigenous mechanisms of female mobilization and cooperation; the second is the historical experiences of colonization, and the third is the present reality of corporate globalization.

Many African countries face economic challenges, stemming primarily from an international economic system that has always undermined African economies, environments, and people. Corporate globalization, the debt burden, and Structural Adjustment Programs (SAPs) continue to increase underdevelopment, poverty, and armed conflicts, in keeping with the legacy of colonialism.[1]

To some extent, women's associations devote much of their time to confronting and trying to solve economic and political challenges. They tend to give priority to social and human-centered goals rather than to narrow feminist preoccupations about gender equality alone. A large part of their effort is not spent on agitating for women's rights or in challenging men. More importantly, they seek to counter exploitative development policies and authoritarian regimes through a number of strategies that include advocating for more democratic institutions and policies; challenging underdevelopment; facilitating access to resources; providing mutual aid in times of hardship; and promoting formal and nonformal education.

In addition, many of these associations have taken a leading role in advocating for peace. An important aspect of female collective action, stemming from indigenous forms of mobilization, is the desire to preserve elements in the culture that safeguard human security. In this regard, these associations tend to advance a type of feminism that is humanistic in orientation and transformative in intention.

The greater part of the study was conducted through fieldwork, and field research projects in Freetown, Sierra Leone over several years and involved 80 associations.

Comparative insights from other countries in Africa as well as at the regional or pan-African level are also included. Despite the diversity of African societies, cultures, and women, there are common themes in their histories, institutions, cultural norms, and values, including the experiences of colonization and corporate globalization that warrant comparative treatment.

The Challenge

Studies of women's associations and movements expressed through collective action challenge our assumptions and explanations. The dominant theoretical expectations and paradigmatic formulations in feminist discourse are derived primarily from Euro-centric experiences that do not necessarily conform to the realities of women in other parts of the world.

In the first place, most feminist movements seek to promote gender equality in a manner that does not necessarily lead to profound social transformations of systems of inequality and entrenched hierarchies. In this trajectory, women can replace men in economic and political positions without necessarily transforming structural inequalities embedded in society.

This line of reasoning is based primarily on approaches that rely on dichotomous models and polarizations to explain gender differences. Simone de Beauvoir, in her classic study, *The Second Sex*, was one of the first cultural critics to point out how dichotomous gender formations worked to women's disadvantage. Later, this problem was given anthropological attention in the anthology, *Women, Culture and Society*.[2]

In this anthology, concerns about sexual asymmetry led to theories about the public/private dichotomy. This attributed the valued public sphere to men and the devalued private sphere to women. Such rigid categorizations do not hold true for all societies and have been contested for failing to show the linkage, overlap, and articulation between these spheres and the potential for social transformation.

For example, African women have historically operated in the public sphere as rulers and political officials, even in patriarchal societies. In addition, women's associations operate in the "public" sphere when they challenge the state, formulate policies, demand change, and lobby for greater female representation in decision-making positions.[3]

As will be discussed later, the unsuitability of some of these paradigms to the African reality has become significant in the politics of representation and domination and in the power struggle within feminist scholarship. Instead of dichotomous models, some African women scholars have chosen to use explanatory models that are more flexible, complementary, overlapping, complex, transformative, and African-centered.[4]

In the second place, skepticism about theory and theorizing makes it impossible to find a single explanation that would not be misleading. It is more appropriate to settle for multiple explanations to better explain the divergent manifestations of female mobilization, agenda setting, boundary crossing, alliance building, and empowerment. Morgan's insistence that "Sisterhood is Global" remains an idealistic vision, blurred by the divergent views of women from different social locations and by local versions of feminism. It is now clear from the academic literature and from international meetings, that deep class, racial, and other social divisions defy universal generalizations about women, gender, and feminism.[5]

In the third place, it is essential to understand the local and cultural context that promotes and sustains collective action. In this regard, indigenous associations, which have existed for centuries, can yield important insights about the gendered nature of women's collective action. For this reason, significant attention is given to "traditional" secret societies and mutual-aid associations in this study.

These associations evolved out of the gender division of labor and the autonomous functioning of "male" and "female" economic and social spheres of activity and influence. They have served as useful mechanisms for mobilization, and continue to act as important power bases for women. In the "modern" context they have also functioned as cultural brokers mediating the process of change between indigenous cultures and Western culture, while at the same time expressing resistance to Western domination.

Finally, external influences and macro-historical factors, with both positive and negative consequences, have left their mark on African societies. The negative factors include the legacy of colonial rule whose leit motif was authoritarianism, brutality, and structural racism. In modern times, colonialism continues through corporate globalization, a factor that reinforces Africa's underdevelopment and one that can fuel wars. Positive external influences include the promotion of a global concensus for democratic values. Another is the opposition to the injustices of the global political economy by transnational grassroots movements, many of which include African women.

Some of the experiences of African women's associations have influenced the international agenda of equality, development, peace, and human rights. They have contributed to the international women's movement by expanding the scope of development and decolonization, and by promoting democratic values that go beyond gender equality. This was particularly evident in their struggle against Apartheid and other forms of racism.

Many of the associations studied have been influenced by the four United Nations World Conferences on Women, starting in 1975, the United Nations Decade for Women and, the Economic Commission for Africa (ECA) by the international women's movement. However, international alliances can be a mixed blessing. On the one hand, they can expand the sphere of women's activities and increase networking across national lines. On the other hand, the international agenda can distract from local priorities and exigencies. Even more significant is the fact that the international community has been remiss in solving global economic problems. It has failed to adequately address the negative effect of international economic policies on countries of the Global South, especially African countries.

Many of these policies promote and sustain underdevelopment, social and gender hierarchies, and poverty as a result of their exploitative, undemocratic, and authoritarian practices. They also contribute to political instability and armed conflicts through the sale of arms and drugs to African countries. Wars in Africa are the often ignored but important dimensions of global capital accumulation.[6]

Review of Relevant Concepts and Theories

Theorizing about women's movements is a challenging endeavor as a review of the literature reveals complexities and contradictions. Nonetheless, studies of women's movements and associations in the Global South have been increasing steadily.[7] In general, they have been concerned with exploring a number of theoretical and explanatory possibilities, but more often with filling the gaps in empirical research.

From the 1970s onward, feminist scholars have emphasized the need to revise and deconstruct the fabricated model of society that reinforced gender polarities, allocated gendered social spaces, and determined gendered destinies. Such intellectual challenges inspired feminist historians to try to write women back into history. Feminist anthropologists gave structural interpretations of gendered sociocultural processes.

Feminist economists focused on studying the household, the gender division of labor, and social reproduction as important aspects of production. Feminist scholars of the Marxist and socialist persuasion have been more inclined to view women's collective action as primarily a function of the general struggle against class oppression. A few scholars have successfully tackled the problem of development and democratization.[8]

Although women's social location in institutions and social networks may vary, this study shows that women can try to determine to some extent whether they are among those that have power and decision-making authority through social engineering and collective action. A review of the various positions and insights into women's collective action can be discussed under the following headings: definition, nature, and origin; collective identities; institutional manifestations and functional roles, and relevance of women's collective action for the social movement theory.

Definition, Nature, and Origin

The sheer numbers, diversity, and proliferation of women's associations make a universal definition of the gendered nature of collective action challenging. This is because women belong to associations whose membership is not exclusively female, and some women's associations have men as members. In addition, these associations share similarities with men's associations in terms of institutional structure and organizational procedures.

However, women's associations tend to display a greater concern about development issues, the welfare of the country, and peace. They are also more inclined to build on primarily group ties of kinship, community, and shared values and display a certain degree of informality in their meetings.

In *Women United, Women Divided*, Caplan and Budjra question the universal application of female solidarity and argue that the origins and objectives of female solidarity are not fixed and can change from one context to another.[9] No single explanation can be given to their origin in Africa. They can be viewed as extensions of the gender division of labor; rights of passage institutions; compensatory mechanisms for the power differentials between men and women; movements for female liberation and as groups promoting political, economic, or religious interests.

While most researchers have stressed their positive aspects, others have emphasized their limitations and weaknesses.[10] For example, Maendaleo Ya Wanawake, the largest women's association in Kenya, earned the reputation of being an adjunct to politically dominant organizations that promote the agenda of the male-dominated political parties.[11] Some associations have been dismissed as possessions of first ladies who maintain a virtual monopoly on women's organizational activities.[12] Others have been criticized for not being feminist enough and for failing to advance feminist issues.[13] Aubrey states:

> There are some instances in which women have resisted openly patriarchal control, yet in most instances women have existed, albeit quietly, on the fringes of the political domain rather than challenge men and the state.[14]

Because of the economic challenges faced by women, one can see how they might define their feminism in more development-oriented and human-centered terms. In so doing one can argue that they are opposing "absolute patriarchy," the kind that is a feature of the colonial heritage and corporate globalization. It can filter down to the national and local levels, influencing and defining institutions in more powerful ways. "Limited patriarchy," on the other hand, is a second line of defense. It is more characteristic of national and local systems that are themselves subordinate and dependent on the superstructure of a global political economy. In "absolute patriarchy," both African men and women are subordinated and oppressed by a global economic system that is anchored in the colonial and similar legacies. Women's struggles for emancipation thus have to be defined in broader terms.[15]

Women's collective action is a phenomenon of the present global economic realities. Groups have emerged from the current momentum for democratization and development, as members of civil society and as nongovernmental organizations (NGOs). Although strictly speaking, not all women's associations are NGOs, almost all of them, rightly or wrongly, now bear that identification. The term NGOs, or increasingly CSOs meaning "Civil Society Organizations," arose out of the need to distinguish between the intergovernmental process of the United Nations and that of groups outside of the governmental structure.

NGOs or CSOs have been effective pressure groups and lobbyists at international deliberations of governments and have helped to make the international agenda more democratic. As a rule, these groups are not popular with many governments and have had to fight for effective representation in the decision-making process of the United Nations. Although they represent various positions, they have become an important feature of the new politics of international partnership.[16]

In the broadest sense, women's associations can be defined as formal interest groups, where membership is primarily voluntary rather than obligatory. They range from local and service-oriented grouping to large and mass-based coalitions dealing with national issues of development, democratization, peace, the environment, and so forth.[17] In the study conducted in Freetown, women's associations are classified primarily according to their functions, which correspond to their political, economic, educational, religious, and "traditional" roles.

Collective Identities

One underlying assumption about female mobilization is that gender asymmetry and gender hierarchies inevitably lead to collective female identities. The anthology, *Transforming Female Identities: Women's Organizational Forms in West Africa*, expresses this line of reasoning. Women's associations are viewed as platforms and spaces for the transformation of female identities. According to Rosander:

> There are important spatial, social, political, economic and religious aspects of women's lives, which may be seen as female ways of coping with the daily constraints with which women live.[18]

Support for this view can be found among cultural feminists, such as Dworkin and Daly. In writing primarily of women in the United States, they assert that there is an ideology of female nature and female essence that has positive qualities, applicable to all women. They advocate a cultural feminist movement that equates women's liberation with the development of a female counter culture.[19]

Other feminist scholars have argued to the contrary. Post-structuralists and post-modernists have challenged these positions as essentialists and dismiss attempts at universal explanations.[20] Following this position, a review of women's movements in the Global South argues for more emphasis on the local and the particular than on the application of universal theories.[21]

While this approach can provide some explanation, it tends to minimize historical and economic factors that can also influence the patterns, trends, and agenda of female mobilization. The history of African women shows many examples of female activism that resulted from challenges to the "universal" and hegemonic pressures of colonialism and underdevelopment. As a result, explanations cannot be limited to the local level alone. To ignore the profound and life-altering historical and ideological legacy of colonialism and corporate globalization would obscure the importance of these associations and their struggles for development and democratization. Essentialism is part and parcel of the global political economy and speaks to the continuing need for essentialist approaches.

Molyneux's distinction between practical and strategic interests, while useful, would also present some problems of application. She argues that women's interests as a group tend to be practical, involving struggle to fulfill their roles as wives and mothers and are inductively derived. Strategic gender interests, on the other hand, seek to change the rules under which women live. These are pursued later, after practical interests have been taken into account.[22]

I propose a third set of interests, namely *sociocentric interests* that could fall between practical and strategic interests. I would argue that women's associations in Africa operate for the most part to advance interests that are related to development and democratization, and oriented toward improving society as a whole. Development issues can thus become feminist issues and can form an integral part of African and other local feminisms.[23]

Given the legacy of colonialism and the current destructive trends in corporate globalization, feminist intentions are intricately bound up with the practical exigencies of economic domination. The gendered nature of collective action is complex, fluid, and influenced by historical events, materialist conditions, international forces, and corporate globalization.[24]

Institutional Manifestations

Some African women scholars have argued that gender systems in Africa have a flexibility that allows for adjustments and modifications of biological systems.[25]

They present alternative views and epistemological challenges to the dominance of Eurocentric interpretations, especially in relation to the concept and institutional manifestation of "motherhood" "daughterhood," and "gender."

In *Male Daughters, Female Husbands*, Amadiume shows how women can functionally play male roles and be socially categorized as "male" within the patriarchal lineage structure, depending on the context. Social motherhood also tends to transcend biological categories and can lead to the empowerment of women. Roles of wives and mothers, Eurocentrically characterized as belonging to the "private" sphere and as devalued, can have political significance and serve as mobilizing forces for development.

For example, Amadiume's study further illustrates how the status of "mother" constructs grouping of wives in Igbo society as opposed to grouping of "daughters" of the lineage. This can lead to jockeying for power between the two groups. Much of this power is derived, not from their affiliation to men, but from women's own economic roles and contributions to the development of their society.

The type of feminism developed within this institutional context was more empowering to women. According to Amadiume, colonialism undermined women through the introduction of male-dominated political, economic, and religious institutions that devalued the empowering traditional institutions of women. Colonialism enforced strict dichotomous divisions of people into biological categories of "male" and "female" that were unequal, without establishing institutional mechanisms for rectifying and transcending these categories.

Amadiume's study also argues that, prior to colonial rule, women's associations among the Igbo represented women's power bases. They were not compensatory mechanisms for women's subordinate positions. Women had important economic roles that granted them important entitlements and privileges and conferred power in social, economic, ritual, and political spheres.[26] Equally important is the study by Okonjo of women's inherent constitutional rights for political participation in Igbo society through a dual system of political representation.[27]

Other studies also show how women can use motherhood in their struggles for equality, especially when it is linked to consumerist and development-oriented issues. This is particularly true of a study of women's self-help groups in Kenya and of women's wings of a political party in Sierra Leone. Steady demonstrates how the concept of motherhood was used to mobilize women for political participation and to emphasize the need for development. Motherhood was a collective concept seen as essential for the advancement of both the society and its women.[28]

In *The Invention of Women: Making an African Sense of Gender Discourses*, Oyewumi argues that the imposition of the socially constructed and unequal Eurocentric concept of "gender" resulted in devaluing Yoruba women. This undermined women's power bases and important roles in promoting socioeconomic development and political participation through collective action. Awe has also shown how institutions such as the Iyalode guaranteed women's political representation and participation in decision making among the Yoruba.[29] Devaluation of women in both Igbo and Yoruba societies resulted from the heightened nature of the "absolute patriarchy" of British colonial rule.

With regard to democratization, feminist scholars have tried to shift the parameters of the democratic debate away from the individual to the group, so that "gender" can become a focal point of democratic expression. They argue that the liberal tradition places too much emphasis on the individual, a factor that undermines the political potential of women as a group, that can benefit from identity politics.[30]

In the process of democratization in South Africa, for example, women were successful in using identity politics to create a powerful lobbying group, the Women's National Commission, which developed a *Women's Charter*. Although not fully implemented, this charter promoted women's rights and equality in politics and decision making and also established institutions that included important development objectives.[31]

Women's Collective Action and Social Movement Theory

The sheer diversity of social movements makes it hazardous to theorize.[32] The concept of "civil society" is an important feature of the social movement theory, concerned with the role of groups outside the formal governmental structure. In particular, civil society groups, of which NGOs are a part, are seen as critical in advancing the democratization project.

The debate over essentialism is echoed in social movement theory from two opposing viewpoints. According to one view, women seek to create "autonomous spaces" to comfortably discuss issues relating to sexuality, child rearing, and home life. The other view challenges attempts to fix women into a single category, since women's groups will act together with a "shared purpose," only by means of conscious and careful processes.[33] The associations discussed in this book fall between these two positions, in that they tend to fuse women's interests and development interests into a "shared purpose."

Despite its popularity and its seemingly important role in advancing the democratization process, "civil society" is a highly contested concept. It does not necessarily represent views that are progressive, democratic, or pro-women. In fact, from a feminist point of view, civil society can be quite conservative and undemocratic and can encourage misogynistic tendencies, including the rise of religious fundamentalism.[34]

In addition, the concept poses specific dilemmas for women, such as when the state exerts control over what is technically the "private" sphere. Alvarez describes this as "the classical social movement dilemma" because what results is that movements composed of women "seek to advance claims, such as women's rights in marriage and the patriarchal family, that are, by definition, outside the legitimate reach of State intervention."[35]

In addition, the civil society argument does not explain the fact that women have and can mobilize with or without considerations of the state. For example, it has been argued that Algerian women's associations do not fit neatly into the category of "civil society groups."

Lloyd asserts a distinctive location for women's associations in Algeria that is outside the influence of civil society concepts. She refers instead to a complex network of social practices and social relations that constitute the sphere of all popular democratic struggles and the rule of law. She contends that this alternative framework is more attuned to the cultural specificities of Algerian society, rather than to the ethnocentrically and politically driven concept of civil structures, defined by reference to the State.[36]

Another dilemma is posed by the role of conflict in the debate, as the relationship between civil society and the state is usually characterized by relations of power and conflict. Tarrow defined contentious politics as occurring when ordinary people, often in league with more influential citizens, join forces in confrontation with elites, authorities, and opponents.[37]

As will be shown later, the confrontations between African women and the state are well known, especially during the colonial period. In the last two decades, confrontations between women's associations and the state have been aimed at resisting economic restructuring and corporate globalization. These widely reported media events have taken the form of anti-Structural Adjustment Riots in Zambia, Senegal, Tanzania, and other countries. In other instances, they have involved protests against complicity of the state with exploitation of oil resources by multinational corporations such as Shell, Chevron, and Texaco in Nigeria.[38]

Anti-state protests have also been evident in the campaigns for democratization. The targets have been authoritarian regimes, especially those involving the military, as in the case of Sierra Leone. Other relationships with the state have been less confrontational, particularly when the goal is to promote development. In this regard, many of these associations have worked in collaboration with the state.

Most women's associations are not merely reacting to gender-based discrimination, gender hierarchies, and women's secondary positions in society. Instead, they are what I have termed *socio-centric* and have, as their priority, humanistic concerns that affect the whole society. They can best be described as *shadow development agencies*. In other words, they take on civic, political, and economic functions that would ordinarily be performed by governments or other agencies in more affluent societies. It is through these activities that they seek to empower themselves and to develop their own brand of feminism that is humanistic in scope.

It is important to recognize, however, that there are both negative and positive aspects of this type of collaboration. On the negative side, governments can become less accountable for development and leave it all to the women.[39] On the positive side women can enhance their potential as important economic and political actors and as contributors to the development of their societies.

The Nexus of Collective Action and Reaction: A Conceptual Model

This book presents two dominant and overlapping themes in seeking to explain the proliferation, functions, and effectiveness of women's associations in Africa. The first can be termed the "development–underdevelopment nexus" and the second the "democratization-authoritarianism nexus." The book argues that, it is in the resulting *nexus of collective action and reaction*, and the dynamic interface of the two themes, that the symbolic and material significance of women's collective action are expressed. This results in a type of empowerment that promotes a socio-centric agenda that aims to advance society and humanity as a whole.

The tensions and contradictions promote a new social consciousness that seeks to both challenge and transform unequal relations of power. These relations are not necessarily determined by gender, but by the position of the majority of African countries within the global economy, which overwhelmingly results in their

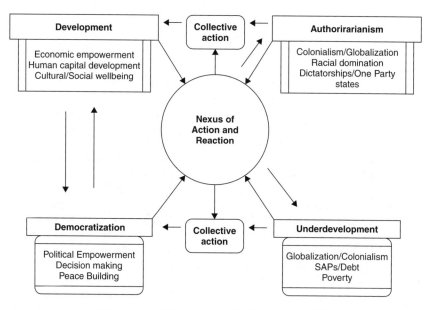

Figure 1.1 Women and collective action model.

underdevelopment. The activities of these associations are simultaneously acting and reacting to these forces, as they try to promote development and democratization (see figure 1.1).

These themes express continuities in the traditions of mobilization, applicable under pre-colonial, colonial, and postcolonial conditions. They serve as building blocks for developing female identities and for promoting a brand of African feminism that transcends narrow concerns about gender equality. In the face of overwhelming economic challenges and threats to the survival and prosperity of women, men, and children, feminist concerns may assume a lower priority to the exigencies of socioeconomic development, human security, and human well-being.

The Development–Underdevelopment Nexus

The dynamic interaction linking the promotion of development objectives while resisting the pressures of underdevelopment is what defines the first theme of this book. Many African countries face severe economic hardship, unemployment, environmental degradation, and deadly epidemics. Most studies of African economies paint a dismal picture and poverty is expected to rise by nearly 90 million by 2008.[40] Of the 49 countries listed by the United Nations as Least Developed Countries, 33 are in Africa. Above all, of the 40 million people worldwide, estimated by UNAIDS to be living with HIV/AIDS, sub-Saharan Africa remains the most affected area representing approximately 8 percent of the population.[41]

The new millennium finds many countries in a protracted state of economic crisis, resulting in political and social instability. Africa is among the regions showing the lowest social indicators for women.[42] Debt servicing burdens, SAPs, the negative impact of globalization and the legacy of colonialism and authoritarian rule, all contribute to the domination and destruction of African economies and to underdevelopment.[43]

The book focuses on how women deal with problems of underdevelopment through collective action. It also examines the dynamics between the challenges of "development" related to improving the standard of living and well-being of people, and the burden of "underdevelopment," which works against these goals. Resisting underdevelopment requires combating global economic forces that include agricultural subsidies, widespread poverty, and protracted economic recession.

The study argues that collective action is rooted in historical struggles against colonialism, structural racism, and the current destructive trends of globalization that sustain and reinforce underdevelopment. As a result, the struggles for development and democratization have become overwhelming priority objectives. Opposition to patriarchal domination and its cultural manifestations is embedded in these primary objectives.

Development, in this context, refers to sustainable economic growth that results in social and human development, and in the equitable and just distribution of the benefits of development. Ake has argued that development is an integral element in democratization.[44] For women, development pertains specifically to their economic and political empowerment, the development of their human capital, and their social, cultural, and personal well-being. In this context, development is not the same as modernization, which was pioneered by colonialism and dominated development thinking in the 1960s and 1970s. For the most part, modernization retarded the integration of women in equitable and beneficial development.[45]

From the 1950s onward, the international community has sponsored a number of development decades. This culminated at the turn of the century in the Millennium Declaration of the United Nations that was endorsed by heads of state at the Millennium Summit in the year 2000. United Nations Declarations and Plans of Action have increasingly stressed the important role of "women in development." Since the mid-1980s, "gender in development" has become the preferred term of the United Nations.

Unfortunately, much of the work of the United Nation System on gender equality and the advancement of women offers only bureaucratic solutions that are embedded in poorly resourced national institutions such as women's bureaus. Furthermore, many women NGOs formed in the last two decades tend to operate within a framework in which agenda setting is dictated by the priorities of the international community. These priorities do not include transformation of the unjust and undemocratic international economic system that destroys many of the economies of the Global South and that in turn lead to social and political instability.[46]

Gender mainstreaming is a major tool used by the United Nations, governments, and women's associations to achieve gender equality. An important aspect of gender mainstreaming is gender budgeting, which employs a gender-blind, needs-based approach to study both the quantitative and qualitative aspects of national budgeting.

The aim is to see whether the allocation of resources follows principles of needs and impact assessment.[47]

Women's associations are increasingly tracking and analyzing national budgets. The result of such efforts was reflected in the South African Budget Initiative of 1993. Women's associations have been particularly critical of military expenditures and image-building projects that tend to receive a disproportionate share of the budget in many African countries.

According to Mama, militarism provides a good base through which to examine the dialectics of public and private life.[48] It can be argued that gender budgeting can help to monitor military expenditures that direct resources away from development and contribute to authoritarianism. It can also reveal the damaging effects of SAPs that impose cutbacks in the social sectors of health and education, both of which are of vital importance to women.

While it has received measured success in promoting gender equality through counting and adjusting budgets, gender mainstreaming *per se* has not led to the advancement of the majority of African women. This is due partly to the fact that it lacks the power to transform the underlying structural impediments in the national and global economy that reinforce gender inequality.

Increasingly, the "poverty eradication" paradigm tends to be giving priority over "gender" concerns in Africa, although the role of women is essential to any effective strategy to combat poverty. This is evident in the emphasis given to poverty by the New Partnership for African Development (NEPAD) and the African Union (AU), and resonates with the goals of the Millennium Declaration to reduce poverty by half by the year 2015.

Anti-poverty declarations of these types would eventually benefit women, since women are disproportionately affected by poverty. Women increasingly head more households, and the majority of these households are of low income.[49] Anti-poverty approaches have been strengthened by the human rights paradigm, which now incorporates the right to development as well as gender equality and recognizes that women's rights are human rights.[50]

The international agenda for development, contained in the *Beijing Platform for Action*, has given priority to the reduction of poverty as one of its thirteen areas of concern. Many women's associations in Africa work for the advancement of women in these areas by taking on the challenge of underdevelopment and by fighting poverty. However, some of the problems of underdevelopment are international in scope and will not be solved by United Nations Declarations and Plans of Action alone. By and large, the United Nations has failed to advance fundamental changes in the international economic system, which at present is mainly responsible for accelerating underdevelopment and poverty.

Underdevelopment is understood as a historical and international process of economic exploitation of countries of the Global South.[51] Used synonymously with the term "dependency," it is a condition with antecedents in the trans-Atlantic Slave Trade, colonialism, and structural racism that continues to be expressed in corporate globalization.

Corporate globalization is the unfettered flow of transnational capital to accumulate wealth for a few industrialized nations and multinational corporations, by dominating and destroying the economies of weak nations, particularly those in Africa. It is based on neoliberal policies that promote privatization and so called free trade, but reserves the right to benefit from such processes to rich and powerful corporations.

Globalization has led to what has been described as a "race to the bottom" for most of the countries of the Global South, particularly those in Africa.[52] Frequently referred to as "the new market imperialism," it is aided by the policies and practices of multinational corporations and their gigantic mergers, as well as financial institutions and other institutions, such as the World Trade Organization (WTO), the World Bank, and the International Monetary Fund (IMF).[53]

According to most studies, there has been a decline or only a slight increase in some development indicators in Africa, primarily as a result of poverty (table 1.1). According to the United Nations New Agenda for the Development of Africa in the 1990s (UN-NADAF), economic growth has been declining, from its highest rate of 5.3 percent in 1997 to 2.0 percent in 2000 (figure 1.2). This is below the minimum of 6.0 percent required by the United Nations. The decline is particularly marked in the area of health (table 1.1). In 2002, Africa has the highest rate of adults and children infected with HIV/AIDS (figure 1.3).

The foreign debt for the continent as a whole stood at 298.3 billion U.S. dollars in the year 2000 (figure 1.4). Meanwhile, aid to Africa has been declining steadily,

Table 1.1 African development indicators*

	1990	2000
Economy		
GDP growth averages**	1	3.7
All-Africa exports as % of world total	2.3	1.6
Total debt service service as % of exports of goods and services	20	11
Manufacturing value added as % of GDP	17.4	14.2
Military spending as % of govt expediture	14.6	9
Population		
Total millions		
All Africa	622	798
Sub-Saharan Africa	508	660
Population growth (annual %)	2.9	2.4
Fertility rate (total births per woman)	6.1	5.2
Health		
Under-5 mortality rate (deaths/1000 live births)	158.5	161.6
Immunization against measles (% of children under 12 months)	64.1	52.8
Life expectancy at birth (years)	52	49
Access to safe drinking water (%)	53	57
Sanitation coverage (%)	54	53
Education		
Net primary school enrolment (%)	56	60
Adult illiteracy (males) (%)	40	30
Adult illiteracy (females) (%)	60	47

Notes: * Figures are for sub-Saharan Africa except where indicated.
** Figures are averages for periods 1990–1995 and 1996–2000.

Source: UN Africa Recovery from UNDP, UNICEF, WHO, World Bank data.

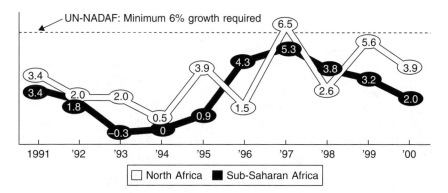

Figure 1.2 Economic growth in Africa UN-NADAF projection and actual.
Source: UN Africa Recovery, from world Bank data.

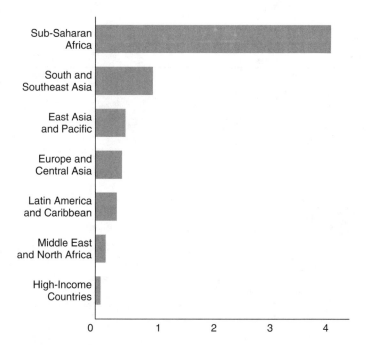

Figure 1.3 Adults and children newly infected with HIV (millions), 2002.
Note: UNAIDS regions differ from world Bank definitions.
Source: UNAIDS 2002.

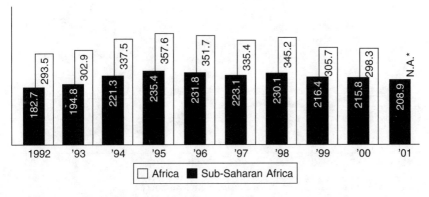

Figure 1.4 Africa's foreign debt (US$ bn).

Note: * Comparable debt figures for North Africa are not yet available for 2001.
Source: UN Africa Recovery, from World Bank data.

and the official net aid in the same year was 15.7 billion U.S. dollars—less than the annual interest paid for debt servicing to the rich nations (figure 1.5). Similarly, foreign direct investment to Africa fell to 2 percent of world flows between 1989 and 1994 and to 0.6 percent in the year 2000 (figure 1.6).

Underdevelopment has also resulted in environmental degradation, wars, conflicts, and faulty economic policies. From the 1970s onward, dependency theorists and Marxist feminists have presented major challenges to modernization theory, which, while predicting economic growth at the national level, fails to take into account the political economy at the international level.[54]

Underdevelopment is reinforced by corporate globalization and presents obstacles to African women that are formidable, structural, and enduring.[55] It has compelled women's associations to place as one of their priorities, combating the economic devastation brought about by the global political economy.

The Democratization–Authoritarianism Nexus

The struggle for democratization is linked to the challenge of authoritarianism and is the second theme of priority. The worldwide momentum for democratization, following the "glasnost" era of the Soviet Union, opened spaces for women in politics, albeit with limitations. This political 'opportunity structure' has led to the proliferation of women's associations demanding an important role in the process of democratization and in government.[56] Prior to this period, political rule had assumed an authoritarian character throughout the colonial and postcolonial periods.

Democratization in this study does not only mean elections. It is a process that, in ideal terms, is homegrown and releases the creative potential of all people in the participation and enhancement of the decision-making process. Homegrown democracy has been a goal for many Africans, and some African scholars have stressed the desirability for democracy as an end in itself.[57] There is now a general consensus that

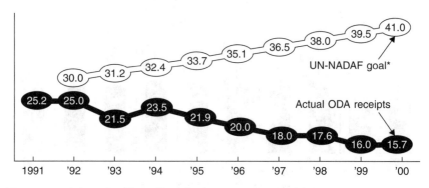

Figure 1.5 Aid to Africa (net official development assistance, $ bn).
Note: * UN-NADAF projected $30 bn ODA in 1992, with a 4 % increase each year after.
Source: UN Africa Recovery, from OECD data.

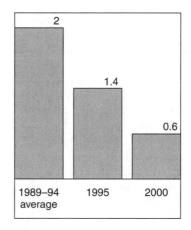

Figure 1.6 Foreign direct investment to Africa as percent of total world flows.
Source: UN Africa Recovery, from Unctad data.

democracy is not reserved for the political and economic elites. It is based on the development and enforcement of fair and equitable laws, and the guaranteeing of equal human rights with justice. It operates at both national and international levels. However, democracy at the national level can be undermined by unjust and undemocratic policies emanating from the global political economy.

Democratization is also understood as promoting a type of development that is sustainable and that ensures an equitable distribution of the benefits of development. One important aspect of democratization is the development of human resources through education, training, as well as economic and political participation. Democratization is best achieved when based on enlightened leadership, accountability, and responsiveness to the needs, concerns, and aspirations of all the people governed.

Authoritarianism, broadly interpreted, represents political and economic domination stemming both from internal and external forces. It destroys human creativity, imprisons the mind, and usurps the rights of the people. It exploits natural resources for the benefit of the few. It violates the constitution, undermines the legal capacity of people, and silences public opinion and dissent. Popular participation, open consultations, and accountability to the people governed are ignored with impunity.

Women have been active in promoting democratization and a democratic culture, and in challenging authoritarianism. Both these elements have important implications for development. In seeking to build the human capacity and social infrastructure essential for development, women's associations are contributing to participatory politics, conflict prevention, and the promotion of human rights with justice. Fourteen countries in Africa have had women serve as prime minister, deputy president or acting president. These are Liberia, Rwanda, SâoTomé and Principé, Senegal, Mozambique Central African Republic, Uganda, Burundi, Lesotho, South Africa, Zimbabwe the Gambia, and Guinea Bissau.

The main argument of the book is that although women continue to struggle to advance their status and modify their socially constructed gender roles, problems related to underdevelopment and authoritarianism take priority over narrow feminist goals of gender equality. Unless these problems are resolved, collective action intended to promote primarily feminist goals will likely remain a luxury.

Women's associations operate on a more pragmatic level. Although they tend to be fragmented, their impact is discreet and incremental. The cumulative nature of their activities, nonetheless, can have cascading effects, similar to large social movements. On some occasions they assume the characteristics of militant and sporadic social movements engaged in contestation with the state. As Tarrow points out, even movements that may appear to be apolitical, and focus on their internal lives or those of their members can have political impact. He states:

> Organizers use contention to exploit political opportunities, create collective identities, bring people together in organizations, and mobilize them against more powerful opponents.[58]

Women's associations have been, and will continue to be, important agents in African development efforts. In this regard, they are political actors in the broadest sense. They influence development policies and programs through formal and informal processes. They are also active in building a strong civil society culture, and in promoting an agenda for democratically oriented political systems, attitudes, and behaviors. This partly explains why Africa has been considered as the continent with the greatest number of women's associations.[59]

In development circles, these associations are constantly referred to as potential vehicles for change. They have worked with governments, donor agencies, and international organizations in delivering important services. They can also be regarded as the real "experts" in articulating and defining women's needs, concerns, and aspirations. They seek to advance social goals through their efforts and a brand of feminism, in which development and democratization are central.

By working within systems that are bureaucratic and closely aligned with institutions of the State, they often run the risk of becoming too close to the government, donor agencies, and international NGOs, and of losing sight of their own priorities.

These are some of the pitfalls of nation building that are different from the nationalism expressed during the colonial era, when men and women were united in the struggle for independence. Nationalism played an important role in defining women's associations, when they came into conflict with the patriarchal colonial system. Independence movements benefited from the activism and daring demonstrated by women. Among well-documented examples are: the Women's War of 1929 in Nigeria; the refusal of Senegalese women to feed French soldiers as captured in Ousman Sembene's *God Bits of Wood*; and the 1951 protest of the Women's Movement in Sierra Leone against the high cost of living. In East Africa, women were active militants in the Mau Mau movement of Kenya, and in the nationalist and liberation movements in Tanzania, Mozambique, Angola, and Guinea Bissau, where they resisted and fought against colonial rule. In South Africa, women were among the most ardent participants in movements that struggled against Apartheid, such as the African National Congress (ANC) and the Pan African Congress of Azania (PAC). They were also active in trade union movements opposed to Apartheid. In post-independent Africa some women's associations have continued to challenge the state, and some have criticized the New Partnership for African Development (NEPAD) for being too closely allied to the neoliberal goals promoting the domination of market forces and dictated by the West.

The African state continues to be defined by the legacy of colonialism and underdevelopment, which is now being vigorously and insidiously propelled by corporate globalization. Globalization aims to create a single integrated market through the unrestricted flow of capital. It also promotes the elimination of trade and other protection of some countries, while reinforcing them in the more affluent countries. It empowers multinational corporations and rich nations by undermining the rights of sovereign states and dominating the economies of weak nations, many of which are in Africa.

Empowerment

Empowerment, broadly defined in this study, is the ability of women to mobilize political, economic, educational, human, social, and cultural resources. The aim is to promote development, democratization, and responsible citizenry. Such empowerment is expressed both inside and outside of formal political processes. It gives priority to challenging the deleterious effects of the global economy on African women. The widespread poverty evident in most countries of the Global South, especially in Africa, is viewed as a crisis of development and a source of disempowerment.

Two associations of the Global South can be cited as pioneers in development of the empowerment approach in response to the crisis of development in the 1970s and 1980s. These are the Association of African Women for Research and Development

(AAWORD) and Development Alternatives for Women in a New Era (DAWN).[60] Both challenged the idea of "integrating women in development" that was popular in international circles at that time. They viewed it as based on an exploitative global political economy that resulted in massive impoverishment of the Global South. Instead, they proposed alternatives in which the following criteria were seen as prerequisites.

The first is mobilizing women through grassroots women's associations that challenge the status quo and demand change and social transformation. The second involves strengthening women's skills, education, training, and legal capacity. The third seeks to promote women's rights and entitlement to strategic resources such as land, finance, facilities, and services. Finally several factors were viewed as essential to promoting women's empowerment, namely the promotion of self-determination, autonomy, options, and self-reliance through women's associations.

As expressed by AAWORD in 1985 and later by DAWN in 1986, feminism has to take into account race, class, colonial history, and locations in the global economy.

> This heterogeneity gives feminism its dynamism and makes it the most potentially powerful challenge to the status quo. It allows the struggle against subordination to be waged in all arenas—from relations in the home to relations between nations—and it necessitates substantial change in cultural, economic and political formations.[61]

Another valuable approach to empowerment is the one proposed by Longwe and Clarke in a study of women and leadership in Africa.[62] Two indices of empowerment are suggested. The first is a Women's Empowerment Index (WEI) that measures levels of representation in political and managerial positions. The second is a Women's Self-Reliance Index (WSI) that measures women's individual capacity to advance in terms of education, training, and access to resources. The authors conclude that there is absolutely no correlation between self-reliance and empowerment. This is borne out by the fact that countries in Africa that show a high representation of women in political positions also have high rates of female illiteracy. This is particularly true of Mozambique, which has 28 percent of women in parliament.[63]

I propose a third index, namely a "Collective Action Index" that is necessary to bridge the gap between the Empowerment index and the Self-reliance index. Women acting collectively can help to accelerate the transition from self-reliance to empowerment. As will be seen, the agenda and strategies of the women's associations in this study aim to achieve both self-reliance and empowerment, broadly defined, by promoting development and democratization.

Chapter Two
The Context and Background

Introduction

The primary context of this study is Freetown, the capital of Sierra Leone. It is selected because of the extensive fieldwork conducted by me on a variety of women's associations over several years. It also has a history of women's collective action that predates the colonial period. Known for its pioneering achievements in education in the nineteenth century, the country earned the title of being "the Athens of Africa" and played a pioneering role in nationalist movements for the decolonization of Africa.

Yet in the 1990s, it shocked the consciousness of the world as a country devastated by a brutal ten-year rebel war. Less known is the fact that fuelling the war were imported weapons and illegal drugs, primarily from the West, used by the rebels to terrorize and wreak havoc on innocent people. Also significant is that the war was fueled by conflicts over diamonds involving Western companies, who benefited most from the pillage and the spoils of war. The rebel war left some areas of the country in ruins and the economy in a state of chaos, a factor that has reinforced the country's underdevelopment and dependency.

Nonetheless, the peace accord signed in 2000 appears to be holding. Some of the leading perpetrators of the war are being tried in a United Nations Special Criminal Court in Freetown, although the rebel leader, Foday Sankoh, is now deceased. The citizens, especially women, have been living in relative calm and look forward to full return to the more peaceful days characteristic of much of the history of the country. They are also hoping for a period of socioeconomic development that appears to have eluded the country since the 1980s.

Sierra Leone is listed among the fifty countries classified by the United Nations as Least Developed Countries (LDCs). The population is projected (for 2000) at 5.4 million with a slightly higher percentage of women, approximately 51 percent (table 2.1). Ironically, despite its poor economic and social indicators (table 2.2), the country is rich in minerals, especially diamonds and other natural resources. It has benefited little from its natural wealth.

The current economic crisis in Sierra Leone stems from both external and internal factors. Since its incorporation into the world economic system, ostensibly through colonial rule in 1808, Sierra Leone's resources have been exploited primarily for export and for the benefit of others residing outside the country. At present, diamond

Table 2.1 Sierra Leone Projected Population

	2000	2005	2010	2015
Population (000)				
Total	5399	6165	7014	7915
Males	2660	3041	3464	3913
Females	2739	3123	3550	4002
Urban	2171	2725	3376	4112
Rural	3228	3440	3637	3802
Percent urban	40.20%	44.20%	48.10%	52.00%
Dependency ratio (/100)				
Total	92.5	91.1	88.3	83.6
Aged 0–14	86.4	85.2	82.5	77.9
Aged 65+	6.1	5.9	5.8	5.7

Source: Sierra Leone in Figures 1998—Central Statistics Office—Sierra Leone Web.

Table 2.2 Wealth, Poverty, and Social Investment

Indicators	1994/1995	1995/1996	1996/1997
Real GDP per capital (PPP$)	790	717	643
GNP per capital (US$)	207.7	180	210
Income Share			
Lowest 40% of households	16.2	—	—
Ratio of highest 20% to lowest 30%	3.22	—	—
People in poverty			
Urban (%)	73.75	—	—
Rural (%)	88.3	—	—
Social security benefit expenditures (as % of GDP)	0.05	0.06	0.06
Public expenditure on			
Education (as % of GDP)	2.72	1.7	1.45
Health (as % of GDP)	1.02	0.49	0.52

Source: Sierra Leone in Figures 1998—Central Statistics Office—Sierra Leone Web.

mining is dominated by Canada, Great Britain, and the United States. The real business of making money from diamonds does not take place in Freetown, but in Antwerp, Toronto, New York, Tel Aviv, London, and Beirut.

Rich countries are habitually given high ratings in the *Human Development Report* of the United Nations Development Program. What this report fails to show is the structural linkage between poor countries and rich countries that has led to the continuous underdevelopment and devastation of the former. In the case of Africa, one is reminded of the continuing relevance of Rodney's thesis, which convincingly shows how the exploitation of Africa's rich subsoil and African labor benefited Europe and contributed to European economic power and imperialistic domination of Africa. The

Aerial view of Freetown. Courtesy: Christopher Greene.

title of his book, *How Europe Underdeveloped Africa*, still resonates with the workings of the global political economy, enacted primarily through corporate globalization.[1]

The modern state of Sierra Leone officially "started" in 1787 with the establishment of a settlement in Freetown for the repatriation of enslaved people from the trans-Atlantic Slave Trade. They came from Britain, the Caribbean, North America, and other parts of West Africa. Those that were freed by British Admiralty Courts en route to trans-Atlantic enslavement, and later settled in Freetown, became known as Liberated Africans. The descendants of the various groups in the settlement were collectively known as Krios. They evolved a culture that is a synthesis of African, European, and New World elements, while at the same time remaining distinctly Sierra Leonean.[2]

In 1808, the Sierra Leone peninsula in hich Freetown is located, as well as a few neighboring islands became a crown colony of Britain. The country's interior, an area of over 27,000 square miles, came under British rule as a protectorate in 1896. In 1951, both areas comprising roughly 28,000 square miles were incorporated into a unitary state, which prepared the way for party rule and independence in 1961.

Ten years later Sierra Leone declared itself a republic, later became a one-party state and today has returned to a multiparty system with a government elected in February 1996. After a peace treaty in 2001, which signaled the end of the ten-year rebel war, it held its second postwar elections in 2002 under relatively peaceful conditions.

Throughout the 1990s, the country was in the throes of a rebel war, started in part by insurgents from the neighboring civil war in Liberia led by Charles Taylor. Both wars were sustained by support from several affluent countries that sold arms and illegal hallucinatory drugs to the rebels. Due to the severe economic hardship that results from war as well as from Structural Adjustment Programs (SAPs) and mismanagement, the country's economy has been in a protracted state of crisis.[3]

In addition, many of the foreign companies operating in the country enjoy elaborate concessions known as "tax holidays" during which the country receives no revenue from their mining operations. The rebel war has also affected the productivity of the mines and increased foreign control through the use of mercenaries to guard the mines. This is bolstered by greater reliance on mining companies from countries like Canada, the United Kingdom, and the United States. The country's resources are therefore controlled by foreign companies and by locally-based Lebanese, Syrian, European, and Asian businesses. Privatization, aggressively promoted by corporate globalization, has increased this control.

The People

It is widely believed that the population of Freetown has quadrupled in the last ten years due to the influx of displaced people and refugees. The growth rate of Freetown due mainly to internal migration and displacement has resulted in high-density living and crowded streets.[4] English is the official language and Krio the lingua franca spoken widely by all ethnic groups, particularly in urban and commercial areas. Temne and Mende are also widely spoken and facilitate inter-ethnic communication.

All ethnic groups are represented in Freetown, even though each group can be identified with a particular geographical region. Considerable inter-ethnic interaction has occurred over the years as a result of several factors. These include internal migration; the diffusion of cultural norms from the two major ethnic groups, namely the Mende and the Temne; the cosmopolitan influence of Freetown; the Krio language and urban culture; intermarriage and the spread of Islam and Christianity.

Although migration has altered the historical settlement patterns, the Mendes traditionally inhabit the southern rain forest region along with smaller ethnic groups such as the Sherbro, Krim, Vai, and Gallinas. The Temne are concentrated in the western savannah area along with the Limbas and the Lokko. In the northern and eastern savannah woodland areas are the Susu, Yalunka, Koranko, Kono, and Kissi. The western coastal area, made up of mangrove swamps, rain forests, and grasslands (riverine and upland), is traditionally inhabited by the Sherbro, Krio, and Kroo. Madingo and Fulani groups are found throughout Sierra Leone.[5]

There are, in addition, a number of non-Sierra Leonean settlers, including in particular other Africans as well as Asians, people from the Middle East, especially Lebanese and Syrians and Europeans. North and South Americans and people from the Caribbean are also represented as well as the multinational contingent of peacekeepers from the United Nations Mission in Sierra Leone (UNAMSIL). Over the years, the country has provided asylum to large numbers of refugees from other African countries such as South Africa and Namibia during the Apartheid period, and to Liberian refugees in recent years.

Sierra Leonean women are exceptional in terms of their achievements in a wide variety of fields, and are represented in all professions. They are reputed for their service in executive positions as chiefs and paramount chiefs in the indigenous political systems. In addition, since the 1950s, a few women have been ministers of government and ambassadors, and have occupied high positions in the judiciary as magistrates and judges.

In the field of medicine and education, a number have held leadership positions as Chief Medical Officers and heads of the department of education. Added to this are the women who hold leadership positions as university professors, teachers, nurses, bureaucrats, managers, military officers, architects, and so on.

Despite these achievements, the majority of men and women in Sierra Leone customarily live in rural areas and are engaged in agriculture for their families' subsistence needs, producing mainly rice, cassava, yams, corn, and vegetables. Although women are the greatest contributors to agricultural labor, in terms of time allocation and intensity of work, agricultural activity usually involves the whole household and can include hired help and work groups.

About 40 percent of all agricultural products are marketed and much of the retail trade in food items, particularly the staple rice, is in the hands of women who play a central role in the food system. Women are also involved in other economic activities that advance the socioeconomic development of their communities. A study by Davies and others shows a wide range of occupations in which they have traditionally provided much needed goods and services.[6]

Social Organization

The traditional political system of most of the ethnic groups in Sierra Leone is based on a large number of small chiefdoms and a few paramount chieftaincies. After independence in 1961, provision was made for the representation of chiefs in the

Freetown Cotton Tree—A historical landmark in Sierra Leone. Courtesy: Azania Steady.

national parliament. The country is divided into four administrative areas, namely the Northern Province, the Southern Province, the Eastern Province, and the Western Area to which Freetown belongs.

Land is communally owned in the Provinces and privately owned in the Western Area. The residents of the Western Area are described as "non Native" and therefore ineligible to ownership of land in the Provinces, except on leasehold basis for a maximum period of seventy-one years.[7] Section 3(1) of the Provinces Land Act states:

> No land in the Protectorate shall be occupied by a non-native unless he has first obtained the consent of the tribal authority to his occupation of such land.

A "native" is defined as a citizen of Sierra Leone who is a member of a race, tribe, or community settled in Sierra Leone, other than a race or tribe or community, which is:

(a) European or Asiatic or American
(b) Whose principal place of settlement is in the Western Area.

Article 29 of 1972 further defined a non-native as "any person who is not entitled by customary law to the rights in the Provinces." Legal scholars and students have been debating the human rights implications of these acts, as they discriminate against so-called non-natives and violate the Constitution of Sierra Leone. As stated by Koroma:

> An analysis of the constitution will reveal that the provisions of the Provinces Land Act are not only discriminatory (Sections 15 and 27), but also contravene the right to property (Section 1), freedom of movement (Section 3 and 4) and do not reflect the position of the present law under the Constitution.[8]

Strictly speaking, most groups, including the largest groups—the Temnes and the Mendes—who migrated to Sierra Leone in the sixteenth and seventeenth centuries are not "native" to Sierra Leone. Only the "Sapes," the majority of whom lived in Bullom, and smaller groups such as the Baga, Nalou, Gola, Krim, and Kissy that inhabited this area before the sixteenth century can claim to be the real "natives" of Sierra Leone.[9] This has interesting implications for the relevance of the Provinces Land Act.

Kinship ties form the basis of social organization, the lineage being the most enduring corporate group. Most groups trace their descent through the patrilineal line, with the exception of the Sherbro and the Krio who have a more bilateral form of kinship organization. Despite these rules of descent, matrilineal principles exist alongside patrilineal ones and are not totally subdued. Maternal kinfolk are important in maintaining the various permutations of kinship that are embedded in social relations and mediated by marriage. During important life cycle events, situations of crisis, and the management of pregnancy and childbirth, the maternal kin assumes prominence.

Members of lineages and their descendants form corporate groups for legal, administrative, religious, and economic purposes. However, succession and inheritance are not automatically designated to males. Among the Mendes and Sherbro, for example, women have traditionally held executive positions as chiefs and paramount chiefs. In

Sunset at Sussex Beach—Sierra Leone peninsula. Courtesy: Azania steady.

addition to groups based on kinship ties, secret societies, such as the Poro for men and the Sande or Bondo for women exist among the major groups. Their primary function is to mark the transition from childhood to adulthood, through initiation rituals and to provide collective socialization and education for young adults. They facilitate opportunities to establish lasting bonds and alliances of fraternity and sorority.

The household is usually "non nuclear" for most groups, although a nucleus consisting of a man and his wife or wives (in polygamous households), can be readily identified. The polygamous household as a production unit creates a matri-focal environment whereby a woman, as mother, is central to her own domestic unit, consisting of a mother, her children, and other minors. Senior wives tend to enjoy a high status in polygamous families.[10]

The division of labor in "traditional" rural society is predominantly organized according to gender, although age can also be a factor. For example, post-menopausal women tend to assume political functions and young men can assist in women-specific tasks, such as fish processing, as needed. Gender-specific tasks include cattle rearing performed primarily by male members of the Fulani ethnic group. Cloth making involves both genders with weaving being primarily a male task and spinning a female task. Both men and women fish in some ethnic groups, with women generally using hand nets and men weirs, dams, and boats.

Religion is extremely important throughout the country. It is expressed in diverse ways and in Freetown represents a rich multireligious mosaic. In addition to Christianity and Islam, whose architectural styles, reflected in churches and mosques dominate the landscape of the city, a number of African religions also operate fully or in part to provide an African flavor to the two imported religions. The city intricately separates and blends the Eastern, Western, and African religious traditions that have interacted for nearly two centuries. The degree of religious tolerance is impressive and probably unmatched anywhere in the world.

The influence of religion on ideology often has a direct bearing on gender relations, as most religions have a strong patriarchal proclivity that tends to place restrictions on women. Religious affiliation is allied to ethnicity in very broad terms and some religious associations tend to reflect this. These associations are inclined to be conservative and seek to maintain the *status quo*. Nonetheless, they have inspired the formation of secular associations that include mutual-aid associations and economic associations.

In addition, some religious associations have been directly active in promoting development and in political participation. This is partly due to the fact that economic factors, such as unemployment, inflation, and general underdevelopment, impose severe hardship on most groups. Religion has inspired many public demonstrations by women on economic issues both in the past and in more recent times. Associations such as the Sierra Leone Women's Movement and an association of Methodist Women have protested against the high cost of living and used religious songs to convey the intensity of their grievances.

Women's Associations in Freetown

This case study of women's groups in Freetown, illustrates the two dominant themes of this book. It shows how collective action responds to the development–underdevelopment nexus and the democratization–authoritarianism nexus. Economic, educational, traditional, mutual-aid, and religious associations reflect the former while political, pro-democracy, and peace-building associations reflect the latter. There is a certain degree of overlap of the themes, which results in a dynamic interplay of action and reaction. Associations reflecting the authoritarianism–democratization nexus will be discussed first because of the impact of the recent war.

Due to the precarious nature of Sierra Leone's economy and widespread economic hardship, the concept of women's liberation has emphasized development above everything else. Most women's associations tend to have social and economic development objectives as their central focus. They have also taken advantage of the political spaces opened up by the process of democratization to push their agenda, which often fuses rights to development with women's rights.

Given the present dependency status of Sierra Leone within the world economic system, women's associations in Freetown have functioned more as development agencies within the context of underdevelopment than as associations with the primary purpose of promoting narrow feminist objectives. Stated another way, women's associations give priority to development in a manner that encompasses democratization and gender concerns.

Women's associations range in type, size, and composition and perform a number of functions. Some of them, like secret societies are precolonial, have their roots in rural communities, and are religious in orientation. Others are more urban-based and secular and some were introduced during the colonial era. During the last twenty years, there has been a proliferation of these associations as a result of economic and political factors, the international women's movement and the four United Nations Conferences on Women.

The fact that the membership is female deserves some reflection. Postmodernism increasingly challenges essentialist assumptions that rely heavily on biological criteria, by insisting on recognizing the different social locations of women based on class, ethnicity, national origin, and so on. In-depth studies of women's associations in Freetown reveal that women's associations can be internally differentiated on the basis of class, religious affiliation, occupation, and ethnic origin. So, while biological categories are important in membership, it is only one aspect of mobilization. Other factors pertain to social location, primary group ties, and affiliations with other social groups.

Nonetheless, a certain amount of identity politics based on gender tends to occur during periods of intense political activity. Women are inclined to mobilize across social divisions in challenging military rule and ushering democratically elected governments. Forming federations of women's groups and national women's associations have also been important strategies.

Identity politics is a term credited to The Combahee River Collective, a Black Women's Feminist Group in the United States. According to this group, identity politics occurs when one's identity is taken as a political point of departure, as a motivation for action and as a delineation of one's politics.[11] In Sierra Leone, while one's gender identity can be significant in a particular context, multiple identities (that include ethnicity, class, and religious affiliation) can exist within the female identity and become the building blocs for forming multiple and compounded alliances.

Trends in Women's Associations

The division of labor required for performing a variety of economic tasks and other prescribed gender roles is partly responsible for the formation of women's groups. Secret societies such as Sande (Bondo), work groups, and peer groups were common among the various ethnic groups and some of these continue to function. Other types of associations result from migration and urbanization. From the 1930s onward a number of voluntary associations served as adaptive mechanisms for migrants to the city from rural areas. Their functions included mutual aid, support for economic activities, political activism, and recreation.[12]

Long-term residents of Freetown also formed associations. Many of these were linked to religious organizations and some had economic and recreational functions. These included Women's Guilds, Ladies Working Bands, and the Sierra Leone Women's Movement. In addition, Western types of associations such as the Young Women's Christian Association (YWCA) and Women's Institutes were introduced during the colonial era.

In the 1950s and 1960s, these associations, particularly the Sierra Leone Women's Movement, were concerned with supporting nationalist movements for independence from colonial rule. In the 1970s, a number of international associations headquartered in Europe or the United States established branches in Sierra Leone, especially in Freetown. One of these was the Zonta International Club, a subsidiary of Zonta International based in the United States. Its overall aims to improve the legal, political, economic, educational, and professional status of women have been successful. It is regarded as having an impressive record in organizational terms and as being one of the first to aspire to high standards of professionalism.[13]

The Sierra Leone Association of University Women (SLAUW) represents a branch of the International Federation of University Women. A number of its members are professional women and hold leadership positions in the society. As a result, it has been effective in networking and advocacy for development, and for the education and professional development of women. It has also played an important role in supporting development projects for women in rural areas and rehabilitation projects for women victims of war.

For the most part, agenda setting was based on local priorities set by the women themselves. However, since the International Women's Year of 1975, the United Nations Decade from 1976 to 1985, and subsequent world conferences on women the agenda and direction of women's associations have become internationalized. Their activities are increasingly being funded by international donors, leading to dependency on external funding and to the priorities and agendas of donors. Given the intergovernmental framework of the United Nations, many of these programs tend to be bureaucratic in nature and function.

In preparation for the Fourth World Conference for Women held in Beijing in 1995, a study profiling Women's associations referred to as Non-Governmental Associations (NGOs) was conducted in 1994.[14] The study supports my findings that there is a tendency for agenda setting to become increasingly dictated by external rather than by internal factors.

The 1980s witnessed the proliferation of women's associations partly as a result of the momentum gained from the United Nations mid-Decade conference in

Copenhagen in 1980. This conference led to the establishment of national machineries or "Women's Bureaus." These represent focal points in governments for policies on equality, development, and peace that would lead to the advancement of women.

The 1980s increased the need for women's contribution to development, as this was a period of economic downturns for many countries of Africa. It has been referred to as "the lost decade of Africa." The neglect of women's roles in the development process was seen as a major factor in the failure of development.[15] Women were under tremendous economic pressures as they witnessed the erosion of household incomes and the urgent need for increasing their contribution to the family budget.

Associations provided opportunities to acquire additional skills, mobilize capital, and gain access to resources such as technology, markets, and information. They also offered material and moral support during times of hardship. In the area of health, one association, the Marie Stopes Society, places emphasis on reproductive health and family planning, along with other family planning associations, such as the Planned Parenthood Association.

The Nairobi World Conference on Women of 1985 stimulated an increase in the number and variety of associations, especially among professional women. These included the Women and National Development Association and the National Organization for Women, which were organized as umbrella associations, reflecting a broad representation of women of Sierra Leone.

The Women's Federation for World Peace and Partners Women's Commission were also formed during this period. Although part of their objectives were to provide a forum for their own interests, especially in promoting adult education, some of their programs were aimed at improving the situation of women in impoverished communities. All of these associations operated at the national level with branches in other parts of the country, enabling them to establish extensive networks. Some of their networking also included partnerships with pan-African and international associations.

The 1990s continued the trend of mobilization and was spurred on by the preparations for the fourth World Conference on Women in Beijing in 1995. A number of smaller associations were either initiated or reinvented during this period. Many women shunted from one to the other or became members of multiple associations.

Large-scale associations were formed at this time and included the Grassroots Gender and Empowerment Movement founded in 1992 to promote involvement of grassroots women in the process of development. Activities included family life planning and the setting up of credit schemes for women. The Federation of Muslim Women in Sierra Leone formed in 1994 focused on promoting the rights of Muslim women and their participation in development.

In 1993, the then first lady, Mrs. Valentine Strasser, established the Sierra Leone Women in Development Movement (SILWODMO) with projects for credit training of women entrepreneurs. It was officially launched as the umbrella organization of women NGOs and reputedly had a membership of several thousands from all over the country. Other first ladies such as Mrs. Hannah Momoh and Mrs. Patricia Kabbah have been influential in forming associations such as the Sisters Unite Association and in supporting existing ones.

In the case of Nigeria, associations of first ladies have tended to co-opt the agenda of other associations and to align themselves too closely with the state, representing what some scholars have referred to as "wifeism." This cannot be said with certainty of Sierra Leone, although first ladies can enhance the profile and resource base of associations with which they are affiliated.[16]

Tensions Between International and Local NGOs

International NGOs have been operating in Sierra Leone for several years. They often consist of foreign personnel who dominate the organization and Sierra Leoneans who are hired to carry out their programs without necessarily having decision-making authority or control of the budget. As a result, tensions have always existed between international and local NGOS. Most international NGOs are more highly paid than their local counterparts and live opulent lifestyles, that are often a source of resentment. They are sometimes perceived as displaying racial attitudes and contempt toward the local people, reminiscent of the colonial era.

Local NGOs are increasingly developing international connections that create problems of coordination, agenda setting, and alliance building. As a result of the overlap of many of their activities, they tend to compete with each other for resources and for membership. In addition, external donors, who are often perceived as representing Western interests, increasingly set their agenda. The so-called Western agenda came into conflict with the government's agenda, whenever it was felt that the NGOs were not promoting the development goals of the country, but were instead, advancing the agenda of international NGOs.

It is difficult to assess the impact of NGOs, due to the general lack of coordination and their tendency to work in an ad hoc manner. Coordination and monitoring, especially of donor aid, have always been a problem. NGOs were also criticized for availing themselves of certain privileges reserved for government officials such as access to duty- free goods.[17]

Federations and Associations of National Unity

There is a long history of women's associations forming alliances in an attempt to develop umbrella associations or federations. Mrs. Hannah Benka-Coker formed the first Federation of Women's Associations in the late 1940s. It was eclipsed in the 1950s by the Sierra Leone Women's Movement.

In 1958, the National Federation of Sierra Leone Women's Organizations was formed in an attempt to unify all women's associations under one umbrella. Its aims were to help develop the country, to help train women for responsible citizenship, and to advance the status of women. Emphasis was placed on education, training, and literacy, as important requirements for responsible citizenship and for developing the country. According to the preamble of the constitution:

> We, the women of Sierra Leone, wishing to contribute to the worthy development of our beloved country, do believe that consultation and joint action on our part, are essential to such a contribution. We therefore come together in a Federation of all women's associations, committing ourselves to the strengthening of relationships within our community; the development of all people therein and the giving of fuller attention and effort to fostering cooperative effort on behalf of the children and women of our country.[18]

Women's wings of ruling political parties have also promoted the idea of an umbrella organization. Some have made claims of representing all the women of Sierra Leone and their associations when their respective parties are in power. Federations that do not wish to become affiliated to any regime and lose their identity and autonomy, however, often contested this trend.

Other attempts to unite all women of the country have included associations that placed priority on development, such as Women and National Development Association (WAND), the National Organization for Women (NOW), and the Women's Commission of the Sierra Leone Adult Education (SLADEA). Some individual women have been outstanding in forming and leading umbrella associations and federations for development and for the advancement of women over the years. Notable among them was Mrs. Constance Cummings-John of the Sierra Leone Women's Movement. Others, such as Mrs. Gracie Williams, former principal of the Annie Walsh Memorial School and former president of the Sierra Leone Association of University Women have evoked the theme of love and unity for national development.[19]

Sierra Leone Association of Non-Governmental Organizations

On the eve of the Fourth World Conference on Women in 1994, a workshop was organized to develop a policy framework for better collaboration between NGOs and the government. An NGO document was launched by the government and a new organization, the Sierra Leone Association of Non-Governmental Organizations (SLANGO), was officially formed to coordinate NGOs. The first director was Mrs. Clarice Davies who in an interview expressed the importance of coordination to achieve the maximum effect and help strengthen the capacity of NGOs:

> There are so many NGOs with overlapping objectives and activities that it has become necessary to co-ordinate them and to have a central registry. This will help to facilitate their work and help to make them more effective. Some of them do not have the capacity to carry out their activities and we try to help them in this regard and to help with networking and alliance building. So far we are having some positive results.[20]

New criteria for eligibility, as well as a date for the commencement of registration of all NGOs, were set up. The policy established a rule that reinforced the shared responsibilities between the government and NGOs, and the need for a Memorandum of Understanding to be established with the government. The government in turn agreed to provide funds for capacity building of NGOs to ensure achievement of development objectives. This ushered in a new phase of NGO relationship with the government that called for greater collaboration, coordination, monitoring, and evaluation of NGO impact on development.

National Policy for Women

In 1991, a National Policy for Women was developed and approved by the government to facilitate the integration of women into development. The policy was implemented by the Women's Bureau but is increasingly being incorporated in the mandate of the Ministry for Gender, Social Welfare and Children's Affairs, headed by its minister, Mrs. Shirley Gbujama.

The Women's Bureau, a feature of the UN Decade for Women was established in 1988 under the Ministry of Rural Development, Social Services, and Youth. It once worked closely with the Women and National Development Association in its capacity as an umbrella organization. The objectives of the National Policy are intended to sensitize members of the public on gender issues and their relation to the overall development of the nation.

The policy is also designed to facilitate the appointment of women to executive and decision-making positions in the civil service and in the government. The establishment of crèches and other facilities for working mothers was also promoted. Nonetheless, progress has been slow in implementing this policy and in accelerating women's representation in the formal sector and in managerial, economic, and political decision making. Some of these associations, have served as pressure groups for policy reform leading to the ratification of the Convention on the Elimination of All Forms of Discrimination Against Women.

Some of the recommendations of the *Nairobi Forward-Looking Strategies for the Advancement of Women* of 1985 and the Beijing Platform for Action of 1995 were implemented through the efforts of women's associations.

Results of a PostWar Survey on Focus Groups

A targeted study of a sample of thirteen women's associations was conducted in 2002 the postwar period, as part of the long-term overall field research that involved 80 associations. This focus group study showed that development is still high in their priority objectives and in their activities.[21] All of these associations were founded between 1984 and 2001. There were six of them operating at the national level and seven at the local level. The number of members ranged from 15 to 10,000 and their ages from 28 to 45.

Five associations listed the most popular occupation of their members as teaching, four as trade, one as agriculture, one as health, and two were miscellaneous. With regard to the level of education, nine associations comprised members who had received secondary school education and the others had members with less education. Nine listed Christianity as the most popular religion among members. Three listed Islam, and one had members of both religions.

In terms of their objectives and activities, education occupied the highest rank at 77 percent, followed by economic and social development at 69 percent. Skills training for self-reliance ranked third at 62 percent; networking and advocacy was 46 percent and promoting peace constituted 31 percent of their activities. Other activities included counseling displaced people, networking, and capacity building

In terms of their relationship to the government, 39 percent considered it to be collaborative, 23 percent as cordial, 15 percent had a tentative relationship and 23 percent noted that they had no relationship at all with the government. In response to the question about the major problems facing the country, many listed economic recession, armed conflict, poor infrastructure, and general underdevelopment. Major problems facing women were poverty, illiteracy, unemployment, poor housing, low health status, male dominance, and domestic violence.

The rebel war was singled out as having had a major impact on the associations. Many of them expressed a decline in membership and decreasing donor support. They have also had to increase activities dealing with trauma healing and counseling of female victims of the war. Other activities included rehabilitation and integration, which incorporate providing micro-credit, education, and training.

The general conclusion from this study is that despite the war most associations still seek to promote development goals while resisting underdevelopment. They also seek to promote democratization while challenging authoritarianism. This continues a trend that has been evident since the 1950s.

The overwhelming challenge to the majority of women is underdevelopment. As a result, a large part of their activities consists of seeking to alleviate economic and social hardship for themselves and for their families. Even when these associations are made up of professional women and women in business, who are more advantaged, their agenda often includes programs for less fortunate women, especially women in depressed areas in rural and urban settings.

Chapter Three
Collective Action for Political Empowerment

Introduction

Women have a long history of political participation in Sierra Leone. This has been manifested in their roles as chiefs, paramount chiefs, mayors, tribal heads, cabinet ministers, ambassadors, and members of parliament. Female chiefs held, and continue to hold executive positions, especially among the Mende and the Sherbro. Some became legends in their time—like Mammy Yoko for building the Kpa Mende confederacy, consolidating alliances with other chiefdoms and trying to resist being outmaneuvered by the British colonial elite. She also used the power base of secret societies, such as the Sande, discussed later, to consolidate her position.[1]

As we shall see, affiliation to a secret society has been a major factor in the mobilization of women in some mutual-aid development associations in Freetown. It has also served as an important symbol of mobilization of women for political participation and for political office. In modern politics, affiliation to a secret society can become important in terms of campaigning for political support. These associations and their spin-off mutual-aid associations are often ushered into the arena of national politics through leadership and direct links with political parties.

Other well-known female chiefs include Mrs. Honoria Bailor Caulker and Madam Ella Koblo Gulama. Madam Ella Koblo Gulama, a college graduate, scored a number of firsts. She was the first woman from the Provinces to obtain a degree from the Teachers Training College; the first superintendent of schools; the first female Paramount Chief in Kaiyamba Chiefdom, Moyamba; the first female legislator and the first female cabinet minister of state in West Africa.[2]

Elected in 1957, she was a member of the government that ushered the country to independence. As a paramount chief, she was exempted from holding a portfolio and was able to have an impact on the work of other ministries. The participation of traditional rulers in parliamentary deliberations and in the electoral process could be crucial in legislation, in formulating policies, and in getting out the vote in their constituencies.

The Congress of British West Africa, in particular the West African Youth League, founded in 1938 by I.T.A. Wallace-Johnson played a seminal role in the decolonization of Africa. Among the few key female members was Mrs. Constance Cummings-John who became a candidate for parliament and was mayor of Freetown in 1966.

For many years, she led the Women's Movement. Mrs. Cummings-John eventually involved the association in politics, which in the 1950s and 1960s was a slight deviation from the norm for women's associations. In general, women achieved positions in politics primarily as individuals, based on their party affiliation and qualification, and not as a result of pressure from women's associations. This, however, set the stage for women's collective action in politics and in the process of democratization. Other female mayors of Freetown include Dr. June Holst-Roness, a physician, and Ms. Florence Dillsworth, an educator and former principal of the St. Joseph's Convent Secondary School for Girls.[3]

For the most part, political activities involving women's associations had been confined to women's wings of political parties. Between 1970 and 1971, the period of the government of the All People's Congress (APC), led by Siaka Stevens, women took center stage as pivotal actors in leading the process toward centralization and the establishment of a one-party state.[4]

The most important charismatic leader and influential female politician during this period was Mrs. Nancy Steele, Secretary General of the Women's Congress, the women's wing of the APC.[5] A trained nurse, Mrs. Steele became politically active while she was a student in England and was among the core leadership of the APC. A major political force in the early 1970s, she introduced the image of female power in politics through her militant, yet maternal approach toward political participation. It was customary for her to invoke the virtues of "motherhood" in demonstrating women's power and capacity for leadership. By linking "motherhood" with militancy, she challenged gender stereotypes about male prowess. She evoked the mysteries of motherhood by infusing it with the ultimate life-giving and life-sustaining force. In one of her political speeches she stated:

> We give birth to men, so in a way we own them. If women were leaders they would put an end to all this political instability. Women act more decisively because for them the welfare of the children, and hence the future of the country is paramount[6]

The type of dynamic activism displayed by Congress through the leadership of Nancy Steele gave women a high profile in political mobilization, and female militancy-Hitherto, women's wings of political parties had been relatively passive within the parties and were content in being silent followers. Unity was also a popular theme in her speeches since ethnic cleavages were threatening the party. Her vision was for a united country under the leadership of the APC. She was willing to use a militant approach to achieve this. She formed a military section in Congress and led militant protest marches against the opposition.

The APC government was in power until 1992 when the government of President Joseph Momoh, the successor to Siaka Stevens, was overthrown in a military coup. This ushered in the National Provisional Ruling Council (NPRC), led by Captain Valentine Strasser. It lasted until 1996 when it yielded to the pressure, especially from women's groups, to return the country to constitutional government.

The 1996 election was historic since for the first time in the country, a woman, Jeredine Williams, ran for president as leader of the Coalition for Progress Party. Although she did not win, she inspired other women to seek the highest position in the 2002 presidential and parliamentary elections. In 2002, Zainab Bangura, whose running mate was also a woman, Deborah Salaam, made a bid for the presidency. In addition, one other political organization, the United National People's Party presented a woman, Haja Memuna Conteh, as the presidential running mate of Dr. John Karefa-Smart. None of these women were elected but they hold an important place as pioneers in the political history of Sierra Leone.

Despite the achievements of a few women, the record of female representation in parliament and in ministerial positions has been dismal. Less than 10 percent have been elected to the legislature. The record is slightly better for appointments to positions in the cabinet. In 1986, under the APC, two women were elected. One became Minister of State for Trade and Industry and the other for Communication.

In 1991, two women held ministerial posts and in 1994 there was only one. Four women have been appointed as members of parliament by Presidents of State. In the 1996 elections, three were elected out of sixty-eight members of Parliament and two were appointed as Ministers of State. Mrs. Amy Smythe became the first Minister of Gender and Children's Affairs and Mrs. Shirley Gbujama was appointed as Minister of State for Tourism and Culture. After a parliamentary shuffle in late 1996, Mrs. Gbujama became the Minister for Foreign Affairs.

That same year, Ms. Kafatu Kabba was made Deputy Minister of Works, Energy and Power. Upon the return of the government in 1998, Ms. Shirley Gbujama was moved to the Ministry of Gender, Social Welfare and Children's Affairs, replacing Ms. Amy Smythe. The Deputy Minister for Transport and Communication was Ms. Susan Lahai.

Between 1999 and 2001, Dr. Kadi Sesay became the Minister of Development and Economic Planning with a female Deputy Minister, Ms. Memunatu Koroma.[7] After the 2002 elections three women were appointed as ministers. Mrs. Shirley Gbujama retained her position as Minister for Gender, Social Welfare and Children's Affairs; Dr. Kadi Sesay became Minister for Trade and Industry and Mrs. Agnes Taylor-Lewis was appointed Minister of Health. In another cabinet reshuffle in 2004, she was replaced by Mrs. Abator Thomas, who was then president of the 50/50 association, an advocacy group for equality of gender representation in government and in parliament.

Modern political systems patterned after the Westminster model remain the preserves of men although a few women have occupied positions as Ministers of State, Members of Parliament as well as Aldermen, Counselors, and Mayors. A few others have held decision-making positions as Judges, including High Court Judges and as Ambassadors, even before the International Women's Movement of the 1970s.

Authoritarianism, War, and Economic Decline

Authoritarianism and political instability bear all the markings of the male-dominated politics of Sierra Leone, starting with colonial rule, through one-party regimes and military juntas. It has culminated in a ten-year rebel war. Added to this fact are the protracted economic difficulties that have been exacerbated by a number of factors, including the loan policies of international financial institutions, such as the World Bank and the International Monetary Fund. The conditions for loans, which included Structural Adjustment Programs (SAPs), have been extremely exploitative as demonstrated by Weeks, and further plunged the country into a deeper economic crisis.[8]

The following is a brief review of the tumultuous events that have come to define Sierra Leone politics since the general elections of 1967. The Sierra Leone People's Party that led the country to independence lost to the APC led by Dr. Siaka Stevens; the army seized power and governed the country through the National Reformation Council. Thirteen months later, in April 1968, this council was overthrown by another military coup and a representative government under Siaka Stevens' APC was restored. The inauguration of a new party in 1970 led to a series of disturbances, which resulted in the declaration of a state of emergency.

In April 1971, Sierra Leone became a Republic with Dr. Siaka Stevens as President. Immediately afterward, there were two assassination attempts on his life. A series of political disturbances preceded the 1973 general elections, and by the time they were held, Sierra Leone was virtually a one-party state. In 1977, student protests led to widespread unrest in the city as the military used force to suppress the peaceful protests. One-party rule under the APC continued for about two decades, and the country saw the presidency pass from Dr. Siaka Stevens to his handpicked successor, General Joseph Saidu Momoh, who had been head of the army.

After a period of seeming political quiet, rebels in the neighboring Liberian war led by Charles Taylor began entering Sierra Leone, joining forces with disaffected groups in Sierra Leone. Shortly afterward, rebel activity against the Sierra Leone government started in 1991. The government tried to suppress the rebels, without much success. During a brief period when the government was in the process of implementing a change to an Executive Presidency with the cabinet chosen solely by the president (from outside the elected Parliament) there was another military coup d'état in 1992. The president fled the country taking refuge in neighboring Guinea.

The young military leaders—mostly in their late 20s and early 30s—ruled the country under NPRC led by Captain Valentine E.M. Strasser from 1992 to 1996. Although, initially, they had support from the people, especially the youth, military rule was considered unconstitutional and the demands for a return to civilian rule grew with each passing year.

The military regime, which was becoming entrenched, was reluctant to give up power and argued that a return to constitutional rule before the attainment of peace would escalate the war. At the same time, there was pressure from civil society, especially women's associations and the international community, to end military rule and return to a democratic system of government. Some civil society groups, however, doubted the wisdom of holding an election during the period of war. The rebels by that time had become constituted as the Revolutionary United Front (RUF) and were gaining ground.

After constitutional talks it was decided to return to civilian rule. In the meantime, a "palace coup" was staged in January 1996 during which the NPRC chairman was ousted and replaced by his deputy who created a new, but short-lived Council and intensified a fruitless call for peace before elections. According to Akhigbe, the British were very much involved in manipulating the process of democratization since the aim was to put the old politicians of the SLPP back in power and hence

Colonel Yvette Gordon, Republic of Sierra Leone Military Forces—Freetown. Courtesy: Sierra Leone Information Services.

"arrogate decision-making power to external forces."[9] The 1996 elections did bring the SLPP back to power.

Women, Democratization, and the Quest for Peace

The male-dominated and authoritarian nature of formal political authority has for the most part marginalized women. Armed conflict, instigated by the ten-year war, reinforced this by creating two polarities in the exercise of force. One was the military and the other was the RUF popularly known as "the rebels." As it was widely believed that some of the military participated in rebel activities, they also became known as "sobels," a combination of "soldier" and "rebel." Regardless of their nomenclature, the impact on women was generally negative.

As a result, women formed associations to advance their political participation and strengthen their opposition to war. They also saw conflict prevention and peace building as important prerequisites for ensuring development. They combined their political and peace-building objectives with development goals designed to improve their material conditions, welfare, and opportunities. This was necessary in view of the rapid economic decline and underdevelopment, which were being exacerbated by the war.

Eighty percent of the refugees generated by the war were women and children. In addition, over half of the population had been internally displaced. The majority of Sierra Leonean refugees were in neighboring countries like Guinea, Liberia, the Gambia, and Ghana. Others escaped to the United Kingdom, the United States, and Canada.

Women and young girls were further victimized by the war through abductions and gender-specific violations. These included rape, sexual assault, physical illness, and deaths from pregnancies and violence. Additional suffering included mental anguish, trauma, and rejection by their communities once they had been sexually abused. Furthermore, women bore the burden of the war in multiple ways. The war left many communities without their menfolk, either as a result of involvement in the fighting, migration, or deaths. They left women with the sole responsibility of protecting and providing for their households.

Women mobilized against the war on many occasions through protest demonstrations and lobbying for peaceful solutions to the conflict. Prayer meetings were organized in the belief in the power of divine intervention. Preventing conflict and promoting peace were viewed as essential to achieving socioeconomic development. Women were also instrumental in lobbying for a return to constitutional rule.[10]

Women had a stake in constitutional rule, because it advanced the process of democratization. They hoped that this would increase the chances for greater female representation in government, in decision making, and in the peace process. Four major landmark events that contributed to a return to constitutional rule and to democratization were spearheaded by women:

- The role of women's associations, especially the Women's Forum;
- The National Commission for Democracy;
- The national conferences on democratization;
- Female activism in the 1996 elections.

The Role of Women's Associations, Especially the Women's Forum

The momentum for democratization and peace was maintained by mobilizing women and lobbying for a return to constitutional rule and greater female representation in government, in decision making and in the peace process. Preventing conflict and promoting peace were viewed as being essential to overall socioeconomic and equitable development. On the eve of the fourth world conference for women held in Beijing, the Sierra Leone Association for University Women (SLAUW) organized women to network, share information, and plan for the conference. The following associations formed the core of what would later be known as The Women's Forum, which played a pivotal role in the return to democratic government and in the peace process.

- Sierra Leone Association of University Young Women's Christian Association (YWCA)
- The Women's Association for National Development (WAND)
- The National Organization for Women (NOW)
- The National Council for Muslim Women
- ZONTA International
- Soroptimist International
- Women's Wing of the Sierra Leone Labor Congress
- The National Displaced Women's Organization.

In all, the Women's Forum is a network of over fifty women's associations. After a number of meetings, discussions, and media events, involving a representative from the United States Information Service, a seminar was organized which culminated in a resolution to take action for peace.

In 1995, The Women's Movement for Peace was formed as a subsidiary of the Women's Forum, to resolve conflicts through peaceful negotiations and to work for peace. This involved consultations with the NPRC that had ousted the APC government in a coup in 1992. The association made good use of the media to widely publicize its activities at home and abroad. It organized the first peace march in January 1995 in Freetown, Bo, Kenema, Makeni, and Kabala, the main cities and towns. The march was a major event in which women from all walks of life joined hands and sang songs in the spirit of solidarity, shouting, "Try peace to end this senseless war." Jusu-Sheriff summarized its impact as follows:

> Peace groups, hitherto viewed with suspicion as "fifth columnists" and rebel sympathizers acquired legitimacy through association with women who had mobilized a mass movement and enjoyed the support of the international community. As a result of the women's intervention, a negotiated peace settlement became a respectable option that offered both government and the rebels the opportunity to climb down from entrenched positions without loss of face.[11]

Another important association at the time was The Women's International League for Peace and Freedom, (WILPF-SL), a subsidiary of an international NGO founded in 1996 as a result of the rebel war. It aimed to create and build a "better integrated, restructured society in which equity is a driving force."[12] The association recognized the relationship between sustainable peace, economic development, and equity and adopted approaches that would be multidimensional.

It promoted harmonious partnerships and the empowerment of women while ensuring the human rights of both men and women. Individuals and communities were trained to acquire skills in conflict prevention, management, resolution, and negotiation with a view to equipping them with relevant creative tools to handle conflicts. It also advocated against the violation of human rights particularly that of women.

The root causes of rebel activities were identified as social and economic injustice; lack of basic necessities of life and lack of effective mechanisms to address these problems. The association hoped to help alleviate the physical, psychological pain suffered by women as a result of violence, including gender-specific violence, as well as economic hardship and to help them resettle in safe communities.

The Women's Forum viewed the peace process as intrinsically linked to the process of democratization. According to Jusu-Sheriff, the insistence on a return to constitutional rule gave new momentum to an otherwise fledgling peace effort. From the point of view of the women, peace would best be pursued through a return to democratic rule.

Since 1991, the country has been in the throes of war. In addition, a relatively popular military coup in the outset, staged by the NPRC in 1992, was later challenged as unconstitutional. The military junta on the other hand was insisting that, based on the state of the escalating insurgency, peace should be secured before a return to constitutional rule.

During the pro-democracy period, which called for an end to military rule, women associations, especially the Women's Forum, took center stage. Their efforts became intensified after the Fourth World Conference for Women in Beijing in 1995. They were in the forefront of the anti-military campaign; in mobilizing women for participation in voter registration and at the National Conferences held in preparation for the 1996 elections.

The Forum worked on various issues related not only to peace and democracy but also to development, party politics, and human rights. By strengthening women's associations, it hoped to develop a dynamic pressure group at a critical time in the country's history. Women hoped that their role in the democratic process would be strengthened to ensure more equitable and meaningful participation in politics, in promoting development, and in ensuring peace. Issues of interest to women were viewed as part of the national interest, especially since most women's issues were oriented toward economic and social development. Peace was viewed a sine qua non for development.

Their solidarity was expressed in the stand against injustice, under-representation of women in decision making, poverty, and underdevelopment. The group sent representatives to the Beijing conference to bring to the attention of the world the devastating effects of the war on the people of Sierra Leone. They also articulated their commitment to ending the war and to building a peaceful and democratic society.[13]

Madam Ella Koblo Gulama, Paramount Chief of Kaiyamba Chiefdom. Courtesy: Obai Kabia.

Although the Women's Forum was an umbrella organization, each member group maintained its autonomy and acted like a caucus responsible for promoting and advocating its own agenda. Task forces were set up for issues of national interest that cut across the specific objectives of each association.

Networking was an important strategy in advocating for women's rights and in presenting a united front. According to Aisha Dyfan, public relations officer and spokesperson for the Women's Movement for Peace of the Women's Forum at that time, the emphasis was on networking on issues and consensus building.

Care was taken to ensure that no one person or personality dominated the association or served as a symbol of unity. Rather, the women were bound together by solidarity in the fight against injustice, under-representation of women in decision-making positions, poverty, and underdevelopment. They also advocated for strengthening security and for boosting the anti-war role of the media.

On February 6, 1996, the association held a conference to advance their agenda and to bring public attention to the importance of women's contribution to the peace process and to political affairs. The main objective was to prevent conflict and influence the peace process. This was particularly important since the rebel war led by the RUF had reached alarming proportions in terms of the death toll, cruelty, and suffering of innocent people. Young boys were being recruited as rebels in large numbers and young girls were being abducted to provide domestic and sexual services to the rebels. Hallucinatory drugs were used to render young boys more susceptible to following orders to commit heinous crimes.

The Forum's strategy was to build alliances horizontally through a consultative and democratic process with women's associations and the women's wings of political parties. This served to enhance their base of solidarity and to strengthen their position as a pressure group and a women's movement for peace and development. In addition to their political objectives, which included mainstreaming gender issues in political parties and ensuring representation of women as candidates for parliament in the forthcoming elections, they also advocated for an improvement in development indicators. These included improvements in health, literacy, and female education in quantitative and qualitative terms.

Unfortunately, women became bogged down with concerns about the process of democratization, with its limited goals of elections, than with securing a foothold and power base in the inner circles of the political machinery. According to Aisha Dyfan, one of the problems with the strategy adopted by women was that they were not tactical enough to play the political game. She adds:

> While women were busy trying to make the system work to ensure a democratic transformation, they were not strategizing on how to get women into the system and into participating effectively in politics as leaders and decision makers.[14]

She feels that women were low on the list of candidates ranked for the election and subsequent positions in the new government. In addition, there was not enough political education, although attempts were made to educate women about the democratic process and elections in public places such as mosques and markets. She also felt that things needed to change so that women could give up the role of cheerleading and parading support for political parties, and seriously seek political positions themselves. Once in office, they should then try to change the system from within.

Mainstreaming Gender Issues in Political Parties
The Women's Forum played a major role in mainstreaming gender issues into the political platforms of political parties. They invited political parties to meetings and encouraged discussions and debate on the issues of primary importance to them. Women's wings of political parties also joined the Forum. They became very active and forceful; incorporated women's issues in their agenda; and forcefully articulated women's needs, concerns; and aspirations in their speeches and during meetings. The Forum also gave advice to women's wings and urged them to include gender issues in their party's manifestos.

Communiqué, Declaration, and Press Statement
The Women's Forum felt that although it had asked political parties to review their platforms in relation to their integration of women's issues, they had not done so in a satisfactory manner. As a result, they issued a communiqué that consolidated their positions at the preparatory meeting for Beijing held in Dakar in November 1994 and at the Beijing conference itself. The communiqué to the government put forward the following demands, most of which were development-oriented:

- Reducing the illiteracy rate especially of women and the girl child. This should include emphasizing the nonformal educational policy at the community level and access to basic formal education.
- Providing accessible basic health care with an emphasis on health education, nutrition, household food security, and reproductive health.
- Promoting entrepreneurship—business management skills at community level and access to credit and savings programs to give the ordinary woman power to determine her own circumstances and reduce poverty.
- Making a speedy reformation of laws inimical to the interest of women, for example, marriage, property, inheritance, and divorce.

The communiqué continued with a declaration emphasizing women's involvement in peace building, rehabilitation, and political decision making. The need for sustainable solutions to peace and to the building of a democratic culture was central to the declaration.

- We know that the government with the support of the international community is committed to reconstruction and rehabilitation following the war. We emphasize that the succeeding government continue these activities. We call for women's involvement in the decision-making processes at all levels regarding these arrangements.
- Peace continues to be a priority but we emphasize sustainable peace following the end of hostilities. We demand that women constitute 50 percent of any peace delegation to all negotiations for peace.

- We want the political parties to know that democracy does not end with elections but that these processes continue. They must therefore be committed to building a democratic culture within Sierra Leone. The strategy must include equal employment opportunities and equal representation in state boards and commissions.
- All these issues have been deliberated at local, national, and international levels. They have come out in the Sierra Leone National Report to the Fourth United Nations World Conference on Women in Beijing in September 1995. We call on political parties to familiarize themselves with the Platform for Action as contained in these reports.
- We therefore demand that the government set up a Monitoring and Evaluation Commission to ensure the implementation of equal opportunities issues in order to reduce the gender gap.[15]

Realizing the reluctance of the government to hold elections by claiming that peace should be secured before elections, the Women's Forum issued the following press statement on March 11, 1996, requesting the government to ensure the safety of its citizens during the elections.

> The Women's Forum, a network of women NGOs and groups has been moved to call on the Head of State, Brigadier-General Julius Maada Bio and the NPRC government to provide the entire citizenry of Sierra Leone with adequate security for the process leading to and during the run-off for the Presidential elections[16]

Throughout the pre-election period for the 1996 elections women staged large-scale campaigns and demonstrations against military rule. They also successfully challenged the attempt of the NPRC military government to postpone the elections on account of the continuing rebel war.

In addition to the Women's Forum, other women's associations were also active in promoting democratization and the peace process. Among the more visible ones were The Women Organized for a Morally Enlightened Nation (WOMEN). The group was motivated by the belief that a democratic Sierra Leone will provide equality, opportunity, and access for all citizens, the absence of which has been among the root causes of the national failure to develop.

Based on its conviction that women are the most disadvantaged people, whose resources were underutilized in the country, the association vowed to improve the situation. Its primary objectives included the establishment and maintenance of democracy and a democratic culture in Sierra Leone through education, training, advocacy, and institution building.

Another association Legal Access Through Women Yearning for Equality, Rights and Social Justice (LAWYERS) aimed to empower women and enhance the democratization process through the promotion and protection of the rights of women and the girl-child. Its objectives included offering professional legal advice, counseling, and assistance to women and girls in Sierra Leone with particular regard to the disadvantaged ones. It also sought to render the legal system and justice more accessible to them. It served as a resource center that aimed to conduct research into the law with particular emphasis on the rights of women and the girl-child and organized workshops and paralegal training for women.

The National Commission for Democracy

Another important development of the process toward democratization was the establishment of the National Commission for Democracy in 1994 by Executive Decree number 15 of the NPRC. The Commission was headed by Dr. Kadi Sesay, former head of the English Department at Fourah Bay College, University of Sierra Leone, who was devoted to gender equality and the advancement of women. Dr. Sesay has also taken her advocacy to the international stage as a frequent speaker on women and peace issues in Sierra Leone.

The Commission's main objective was to promote civic education and patriotism as well as to reduce inequalities and prepare the country for democratic rule through building democratic institutions. It helped to create a climate and momentum for mobilizing civil society groups, especially women. The Commission's plan was "to assess for the information of government, the limitation to the achievement of true democracy arising from the existing inequalities between different strata of the population and make recommendations for redressing these inequalities."[17]

Educating the public about the constitution and ensuring legal literacy about their rights under the law were important strategies in the work of the Commission. This resonated well with women who still faced gender-based discrimination in the legal and political systems and in participation in decision making. The work of the Commission is well known and included mobilizing civil society through the media and reaching out to grassroots movements and people from all walks of life. The central focus of its work was to develop and implement a realistic agenda for the democratization process.

It held a Consultative Conference in May 1995 with representatives from a wide variety of grassroots organizations as well as traditional leaders and chiefs. A national pledge of allegiance was launched which developed into a process of citizenry and patriotism that gave visibility to civil society, especially women's groups, and its important role in the democratization process. The country prepared for elections in 1996, under the auspices of the United Nations. It was headed by Dr. James Jonah, the Chairman of the Independent National Electoral Commission (INEC) and a former under secretary-general of the United Nations.

National Conferences on Democratization

The first national conference held at the Bintumani Hotel in preparation for a return to constitutional rule was held from August 15 to 17, 1995. Of the sixty delegates invited only three were women. This led to a protest by the Women's Forum resulting in an increase in the number of female participants to sixteen. Although representing different groups, they presented a unified position. This included the urgency for peace, ensuring fairness in the electoral process and in women's representation, and containing the army in the electoral process. They also insisted on adhering to the timetable already set for the elections, which were being delayed by the military government.

The second Bintumani conference chaired by Mrs. Shirley Gbujama of the Sierra Leone People's Party (SLPP) was held on February 12, 1996 to discuss the status of the peace process and its implications for the elections. The timing of the elections in relation to securing peace was again central. Women were active, once again, at the second Bintumani conference and lobbied delegates for support of their position. The following united position was presented by the women attending the conference:

- Women reaffirmed their wholehearted commitment and support for every genuine effort being made to bring a sustainable peace to Sierra Leone as speedily as possible.
- Women therefore demanded that the elections (an essential and fundamental part of the peace process) go ahead on February 26, 1996 as agreed at the National Consultative Conference by the NPRC government, the Political Parties, Civil Society, and the INEC.

The women's position won the vote at the conference. The elections took place as planned. This was considered a crucial turning point. The Forum had demonstrated its ability to mobilize women through its links to many associations and by its process of consultation and discussion. It established its credibility as an umbrella association of women with ideals and principles that were closely linked to democratization and development objectives. It attributed its success to sound organization, the clarity of their message, and absolute persistence on the right to be heard on all issues.

Female Activism in the 1996 Elections—A Test of Unity

The elections were held on February 26, 1996 in a limited capacity, because of rebel activity in parts of the country. Constitutional rule was reinstated, with Ahmad Tejan Kabbah as president, and a multiparty legislature was sworn in. Though there was some accommodation of the RUF, and a "peace accord" signed in November 1996 in the Ivory Coast, there was no clearly defined role for the RUF leader, Corporal Foday Sankoh, who, many years previously, had been a member of the country's armed forces.

The Women's Forum and civil society groups can take credit for the eventual elections. Women's groups had sponsored and initiated debates about the war, the peace process, and the democratization process, and stressed the importance of the people choosing their leader. Associations such as the Women's Forum and the WOMEN also gained prominence as leaders mobilizing women for participation in democratic politics and for promoting development objectives.

On the day of the elections, women voted in large numbers and defied the military government's threats, chanting "we want to vote." They got out the women's vote and participated in providing logistical support for the electoral offices. They also demanded security before and during the elections, a period, which was marked by sporadic violence, sometimes instigated by the military government.

The election was a welcome relief from one-party rule and military rule after a period of almost twenty years. Repeated threats from the rebels and the reluctance of the military government to proceed with elections did not prevent the outcome.

Dr. June Holst-Roness, former Mayor of Freetown. Courtesy: Sierra Leone Information Services.

Law Courts Building—Freetown. Courtesy: Azania Steady.

Women were credited for the return to constitutional rule by several prominent politicians, including the Head of State, Alhaji Ahmad Tejan Kabbah, in a number of speeches after the elections.

The unity achieved by women during the period of electioneering and democratization faced challenges after the elections. It became publicly known that some members of the Women's Forum objected to the procedure of nomination used to select the Minister of Gender and Children's Affairs, Mrs. Amy Smythe. They allegedly felt that she would not be in a position to represent women of Sierra Leone, not having been initiated into the Sande/Bondo secret society. However, the public statement issued mentioned opposition to the selection process, rather than Mrs. Smythe's cultural affiliations.[18] After much deliberation at high levels, the appointment was confirmed.

Some pro-democracy movements comprising both men and women were established after the elections. One of these was the Campaign for Good Governance (CGG) established in July 1996 after the general elections to promote democratic awareness and popular participation in political affairs. It also had on its agenda the strengthening of democracy both inside and outside the government. It was based on the premise that "civil society had a role to play in ensuring good governance and in building a strong, viable and well-informed civil society to act as a counter-balance to the powers of government."[19] The objectives included strengthening democratic institutions of government by facilitating training of personnel within specific institutions of government.

Activities in this regard have included workshops for cabinet ministers, members of parliament, members of the judiciary, and members of the media. An international experts panel consisting of cabinet ministers, judges, and journalists from South Africa, the United States, and Ghana was also invited to attend.

Gender Bias in the Body Politic

Despite internal differences, women shared a common marginalization in the formal political structures for over three decades. According to Jusu-Sheriff, all the major parties in the conflict ignored their demands for direct consultations and for a place at the peace table. This was because their idealism was threatening to traditional male-dominated politics. They were also blamed for failing to translate their political capital, gained through the peace process, into political gains. In Jusu-Sheriff's view, they tended to surrender to a civilian government that promised to take over responsibility for the peace process, rather than secure a critical mass of political positions for themselves in the new government.[20]

Though valid to some extent, one cannot but help call into question the reason given for this, as it was not borne out in my interviews with members of the Women's Forum. According to Jusu-Sheriff:

> The peace process was a sufficient achievement for many of the women's groups who were not comfortable in the limelight.[21]

The women I interviewed did not think that shyness was a factor. On the contrary, they were highly visible, articulate, and demanding. Many of them gave another reason for their marginalization in the resulting political structures. It had more to do with their failure to anticipate and plan well for the post-election period. They felt that they should have had a better and more concrete strategy for securing political positions and consolidating their power base after elections. They also felt that by being intensely involved with the process, they lost sight of the content and outcome of the elections. This did not necessarily signify their contentment with invisibility or their satisfaction with being left out of the limelight of formal politics. They were simply out-maneuvered by the male-dominated political machinery.

1997 Coup d'Etat and Escalation of the War

The elections did not bring about the stability that some had expected and a few months after the first year of its rule, the government was overthrown by a coup on May 25, 1997. President Tejan Kabbah fled the country to Guinea. The military took over and formed the Armed Forces Ruling Council (AFRC) with support from the RUF with whom this faction of the army had become allied.

The rebel war eventually reached Freetown. As Freetown was the national capital and densely populated, the fighting, which eventually ensued, resulted in parts of the city being severely damaged or destroyed and about 2,000 or more people losing their lives. Armed conflicts pitted the AFRC and its allies on the one hand against civil defense forces (Kamajors and others) and ECOMOG forces on the other. ECOMOG comprised mostly of Nigerian soldiers sent as a monitoring group for the Economic Community of West African States (ECOWAS). Constitutional rule was ultimately restored in 1998, with the return of the president and other ministers from exile. Many parts of the country and the capital were devastated. The rebels still controlled large sections of the country, including the diamond-rich areas at that time.

The government captured several leaders and supporters of the rebels, charged some with treason and put them on trial. A number of them, primarily military officers were found guilty and put to death. The rebel leader Corporal Foday Sankoh was later found guilty of treason and sentenced to death.

While his case was on appeal, rebel protests and anger at his imprisonment took the form of renewed attacks on Freetown and its environs in January 1999. Military engagement with ECOMOG troops resulted in extensive destruction by fire, looting, and bombings. Many private homes and public buildings, including churches, were destroyed. ECOMAG forces restored order by driving the rebels out of Freetown. By 1999, Sierra Leone had 400,000 refugees, the largest number in Africa, the majority being women and children.

The Peace Process Renewed

Both the government and the RUF agreed to negotiations and serious talks about a lasting peace. Foday Sankoh was released to United Nations representatives for consultations in Togo with other RUF leaders. Talks between the government and the RUF began in May and June of 1999 under the auspices of the President of Togo, the ECOWAS chairman at that time.

Three attempts had been made to secure peace namely the Abidjan Accord of November 10, 1996; the Conakry Accord of October 23, 1997 and the Cease-Fire Agreement of April 17, 1999. All of them were broken. In November 2000, the government and the RUF agreed to a cease-fire for thirty days, which was later extended. This was followed by the setting up of a United Nations peacekeeping mission, UNAMSIL, and the disarming of all combatants. A Joint United Nations-Sierra Leone Government Special Court for Sierra Leone was established to try those with the greatest responsibilities for the atrocities committed during the rebel war. Foday Sankoh, the leader of the RUF, who was one of those indicted by the Special Court died while in hospital awaiting trial on July 29, 2003.

One association that played an important role in the peace process in both Sierra Leone and Liberia is the Manor River Women's Peace Network (MARWOPNET). Formed in 2000, with the support of Femmes Africa Solidarité, a pan-African women's association, it was awarded the 2003 United Nations Prize in the field of human rights.

It is made up of networks of women politicians, journalists, lawyers, academics, and representatives from the private sector in the fields of peace, human rights, and development. Its members are from the three Manor River states—Liberia, Sierra Leone, and Guinea and it has played an active role in advocating for peace and in getting women involved in peace talks.

Its founders share the conviction that women can contribute meaningfully to the quest for regional peace and security and that the lasting absence of conflict is a necessary condition for fulfilling the human rights for all. The work of MARWOPNET is wide ranging and includes advocacy and training of women in conflict prevention, conflict resolution, peace building, and techniques of negotiation.[22]

Militarism, Peacekeepers, and Gender Vulnerabilities

It is estimated that between 50,000 and 64,000 women were victims of sexual abuse during the war. In addition to being abducted, women and girls were raped and forced into so-called marriage and sexual servitude by rebel forces.[23]

The presence of peacekeepers has been a mixed blessing for women because of its gender implications. Women's vulnerabilities to sexual exploitation, violence, and to sex work have reputedly increased as a result of militarism, the war, and the United Nations peacekeeping mission in Sierra Leone: (UNAMSIL). Prostitution has increased due to the large contingent of unattached men that make up the peacekeeping force and the military itself. The main beaches are full of prostitutes and peacekeepers and no longer offer the pleasant family recreational venue for local residents.

Other forms of sexual relations between local women and peacekeeping forces range from casual sex to more stable relationships of cohabitation that have resulted in the birth of children. Peacekeepers represent a source of income in a country ravaged by war and grave economic problems but this has been a mixed blessing for women. It has been reported that some of these relationships have resulted in violence against women and the spread of the HIV virus.

The military in general, and peacekeeping forces in particular, are reputed to be groups at high risk for HIV/AIDS. The problem posed by peacekeepers in the spread of HIV/AIDS has been the focus of discussion at the United Nations and is the subject of a Security Council resolution.

According to health workers interviewed in Sierra Leone between 2000 and 2002, the incidence of HIV/AIDS has been increasing since the presence of ECOMOG forces and UNAMSIL. This is confirmed by estimates of the United Nations and other sources.[24] A number of programs have been set up to deal with the threat of HIV/AIDS. These include the Sierra Leone HIV/AIDS Rapid Response Project (SHARP).

The experience of Sierra Leone shows that gender vulnerabilities are intensified in times of militarism and war. Women and girls are affected by war in a manner that demoralizes them, puts them at risk physically, and destabilizes the social fabric.

End of the War, the 50/50 Group and the 2002 Election

In January 2002, the president announced that the war was finally over and presidential and parliamentary elections were held on May 14, 2002. A total of ten parties contested the elections. Women's Associations, including the Women's Forum, mobilized for the elections and welcomed the re-opening of the political space and the declaration of an end to the war.

One association, the 50/50 Group, stood out for making the boldest demands yet for gender equality in government and for maintaining a high public profile. It aimed to have equal representation of men and women at every level of decision making. Its activities included recruiting, training, and supporting women seeking elected office in order to remove some of the barriers women face. The association realizes that women have to have substantial resources if they are going to achieve the objectives of gaining political office. As stated by Dr. Nemaia Eshun-Badan, a leader of the group and an education and training consultant, in a fund-raising letter:

> More than half of our capable trained aspirants will lose the opportunity to stand in the election because they are unable to raise sufficient funds to pay their party and the National Election Commission candidate fees. We stand to lose all the ground we have gained if we don't raise money to be distributed among our women candidates.

The association is seeking to transform politics by having a fresh parliament of new ideas and approaches to peace and progress. It hopes to work for a parliament that is full of integrity and genuine multi-partisan cooperation and where women are present and fully represented in force. The association also aims to influence leaders of political parties and the President by having their endorsement. The letter also states that:

> In addition to training the women, we have asked for, and received, commitments from political party leaders to include women in prominent positions on their candidate lists. We have asked for and received an endorsement of our work by his Excellency, the President of the Republic, Alhaji Ahmad Tejan Kabbah.

The 2002 elections were peaceful and the ruling party, the SLPP, won by a landslide with the President being returned to power for a second term. Three women were appointed as ministers (health, trade and industry, and gender, children and social welfare) and one was appointed as deputy minister.

For the second time in Sierra Leone's history, the presidential candidates included a woman, Zainab Bangura, chair and leader of the Movement for Progress Party (MOP) whose running mate was also a woman, Deborah Salaam. The party issued one of the most comprehensive populist manifestos. The manifesto opens with the following words:

> In this document we offer a vision for the future of our country. This vision is not the property of our movement: it is the birthright of all Sierra Leoneans, if they should use their freedom and our precious democracy to choose it. We as a movement offer only our commitment, our humility and willingness to learn and improve, and our profound determination to conduct ourselves with courage, generosity and integrity in public office.

We have watched our country, our families, our children, ourselves, suffer from the inhumanity of war, a war that has stolen something from all of us. We know that this war was ours, that it was a war which all of us bear some responsibility for and we have resolved to learn its lessons and rebuild our country, by constructing strong, loyal, transparent and patriotic institutions, promoting prosperity for all our citizens, building a just and moral society and cooperating with the community of nations. If we succeed in these things, we shall not only ensure that the horrors of war never ravage our beautiful country again, but that for the first time in our history, all of our people shall have the opportunity to live a life of real freedom.[25]

The Impact of Women on Democratization and the Peace Process

The role of women's collective action was pivotal and historic in the return of Sierra Leone to constitutional rule. It marked a critical point in the struggle against authoritarian rule and the quest for democratization. However, it did not significantly change the authoritarian nature of government nor did it alter the reality of politics as a male preserve. Women were the major force willing to confront a military government for a return to civilian rule, but they did not make significant gains in terms of being elected and appointed to political office.

During my research, I encountered people with skeptical views that included the suggestion that women were used to advance the agenda of male-dominated political parties. This is true but only in part. It is not fully substantiated by my interviews or the stated objectives of women's associations. They advocated intensely for women's involvement at high levels of political participation, in political office at the level of decision making. This is particularly true of the 50/50 Group.

Another source of criticism included the view that women lacked an ideological guiding framework, which blunted their effectiveness.[26] In my view, the Women's Forum did have an ideological framework in their commitment to democratization, peace, development, and the advancement of women. What they lacked was a strategic plan of action to implement their vision and ideological position. Problems of resource constraints also contributed to derailing prompt and effective action toward a realistic and meaningful process of democratization.

Despite the disappointments in failing to gain more political positions, women had moved the process of democratization and development for the whole society in a forward direction. By acting collectively, they were able to advance a vision with concrete suggestions some of their efforts are commendable and could lead to a people-oriented and democratically peaceful, stable, and economically viable society that included and transcended gender equality.

Chapter Four
Collective Action for Economic Empowerment

Economic Profile of Women in Freetown

Freetown's female population is diverse, in terms of economic activity, geographic background, ethnic affiliation, and so on. The majority operates primarily in the informal sector and faces economic difficulties due to high rates of unemployment, inflation, Structural Adjustment Programs (SAPs), and increasing poverty. Many women in the informal sector also have to face the vulnerabilities of this sector to recessions and economic downturn.

In the absence of adequate surveys on women's economic activities and of detailed up-to-date statistics, an attempt will be made to further construct an economic profile of women in Freetown based on the available data and on interviews and observations derived from fieldwork.

According to figures available from two household surveys and the latest census published, the largest percentage of working women in the formal sector is in sales, representing 49 percent. Women in the service sector constitute 16 percent followed by 14 percent in clerical and related work.[1]

Women have a long-standing tradition of trading in Sierra Leone.[2] A number of those in Freetown also own shops, stores, and bars retailing imported goods, food items, clothing, and beverages. A few conduct private sales to friends and relatives from their homes usually of imported ready-made clothes, hats, and shoes. It is also quite common to see small food stalls in front of private homes with family members taking turns as vendors. Small restaurants and cooking facilities can be found everywhere, including private homes.

Most of the trading activities of the majority of women can be described as subsistence trading. The surplus generated is minimal and is used primarily for purchasing food and for the basic consumption needs of the household. Based on my interviews, many of these women barely break even and some operate at a loss. Processed food items such as cooking oil and major staples like rice usually generate more profit, which can be used for more expensive items such as clothing and school fees.

A number of services are provided in the informal sector, such as hairdressing, childcare, home nursing, domestic work, prostitution, and so on. Unlike most regions of the world, the majority of domestic workers in Freetown are men, not women.

Women are also involved in entrepreneurial small-scale cottage industries in textiles and garment manufacture, particularly tie dye, soap making, oil extraction, basket making, and so on. Most of the enterprises of women are home based and somewhat "invisible." Men in manufacturing, on the other hand, are more visible and can often be seen at work in the many roadside industries all over Freetown, making furniture, building materials, garments, and the like.

Industrialization has proceeded at a slow pace and takes place on a small scale. The few industries that exist are capital-intensive and foreign owned, employing only a handful of local citizens, primarily men. One exception is the labor-intensive fisheries industry involving food processing and packing, which employs more women than men.

In the formal labor market, women constitute the majority of elementary school teachers and are predominant in sales and secretarial work. There is also a good representation of women in the professions of teaching, nursing, the civil service, the police force, the military, and the legal profession. A number of women have held top managerial positions as chief medical officers, ranking officers in the army, and permanent secretaries or directors-general in the ministries of government. A few women have also served as cabinet ministers and ambassadors.

According to the United Nations, women's economic activity rate is increasing all over the world. The figures for Sierra Leone show a rate of 44 percent for 1995/1997, which is up from 42 percent in 1990.[3] The percentage of women in administrative posts or positions as managers was 8 percent, and for sub-Saharan Africa as a whole, it was 14 percent in 1997. In the professional, technical, and related fields, it was 30 percent.[4]

Management and Decision-Making Positions

A study of women in management and decision-making positions in Freetown found that, despite the slow progress, some highly qualified women have achieved a high degree of prominence in top positions in the academic and professional fields.[5] At least 50 percent were involved in some form of decision-making activity, although there was little evidence of goal setting or career planning in their work experiences.

The majority of women interviewed for this study did not perceive barriers to their professional advancement. Seventy-two percent stated that there were no barriers and 78 percent felt the promotion criteria were fair. Only 28 percent felt that there were barriers, which they identified as a male-influenced culture and the institutional environment. The following were viewed as obstacles to career advancement: gender-based discrimination, sexual harassment, conflicts with domestic responsibilities, low pay, poor communication, and interpersonal relations. Despite these problems, only 32 percent of the women said they preferred a male to a female supervisor and 29 percent had no preference for either gender.[6]

According to the study, most women had difficulty reporting gender-based offences and seeking redress although they were aware about the necessary action to be taken. They expressed the need for more progress in the following areas: building gender awareness, constructive dialogue, more flexible working conditions and provisions, clearly defined roles and job descriptions as well as strict regulations and disciplinary action for insubordination.

With regard to manufacturing, no women were found in top executive positions although they made up 25 percent of the senior staff. The insurance industry had the highest number of women, followed by NGOs, professionals, and the hotel industry. In the administrative section of the Civil Service, which is the largest employer in the formal sector, women held 15 percent of the top management and senior staff positions. In terms of political office, the election and nomination of women to political positions have been minimal in all political regimes.

About 75 percent of the women surveyed tended to have a long record of service or more than seven years with the same institution. Sixty-one percent were married and 50 percent had professional degrees, of which 14 percent were Masters degrees. In addition, the study also sought to identify strengths and gender-related problems of the woman manager. Conflicting information was given regarding the negative aspects of gender relations and the dynamics in the work place. Equally unresolved was the information on the degree of conflict between women's roles in the workplace and in the home.[7]

Women constitute more than 50 percent of the population and are poorly represented in the formal sectors of the economy especially in executive, managerial, and decision-making positions. As a result, an attempt was made to find out the degree of their participation and the gender-related problems serving as obstacles to their advancement.

One important finding was that women lacked a critical mass of role models in managerial and decision-making positions. The study concluded that more women should be included in top management positions, so that they could also serve as role models to help break down restraining cultural and traditional gender-based beliefs and practices.

Women's Economic Associations

The development–underdevelopment nexus is best exemplified by women's economic associations. While seeking to promote development through their economic activities, they are also resisting underdevelopment through boycotts and protests against economic exploitation by foreign enterprises and institutions.

The tradition of mobilizing for economic reasons has been in existence for a long time. In most rural areas, it was customary for women to organize into work groups and collectives as an important feature of production and socioeconomic development. This followed the customary lines of the division of labor by gender. In addition, women traders, especially in urban areas like Freetown, have traditionally established associations to facilitate their commercial activities.

These included mobilizing capital, securing markets, defending their interests, and withstanding economic pressures from middlemen and large foreign-dominated commercial enterprises. Some of the most important economic associations among women are the Women's Movement, rotating credit associations, cooperatives, professional associations, and guilds.

The Women's Movement

The historical impact of the Women's Movement has a unique historical position Sierra Leone. It provided a model for female economic activism that is still relevant today. In February 2004, some women from religious associations attempted to demonstrate peacefully against the high cost of living, but were not issued permits and were prevented from doing so by the police. According to one report, this planned protest was not instigated by the opposition parties, but by the anger of the women, who felt that the government had let them down.[8]

The Sierra Leone Women's Movement was the first association to mobilize women against foreign economic exploitation on an unprecedented scale. As a result of the alarming rising cost of food, caused by the involvement of Lebanese traders in wholesale food distribution, a group of ten thousand women staged a massive demonstration of protest in Freetown in 1951. Dressed in colorful African clothes, they marched in procession, singing songs and hymns and carrying banners protesting the high cost of living to the secretariat building, the country's administrative headquarters at that time. The women were also protesting the proposed increase in market dues.[9]

> We are hungry
> We are tired of buying head ties for four shillings
> We need milk for our babies
> Governor, give us farmers banks
> Peg Prices
> Governor, do please subsidize rice
> Cost of living is too much for us
> Government action please
> Big firms, hands off our staple food.[10]

They were led by Mrs. Mabel Dove Danquah who had worked with women's movements in Ghana and Mrs. Hannah Benka-Coker who founded the first federation of women's associations in Sierra Leone. The women handed over a petition requesting that women be given a monopoly to trade in palm oil and rice. Since the end of the Second World War, these commodities had been taken over by Lebanese traders who were demanding exorbitant prices. European firms had large trading establishments linking the economy of Sierra Leone to the global trading system.

As a sequel to the demonstration of female solidarity, the women not only secured the monopoly of buying directly from the governmental agricultural station at Newton, but also formed the Sierra Leone Women's Movement, a strong association comprised of women from all occupational groups. They included dressmakers, petty traders, businesswomen, teachers, nurses, and so on, who came together primarily to forestall the rise in the cost of living.

In 1952, branches were established in the Provinces and this brought together thousands of women into a network of trade linking all the important centers of the country. The leaders enlightened farmers about the importance of controlling prices. As a result, members were able to purchase food commodities at reduced wholesale prices, directly from the producers in the Provinces, and to retail them at fairer prices in Freetown, thus bringing down the cost of living.

Some of the members who went to the Provinces, made contacts with women traders and bought goods at cheaper prices so that they in turn could retail them reasonably cheaply. The net result was lower prices all round. Branches were set up in key towns in the Provinces such as Kenema, Makeni, and Bo.

The involvement of women living in the Provinces in the movement was due primarily to the efforts and influence of Mrs. Constance Cummings-John, a charismatic leader, who was the organizing secretary of the movement for many years. She was remarkable at mobilizing women and embodied the spirit of collaboration and alliance building necessary to keep a large-scale movement going. She was able to gain support from the leaders of the Sande/Bondo societies who commanded the loyalties of the vast majority of the women in the Provinces.[11]

The movement is reputed to have made tremendous profits for its members and for itself. A few of the members participated in foreign trade, exporting bananas to the Gambia. In 1953, the movement acquired its own headquarters building at Charlotte Street. The Women's Movement song reinforced their commitment to help promote socioeconomic development and unity.

> We work to make Sierra Leone noble and great,
> A country where God has command,
> An end shall be put to division and hate,
> New unity sweeps through the land.
> No more shall a section all selfishness reign,
> No more shall the jobless seek labor in vain,
> We'll work as a team with one goal,
> Our country, with God in control.
>
> We work for our country to be healthy and strong, New fitness of body and soul,
> To raise up a force to which all can belong,
> With each giving all for a whole.
> So forward to action, no coward delay,
> Our leader unseen preparing the way
> We work with a team with one goal,
> A new world with God in control.[12]

After 1953, the association was able to concentrate on other activities. These included literacy classes for illiterate members and a printing press that published a women's newspaper titled *Ten Daily News*. Fund-raising became a regular activity and special community projects involving the promotion of public health and sanitation were established.

In 1954, two specialized groups were created within the movement. One was the Friendly Society of the Sierra Leone Women's Movement whose main function was to promote charity work for the welfare for the less fortunate members. The Society also provided artificial limbs to disabled people and donated medical supplies and money to children's clinics and to the maternity hospital. Funds were also raised for scholarships from the government and other sources to promote the education of girls.

The second specialized group to be formed within the movement was the Freetown Women's Traders Co-operative Society. It provided training in bookkeeping and commerce to members and was responsible for running the cafeterias set up at the headquarters.

In 1955, there were major strikes and riots all over the country by trade unions and the Youth League, a worker-oriented political organization with a nationalist agenda for an end to colonial rule. Some members of the movement, including Mrs. Cummings-John were also active in the struggle for independence from British colonialism.

The association also forged alliances with women's groups in other parts of Africa and internationally. It became affiliated to the Women's International Democratic Federation based in the Soviet Union and the International Alliance of Women, based in England.[13]

In 1966, an award-giving ceremony was held to recognize the contributions of women. It coincided with a major seminar on women, attended by the Prime Minister, other Ministers of State, dignitaries from the diplomatic community, and religious leaders. "The Women's Movement Honors" were bestowed on women for their various roles as businesswomen, career women, community leaders, and housewives. The following is a transcript of one of the certificates issued:

> The National Council of Sierra Leone Women of the Sierra Leone's Women's Movement offers this certificate of good citizenship on behalf of the women of Sierra Leone. The Sierra Leone Women's Movement hereby confers on Dora Wright the honor of the Rank of Ladyship for long and faithful service in the field of small industries in Sierra Leone. (Signed: Constance Cummings-John, J.P. President. Clarice P. Norman, J.P. Secretary, April 5, 1966)

Political Role

As its status became elevated, the association took on a political role and became involved in the national movement for independence from colonial rule. A few of the leaders, including Mrs. Constance Cummings-John became active members of the Sierra Leone's People's Party (SLPP). The party later formed the first government after independence in 1961.

Mrs. Cummings-John won a seat in Parliament, but lost it after a successful election petition. She eventually became Mayor of Freetown in 1966. When the SLPP lost the 1967 elections, she went into exile in England along with some other SLPP leaders. Since then the activities of the association have declined substantially. Mrs. Cummings-John later returned to Sierra Leone and traveled to England frequently where she died in 2001.[14]

Over the years, the association has continued many of its functions, albeit in a limited capacity and with a lower profile. The nursery school and cafeteria are still in operation. The trading network among women continues to operate as an informal network. Its role in mobilizing women at the grass-roots level for economic reasons was a significant development that inspired subsequent women's associations and promoted a high level of political consciousness among women.

The award-giving ceremony to celebrate women's achievements gave women high visibility and led to programs and activities that contributed to social and economic development. The Women's Movement was also the first association to advocate for greater participation of women in national politics and public life. In many ways it was the first truly African feminist association in Freetown, for it combined the historical struggle against colonialism with the quest for socioeconomic development, independence from colonial rule, and the advancement of women.

Other associations have followed the leadership of the Sierra Leone Women's Movement and have sought to protect women's economic interests. They have functioned as economic pressure groups against foreign encroachment in women's commercial activities. Problems of foreign economic domination are chronic and pertain expressly to the Lebanese and Syrian traders who not only dominate the wholesale trade, but have also been steadily moving in the professions. Some have tried to build good public relations by making periodic donations to charity and to educational institutions, but this has not diminished the general feeling of resentment against them.

Lebanese and Syrian speculators are heavily involved in the mining and trading of diamonds, the country's chief natural resource. The general view is that these groups have consistently exploited the resources of the country at the expense of the people. They are often blamed for charging high prices, maintaining trading monopolies on some goods, hoarding large sums of money, channeling profits outside the country to develop their own countries, and engaging in ruthless and corrupt business deals. During the ten-year war and even before, Lebanese traders were among the most targeted by the rebels, and were often admonished for having exploited the country and drained it of its wealth.

The Lebanese and Syrians have become entrenched as a formidable economic ruling class in Sierra Leone and are often perceived as being linked to the political elite through ties of economic alliance, bribery, and corruption. Women's associations mobilized against the economic exploitation from these groups, serving as pressure groups and raising consciousness among women about the destructive effects of foreign economic exploitation. They take collective action to resist such domination that results in underdevelopment, and to develop constructive women-centered alternatives, following the model of the Women's Movement.

The Gara Thrift and Credit Society and Similar Associations

Following the initiative of the Sierra Leone Women's Movement, a number of women's associations have been formed to protest foreign involvement in trading activities, particularly where these affected women's economic activities adversely. The Gara Thrift and Credit Society was formed to protest the high cost of materials used in the production of gara (tie-dyed fabrics) goods and the involvement of foreign men in trading activities traditionally carried out by women.

These protests were directed not only against the Lebanese, but also against the Marakas (from Guinea) and the Yoruba from Nigeria. The Lebanese retailed the materials needed for gara production at very high prices and the Marakas and Yorubas produced gara fabric, an activity traditionally carried out by local women.

Gara is a specially dyed fabric, similar to tie dye, which has become a popular national fabric. The main items required are imported and are consequently expensive. The traditional dyes, like indigo and cola nuts, are less costly but have certain limitations in terms of their availability in large quantities and the fabric would require careful maintenance. The imported dyes have a number of advantages, including wider variety of colors and better retention of the colors.

The original importing company priced these products reasonably. This changed when the Lebanese took over and inflated the price. Women gara manufacturers were outraged. The Lebanese merchants were viewed as saboteurs. As one woman from the Gara Thrift and Credit Society put it:

> The Lebanese take all the money out of the country and send it to their own country. Most of their children go to school in Lebanon and only come to Sierra Leone on vacation.

Women claim that other countries such as Nigeria, Guinea, and Ghana have taken strong measures to protect their people's interests against Lebanese traders and feel that the Sierra Leone authorities should act in a similar manner.

Another association, the Gara Women's Association lobbied to prevent foreign companies, particularly Lebanese merchants from setting up gara production in Freetown. A delegation was sent to the Minister of Trade and Industry to protest the setting up of these industries that would seriously undercut a vital female industry.

Although the Sierra Leone government took steps to expel the Marakas from time to time, it has not made a definitive move toward seriously controlling the exploitative activities of the Lebanese, nor has any effective assistance been given to women in terms of keeping down production costs. The Sierra Leone Co-operative Department offers loans and limited training but no major help has been provided in terms of reducing the cost of production.

The problem of immigrant traders in competition with women is not limited to Sierra Leone but became a feature of female economic exploitation elsewhere. The following is a resolution adopted by the Economic Commission for Africa as early as 1963:

> Access to the market place should be opened to the women of all African countries and where certain market places are monopolized by men, mostly foreigners, governments and local authorities should not hesitate to break this *de facto* monopoly by means of legal and fiscal measures in favor of indigenous women.

In spite of growing regional and national awareness and action, the problem has worsened due to the general impoverishment of African countries as a whole. It has been aggravated by the competitively driven ideology of privatization and liberalization of trade, which are features of corporate globalization. Most enterprises now face the fierce competition of cheaper goods and foodstuffs being imported from Europe, North America, and Asia. Trade liberalization is driving many small enterprises out of business. These are posing serious barriers to economic development in Sierra Leone. It is no surprise then that resisting the forces of globalization which are resulting in underdevelopment, has been high on the agenda of women's economic associations.

Rotating Credit Associations—Indigenous Mechanisms for Micro-Credit

The rotating credit association is an indigenous mechanism that women use for savings and to buttress their financial vulnerabilities. It operates on the principle of contributions by a group at regular intervals to a collective fund, and withdrawals by individuals in rotation. It is an adaptable institution for mobilizing capital and has wide geographical distribution, particularly in West Africa.[15] It is known in Sierra Leone as Osusu.

Access to credit is a major constraint for women who are self-employed, especially in the informal sector. Credit institutions set up during the colonial period functioned primarily for the benefit of European companies and firms. Although a few African men who were qualified for loans were granted credit, women were subject to gender-biased restrictions endemic in European banking practices. Some of these practices are still continued by a number of European banks that still operate in the country.

Some progress has been made in terms of providing alternatives through micro-credit alternatives, similar to the Grameen Bank model, but these are few. Requirements for collateral and male guarantors still exist in some financial institutions. The requirement for collateral such as a house is particularly difficult for most Sierra Leoneans, both male and female.

Rotating credit associations such as Osusu are the only alternative for most women. The Osusu appeal is both economic and social and rooted in traditional forms of mobilization of capital. On the basis of economic rationality, Osusus provide an opportunity for raising capital without too much red tape. Some Osusus operate a separate emergency fund which allows members to make loans in an emergency. In Freetown, the Osusu is particularly popular among market women and petty traders.

Most Osusu clubs are also social institutions. Meetings have a recreational aspect, a relaxed atmosphere, and a sense of common obligation as members collectively pool resources at regular intervals for each other's benefit. Members often share one or more of the following characteristics: relationship through kinship and friendship ties, working in the same establishment, and living in the same neighborhood. This tends to enhance the relationship of mutual trust and to serve as a deterrent against fraud or default in payment. There is also an inherent social obligation that adds to its continuing appeal. Osusus can only work if all its members cooperate. Geertz's well-known analysis of rotating credit associations reveals the complexity of these associations when viewed from a cross-cultural perspective. He states that:

> In Africa, the tendency is towards the development of more complex leadership patterns and more differentiated internal organization and consequently towards increasing administrative costs.[16]

This is particularly true in situations where associations are organized on a large scale. Most of the Osusu associations in Freetown which serve women are small, and do not constitute an elaborate and complex administration even though they adhere to the basic Osusu principle of regularity in contribution, withdrawal, and rotation.

Most Osusus are not readily apparent since meetings are held in private homes. They usually take place toward the end of the month to coincide with "pay day" and the period when business activities are brisk. The size can range from six members to fifty with the smaller ones operating with little or no formality.

There is greater formality in the larger Osusus, which also tend to have a permanent leader. Meetings last longer because of their recreational aspects but the session devoted to business is usually quite brief. The leader is responsible for conducting the meetings and for recording the transactions to ensure that the proper contributions and allocations are made. At the appointed time and day of each week or month, members meet to "throw in" their subscriptions. The person scheduled to "draw" is given the money collected by the leader. She checks it, signs for it, or has her verbal acceptance recorded and the meeting is over within a short period.

No specific name is attached to an Osusu as a rule and members simply refer to their Osusu as "mi Osusus" or "we Osusus" (my Osusu, our Osusu). It is difficult to estimate the number of Osusus in Freetown but three types can be identified: The formal ones such as the Osusu Club, with membership of 12 or more, the collector type, and the targeted type. Informal ones usually have less than 12 members and are generally made up of kinfolk, friends, and workmates.

Formal Osusus tend to be more structured, have stricter rules, and more elaborate bookkeeping methods. Due to limitation of space, only the Osusu club will be discussed at length as most of the others follow a similar pattern. Significant variations are presented by the collector Osusu and the targeted Osusu, which will also be discussed as variants of the basic Osusu.

The Osusu Club

The Osusu Club, used as a case study, is located in the East end of Freetown. It began with a membership of 17 and grew to 52. It is well organized, successful, and is made up of people who live in the same neighborhood around the Kissy Road and Fourah Bay Road where a major market, the Bombay Street Market, is located.

Some of the members are wage earners but most are traders. Although it is a women's association, three of the members are men: a bank clerk, a third grade clerk in the civil service, and a hospital nurse. The leader is a successful female trader in both wholesale and retail marketing of staple foodstuffs—rice and palm oil—and various household goods. Most of the members retail a range of commodities that include fabrics, groceries, household utensils, and goods. Six of the women are dressmakers and another combines a clerical job with the sale of prepared food, known locally as "cookery."

As the majority of members are in commerce, the money is usually used as capital to expand their business. Two market women who now rent stalls hope to establish proper shops instead of stalls. Three current shop owners used their money for renovations and investments to enable them to market a new commodity. One dressmaker purchased an electric sewing machine. Another woman who sold vegetables embarked on a new and more profitable trade in rice and palm oil. The money was used to make her first wholesale purchase. Some of the members use their capital to purchase expensive household items for their families such as refrigerators, stereos, and television sets and for expenses incurred in the celebration of life cycle events. About one third indicated that their intention was to use the money for the education of their children.

The Osusu club is not the only means of savings used by these women. Twelve of them had personal bank accounts in addition to the Osusu, because of the desire to earn interest, which is not possible in the Osusu system.

Collector's Osusu

A "collector Osusu" is a type of roving bank, a variant of the rotating credit association. In the case studied, the collector, a man, was operating the Osusu primarily for market women. Some of the earliest collectors in Freetown were from Nigeria, which indicates the possibility of it being imported into Freetown. The relationship is between two people, namely the collector and the client. In this particular case, the collector collects money from 150 subscribers, mostly market women who trade in a variety of goods.

The sum of money paid is not fixed, as each woman pays whatever she can afford. The collector records each subscription daily in a special records book. At the end of the month, a subscriber receives the total sum of her contribution, less one day's subscription, which the collector keeps as his commission.

There were no serious problems reported for this particular association. The collector was reputed to be a man of integrity and deposited the money in a bank everyday. However, problems with similar types of collector's Osusu have existed. One collector reputedly absconded to Nigeria with all of the money collected.

In spite of the potential risks involved in this type of Osusu, a large number of market women still use it for convenience. It also imposes a kind of discipline, which obligates members to save regularly. Many women expressed their belief that this made for good business sense. Some indicated that they would not have such regular savings with a bank or the post office, because of lack of time to make a trip to the bank. Others expressed dislike for the impersonal nature of these financial institutions.

Osusu for Mecca (Targeted Osusu)
Another variant is the targeted Osusu, such as the Osusu for Mecca found among Muslims in the East end of Freetown. It is formally organized and the objective is to pay for a trip to Mecca for five of their fifty members each year. Selections are made on the basis of seniority. It has been difficult to sustain this association because of the economic recession and the rebel war.

Osusu—A Socially Valid Savings Institution
The Osusu continues to be an important institution for savings and credit for a large number of women engaged in commercial activities, especially in the informal sector. It is likely that this institution will continue for many years to come since it provides a safety valve for many groups who are marginalized by the formal financial institutions and banks.

Micro-credit has become an important strategy for integrating women in development in international circles, but it is not widely available and has its own limitations. It is also a mechanism for integrating women into the global marketing system, albeit at a low and often exploitative level. Although some women have benefited from it, they have not been able to move to higher levels of economic activity.

Micro-credit does not adequately serve the African woman entrepreneur, if she is poised to leverage larger resources and move into the level of wholesale commercial activities. At the moment the most practical form of mobilizing capital for the majority of women traders and service providers is the Osusu.

Cooperatives

The Cooperative Department, usually located in the Ministry of Trade and Industry, is responsible for promoting the development of cooperatives. It is headed by the

Registrar of Cooperative Societies, who is assisted by a staff of civil servants and clerical workers. It helps groups to organize cooperative societies; improve the economic and social lives of their members; audit their records; conduct educational and training programs for the staff of cooperative societies and supervise the operations of cooperative societies.

Cooperatives in Freetown offer opportunities for developing economic mechanisms that will promote self-help and collective autonomy. They are often developed along commonalities based on economic activity and community ties. They aim to protect the producer from exploitation by ensuring a fair price for their product and to cut down on production costs. According to one official of the cooperative:

> A cooperative offers a practical alternative to the capitalist or profit-motivated system. . . . In a co-operative society, people are the most important asset whereas in private enterprise, money is what comes first. Co-operatives provide protection against monopolies operating at the expense of the public. In any field where a strong co-operative exists, there is less likelihood of exploitation. Through co-operatives, people learn to do things for themselves. They develop the habit of self-help.

The majority of cooperative societies, which number over one thousand, are thrift and credit societies. First established in the 1950s, these societies have savings schemes for members from which they are given loans at very low interest rates, around 2 percent. In 1970, these societies collectively formed the Cooperative Savings and Credit League of Sierra Leone.

Other cooperatives include Marketing Cooperatives, Handicraft Cooperatives, and Fishermen's Cooperatives. The majority of them are based in the rural areas. Cooperatives are run by trustees or committees of management that elect the president, vice-president, treasurer, and secretary.

The Sierra Leone Cooperative Society Movement is organized nationally and internationally through the National Cooperative Congress. The raison d'être for a national movement, was presented by one official as follows:

> The big merchant houses, traders, and banks on which we depend have companies in Freetown which are well financed. These are owned, for the most part, by foreign companies. As long as we are dependent economically we are not really free. To replace these institutions, we cannot just continue to work at the local level. We must organize on a nation-wide basis because this is what our competitors are doing.

One of the ways in which cooperatives develop a strong economic base is to pool their financial resources within a centralized banking system. To achieve this the National Cooperative Development Bank was established in 1970. It is now a Registered Cooperative Society. Members of cooperative societies are encouraged to make regular savings at the bank through their local societies to be used as share capital. Societies with bank membership have voting rights and earn dividends from the bank. It can also take big loans on behalf of its members for a period of one year.

The Cooperative Department has been actively encouraging the development of cooperatives among women since they often constitute substantial segments of the productive and marketing communities in the country. They have also encouraged women's associations to form cooperatives within their associations and often attend their meetings to provide advice and disseminate information.

In addition, the department offers advice and professional expertise on accounting and bookkeeping. Women's cooperatives have been able to secure special concessions on wholesale goods purchased from the Produce Marketing Board for retail by their members.

Marketing women's cooperatives are more common than production cooperatives since they seek to safeguard women's interest in trade; raise capital through the establishment of credit facilities; and provide access to markets. In Freetown, the Congo Market Women's Cooperative that has a membership of 42, and the Women's Cooperative of the All Peoples' Congress Association in a membership of 150 have had some success in improving the economic prospects of their members. They have encouraged regular savings by their members and have been able to secure loans from banks.

The Gara Women's Union is an example of a Marketing Cooperative whose aims are to achieve unity among all Gara women and to improve the gara industry. Haja Dean, the President and Founder explained the need to organize women in this way:

> Because of the difficult economic situation of women in our country, most women are unemployed and depend on petty trading. This makes it difficult for women. Our profits are very small.

The first step taken to ease the problems of trading was to ensure reliable access to a market already saturated with gara goods. The Gara Union consequently rented a shop and established it as the main retail store for the finished goods. Members were entitled to use this outlet for selling their manufactured gara materials. This helped to solve some of the problems of access to markets for their products. In addition to market access, the association also secured rice quotas for its members from the Produce Marketing Board and utilized the credit facilities it offered for securing loans for investments.

Attempts at forming production cooperatives have not been as successful as marketing cooperatives mainly because women in Freetown are more involved in trade than in agricultural and craft production. Even though the gara industry entails a certain degree of manufacturing, the emphasis of the Gara Women's Union is on marketing. Limited agricultural activity is pursued in the peninsula villages around Freetown. Some of the food grown in these areas is for domestic consumption but the bulk of it is sold for cash. These are mostly vegetables, greens, and fruits.

One production cooperative, the Women's Farmers Cooperative was established in 1971 as a subgroup of the National Congress of Sierra Leone Women to enable a number of women with the tradition of farming to participate in this activity on the outskirts of Freetown as well as in the Provinces. Some cooperatives operated by the YWCA in peninsula villages such as Gloucester have been very successful.

There are a few smaller cooperatives run by individual women or by groups of friends, relatives, and neighbors. They are usually not registered with the Cooperative Department and are less formally organized. They tend to proliferate during periods of brisk economic activity but in general have a very short life span, usually about one year or less.

Cooperatives as Mechanisms for Collective Action

Although the idea of group cooperation and collaborative work can be regarded as indigenous, the model of cooperatives is relatively new and perceived by some women as alien. A number of women expressed their preference for the more traditional credit and savings institutions such as the Osusus. As one member of a cooperative remarked:

> At least I know that in my Osusu my turn would come every twelve months because there are only twelve of us. I am still not sure what I have benefited from joining the co-operative. You wait for ages to receive something and when you do, it is only a few worthless Leones [local currency] because the interest has to be shared among all the members.

A number of successful businesswomen also expressed cynicism at joining cooperatives. Many of them have been fairly successful and can function on an individual basis. These women tend to invest some of their profits in the business and can secure conventional loans from banks.

One successful businesswoman expressed her views as follows:

> I have been in business for a long time and at this stage my business is fully organized and doing well. I have a part-time accountant and an auditor. I also do regular banking with the neighborhood branch of Barclays Bank and can raise a bank loan on the strength of my business and my property. I have therefore not felt the need to join a co-operative but I am interested in the idea and think it is a good thing for women.

Cooperatives offer services to women who have difficulties meeting production costs or accessing markets on their own. These are women who also need the security and support of other women in similar economic conditions and can pool their resources to resist the forces of underdevelopment.

One major problem faced by cooperatives centers on controversies resulting from the fact that one individual, usually the President, is able to obtain loans from the cooperative bank in the name of her cooperative society. Another controversial issue concerns the management and allocation of members' quotas of wholesale purchases, particularly rice from the Produce Marketing Board.

Some women view cooperatives run by individual women with suspicion. There are at least two instances in which major accusations were made of embezzlement. In one of the cases, court prosecution was threatened but never pursued. There was a general feeling among the members that some leaders were apt to misuse cooperative credit facilities. Consequently, by far the most common fear expressed by women traders reluctant to join cooperatives was the possibility of the leader usurping the rights of the cooperative. This could include raising credit and using quotas for their own benefit.

Although cooperatives are not well organized and are relatively new to many women, in some ways they offer a new concept of credit and savings. Women may in time come to appreciate them both for their immediate and long-term benefits. It is also likely that production cooperatives, especially for the gara industry, would be more beneficial in reducing production costs. Some women have seen production cooperatives work in other areas and often cite the examples of successful craft production cooperatives among Gara women, particularly in neighboring countries like Guinea.

Women cooperatives reveal a process depicted by the constantly changing patterns of association, disassociation, and re-association. The person most likely to be involved in this process is the woman struggling in commerce with very little or no capital. This woman is typical of most women traders in Freetown. The well-established and successful entrepreneur is likely to follow a more independent path.

Sierra Leone is a developing country with major financial problems of underdevelopment. As a result, most women face economic difficulties and have few options besides trying to eke out a living in the informal sector. Unable to secure credit from financial institutions and access to markets, a woman would likely join a cooperative. She may become disillusioned and would then cease her membership and try to go it alone. In time she is likely to become frustrated and once again associate with another group, usually a smaller one.

The process is likely to go on as long as she remains in the informal sector. This reflects the unstable nature of the national economy, which presents challenges for women in this sector. It also provides a good example of the process of action and reaction characteristic of the tension between promoting development while resisting underdevelopment.

Chapter Five
Collective Action for Educational and Occupational Empowerment

Introduction

Historically, formal education and religion were closely linked in Freetown. This was due in part to the role of the Christianized Krios in promoting Western education and to the activities of Christian missionaries in the eighteenth and nineteenth centuries. Education was an important policy of the Anglican Christian Missionary Society (CMS), which played an important role in the early development of Freetown.

Religion and Western education were important by-products of the period of settlement, and repatriation of freed slaves and "liberated Africans" to Freetown. They were also important in developing an African elite to serve as administrators in the project of colonial penetration of Africa. Ironically, at the same time, this elite was able to use its Western education to challenge colonial rule and organize nationalist movements for independence.

The Church Missionary Society played a seminal role in promoting Western education in Freetown. In time, they were joined by other Christian denominations, as well as the Muslim religion, in establishing schools among all ethnic groups, both within and outside of Freetown. Muslim schools emphasized education in Arabic and in the Islamic religion.

The parochialism of most schools, however, declined as the government increased its management of the educational system. The role of the government in administering and coordinating the educational system, in keeping with standardized requirements for certification, has remained central. This role includes developing an educational policy, of which the Basic Education Reform is apparently the most significant, in terms of promoting human resource development.

Gendered Education

Educational systems have never been perfect as far as gender equality is concerned. It is widely acknowledged in the sociology of education and in women's studies that educational institutions are among the primary agents of socialization of boys and girls into gender roles. For the most part, these roles reinforce the evaluative norms and values given to gender differences.[1]

On the surface, the educational policy of the Church Missionary Society (CMS) was gender neutral. The earliest proposals sent to the British government were for the establishment of educational institutions for "children of both sexes."[2] Nonetheless, gender biases soon became evident in the implementation of the policy. The curriculum for girls gave priority to practical subjects such as home economics and needlework, in keeping with their projected future roles as wives and mothers.

In addition to missionary-sponsored schools, private schools for girls were subsequently opened through the individual efforts of women like Miss Phyllis Hazeley, who had been educated in England.[3] The curriculum for women was expanded to include reading and arithmetic. Despite these changes, women's education continued to lag behind that of men in the twentieth century. For example, in the registers for property transactions and marriage contracts, the majority of women made representative marks in the registers in lieu of proper signatures.[4]

Islamic education was available in Arabic but primarily for boys, although separate but limited classes in Arabic were arranged for girls. Islamic education later became intensified by the Ahmaddiya sect, which promoted a dynamic program reflecting a blend of both Arabic and Western education. Unlike earlier male-focused efforts, it targeted both boys and girls for education in both Arabic and English.

Few opportunities existed for women in the first half of the twentieth century for professional work, regardless of religious affiliation. The kinds of remunerative employment open to women, such as dress making, nursing, and commerce did not require formal education. For those in the more affluent socioeconomic groups, girls were sent to expensive finishing schools in England to maintain their high status and to marry well, rather than for professional reasons. According to one elderly informant:

> It was considered undignified for an educated woman to work. Her education was intended to make her into a lady and nothing more. Women had to be respectable, behave properly, and speak properly. Those who attended school in England were expected to acquire the social graces of ladies and to set an example for others to follow.

When the Church Missionary Society decided to train Africans as missionaries to remedy the difficulties of recruiting European missionaries, it was men who received this training, not women. The first principal of the Christian Institution, established in 1816, was Adjai Crowther. The Christian Institution became known as Fourah Bay College in 1827, and was later elevated to the status of the University of Sierra Leone.

When secondary school education became fully established, it also revealed a male bias. In 1845, the Church Missionary Society opened a Grammar School in Freetown, to provide secondary education for boys. At this point the educational disadvantages of women became even more marked. As a remedial measure Miss Sophia Hehlen, a German C.M.S. missionary, founded The Female Institution in 1849 to "bridge the gap." The remedy, however, was only partly successful, as the school's objectives continued to emphasize the subsidiary role of female education.

Women were educated to become suitable companions for husbands from the Grammar School or the Christian Institution. It was customary at this time for prospective husbands, trained as teachers or missionaries, to request a recommendation for a suitable wife from the principal. In 1865, when its new building was opened,

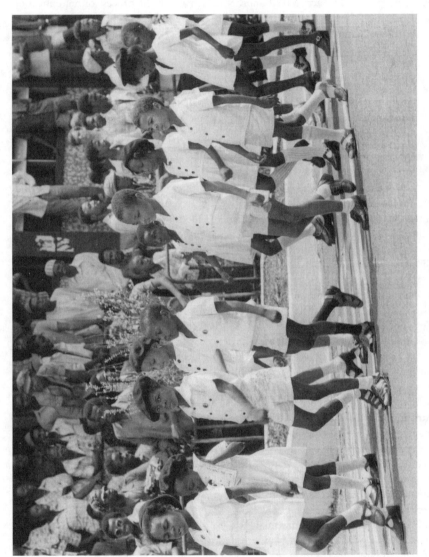

An association's school marching in a Freetown parade. Courtesy: Sierra Leone Information Services.

the Female Institution bore the name Annie Walsh Memorial School, and had as it first African teacher, Miss Kezia Grant who had been educated in England.

By this time it was more acceptable for the well-to-do Krios to send their daughters to England to receive training in the professions of nursing or teaching as well as in the social graces. The account of another informant of the process of her education best illustrates the long-term effects of this early emphasis on the domestic arts in women's education:

> I went to Bethel Infants School, Ebenezer Primary School, and the Annie Walsh Memorial School. I went up to what was then Form II and left at the age of seventeen. Most girls were taught domestic science at school and then went to a seamstress to learn how to sew. The tape measure and scissors were the most important tools for women then. Women were taught canvas work and embroidery. It was also important for young girls to learn how to shop at the market by themselves and to buy sensibly. The market then, was a woman's world. I was married a year after I became apprenticed to a seamstress at the age of eighteen.

The interest in women's education took on renewed significance as opportunities for careers in primary and secondary school teaching became open to women. A teacher-training college was subsequently inaugurated. This new status of women's education helped to open up other career options for their advancement. Mrs. Ingham persuaded the colonial Bishop's Court Fund to give up part of the large Bishop's Court grounds for a cottage hospital where young women could be trained in nursing, which was as yet an unqualified profession in Sierra Leone.[5]

She raised funds for the hospital and in 1872 the Princess Christian Cottage Hospital was opened. The staff consisted of one doctor and three nurses from England. This hospital served the poor in the largely Muslim neighborhood as a mission hospital. It also ran a nurses' training program.

As various missions became established in Freetown other schools were founded for boys as well as for girls. These included the Methodist Boys' High School and its counterpart—the Methodist Girls' High School. The Saint Edwards Boys Secondary School (Roman Catholic) had a female equivalent—Saint Joseph's Convent. The African Methodist Episcopal mission, headed by the Reverend Henry Metcalfe Steady, inaugurated the Girls' Industrial School, as the first vocational school in Freetown.

The influence of female missionaries on the lives of women of Freetown was quite marked. Most of those who went to missionary schools, especially to the Annie Walsh Memorial School, revealed that they emulated their missionary teachers, and a few of them at one time had aspirations of becoming missionaries themselves. This partly explains the popularity of the teaching profession among women, and the philanthropic objectives of several Christian women's associations. The system of "prefects" introduced at the Annie Walsh by Miss Hamlette[6] encouraged the development of leadership, which was later utilized for organizing women's associations.

Contributions of Individual Women to Female Education

Individual women also made outstanding contributions to the education of girls both in the founding of schools and in their successful operation. Among the most outstanding was Mrs. Lydia Reuben Johnson, who founded the coeducational Reuben Johnson Memorial School in 1898. Others include Mrs. Rice (nee Dove) and Mrs. Adelaide Caseley-Hayford, who founded the Girls' Vocational School at Gloucester Street.[7]

Mrs. Hannah Benka-Coker became a legend for her outstanding contribution to female education in Sierra Leone, especially in relation to the Freetown Secondary School for Girls. The names of Constance Cummings-John, founder of the Roosevelt School, Lati Hyde-Forster, former principal of the Annie Walsh Memorial School, and Hajah Salimatu Sesay of the Kankalay School, are associated with female education in a most salutary way.

Freetown schools were pioneer schools and attracted a large number of students from West Africa, particularly from Nigeria and Fernando Po from the nineteenth to the mid-twentieth centuries. The Annie Walsh Memorial School has alumnae from several West African countries, some of whom still attend the annual reunion in December.

Despite the achievements of women in the field of education, several obstacles had to be surmounted, due to gender-based discrimination in the formal educational system. As Lati Hyde-Forster, the first Sierra Leonean female college graduate once recalled to me:

> I was a real pioneer and it took courage because society then was against women receiving higher education. I will never forget my first year at college. This was in 1934 when I was the only female student. I was 23 years old then. The authorities thought it safer for me to share a table with the theologians. On one occasion during service at the cathedral church, a man remarked that I had gone to College to lead a loose life. This was the kind of attitude I was up against. I did not receive much support from society but my father supported me morally and financially, and encouraged me. He was the greatest influence in my life. He was a minister of religion and a barrister, and always discussed intellectual subjects with me from an early age.

Education and Female Professional Advancement

In the current labor market, formal education is a valued asset that greatly increases one's chances for gainful employment and social mobility. It also promotes a system of meritocracy that has been essential for democratization and development. For women, education enhances the opportunity for achieving economic independence. In a society where highly skilled labor is in short supply, gender-based barriers to social mobility have had to become more constrained. As a result, women who receive advanced education have been able to achieve professional mobility.

It is not unusual for women to hold high-level positions in medicine, education, law, the military, and so forth. For example, the chief medical officer, as well as the chief education officer, and the head of the social welfare services have usually been women. In addition, women have served as cabinet ministers, members of Parliament, high court judges, colonels in the military, architects, engineers, professors, managers, and ambassadors.

Their contributions have also been significant at the level of administration. During the decades of the 1980s and 1990s some of the highest professional posts in education were held by women. Among these were Oredola Fewry as Permanent Secretary in the ministry of education and Christiana Thorpe as Cabinet Minister of Education. Both were highly regarded for their significant achievements in the field of education.

Educational Challenges for Women

Despite these achievements, the record for the majority of women in Sierra Leone still leaves much to be desired. The illiteracy rate for the population as a whole, based on figures from the *Human Development Report*, can be as high as 80 percent in the rural areas.[8] As most schools are fee-paying, formal education is expensive. Although educational opportunities increased soon after independence, the government has not been able to provide adequate educational opportunities and facilities for many years. Among the reasons for this failure are the protracted recession and the ten-year rebel war from 1991 to 2001.

Educational deficiencies are a reflection of the pattern of underdevelopment prevalent in many countries of Africa. In addition, policies operating at the international level have resulted in major setbacks to educational opportunities. Among these are Structural Adjustment Programs (SAPs), which impose conditionalities that include cutbacks in the social sector, such as health and education. The impact of SAPs on women's education in Africa as a whole has been profound.

According to UNESCO and the United Nations, the gender gap is closing at the primary and secondary school levels in almost all regions of the world, except in South Asia and sub-Saharan Africa where it appears to be stalling or widening (table 5.1).[9] This is despite the relatively high degree of resource allocation to public education by sub-Saharan countries compared to other regions, as percentage of GDP (table 5.2). In Sierra Leone, gender disparities in education are significant at all levels (table 5.3).

This is partly due to economic adjustment or cost shifting for education from governments to families as a result of SAPs. At the same time, SAPs are also imposing retrenchment in the public sector, the largest employer, especially of women.

Table 5.1 Enrolment rates at basic levels of education in developing countries

Region	Net enrollment rate		Share of girls in primary school %	
	Male (%)	*Female (%)*	*1990*	*1995*
Sub-Saharan Africa	65.7	56.1	45.6	45.1
Arab states	78.6	69.7	45.4	43.6
South Asia	75.6	55.8	42.2	42.1
Latin America/Caribbean	91.9	85.7	49.7	48.6
East Asia/Pacific	91.1	82.1	47.6	47.1

Source: UNESCO 1996: Working Documents, Conference on Empowerment of Women through Functional Literacy and the Education of the Girl Child, Kampala, September 8–13, 1996.

Table 5.2 Unit costs of public education at the various levels as a percentage of per capita GNP in selected country groups

Country group	Primary	Secondary	Higher
Sub-Saharan Africa	15	62	800
Francophone	23	86	1000
Anglophone	12	51	600
South East Asia & Pacific	11	20	118
South Asia	8	18	119
Latin America	9	26	88
All developing countries	14	41	370
Industrial countries	22	24	49

Source: Appendix tables, Mingnat and Psacharopoulos (1985). World Bank Policy Study on sub-Saharan Africa, 1988, p. 75.

Table 5.3 Education in Sierra Leone

(A) Projected Enrollment (000)

Years	Primary			Secondary		
	Male	Female	Total	Male	Female	Total
1992/93	340.7	239.7	580.4	328.1	195.6	523.6
1993/94	350.5	246.6	597.1	340.3	103.1	543.4
1994/95	360.2	253.5	613.7	353.1	210.8	563.9
1995/96	370.6	260.7	631.4	366.4	218.9	585.3
1996/97	381.6	268.5	650.2	380.2	227.3	607.4
1997/98	393.1	276.6	669.7	394.5	236.1	630.6
1998/99	404.8	284.8	689.6	409.4	245.1	654.5
1999/00	418.6	294.5	713.1	424.9	254.5	679.4
2000/01	428.3	301.3	729.6	434.7	260.4	695.1
2001/02	438.2	308.3	746.5	444.7	266.4	695.1
2002/03	448.3	315.4	763.7	455.1	272.6	727.7
2003/04	458.7	322.7	781.4	465.5	278.9	744.3
2004/05	469.3	330.2	799.5	476.2	285.3	761.5

Source: Planning Division, Ministry of Youth, Education and Sports.

(B) Undergraduate courses, University of Sierra Leone*

	1992	1993	1994	1995
New admissions				
Total	438	490	383	398
Males	331	336	300	320
Females	107	124	83	78
Full time students				
Total	1566	1626	1369	1463
Males	1216	1303	1076	1163
Females	350	323	293	293

Note: * Data for Fourah Bay College only.

Source: University of Sierra Leone.
Sierra Leone in Figures 1998—Central Statistics Office—Sierra Leone Web.

An old girls (alumnae) association Thanksgiving Service. Courtesy: Sierra Leone Information Services.

This results in the elimination of incomes and the destruction of the educational and employment prospects of many young people, especially women and girls.

Africa shows a trend in university education for both men and women that is the reverse in many parts of the world. For every 1,000 women, only two will attend university and for every 1,000 men, only four will. Male enrollment is dominant in the fields of science, engineering, and agriculture although there has been a slight improvement for women. As can be expected, women are also under-represented in teaching at the higher levels and at decision-making levels. Only 5 percent of the heads of universities in Africa are women.[10]

Among the reasons for the lingering gender gap in education is the fact that the education system introduced during the colonial era was inherently biased in favor of men. Its limited goal was to train a few men for administrative positions in the colonial government and for providing the services necessary to reproduce and maintain the system.

Before the introduction of government subsidies, payment of fees was required, thereby limiting education to those who could afford to pay. Access to education for girls is even more limited based on a number of factors. These include preference for boys' education as a safer investment, the domestic responsibilities of girls, pregnancy, and early marriage. Attrition affects both boys and girls. However, whereas economic reasons are largely responsible for boys dropping out, social, cultural, and biological reasons are additional factors that feature more prominently among the reasons for girls abandoning their education and restricting their prospects for employment.[11]

In the field of technical education, the enrollment of girls is only 0.5 percent, a factor that results in a low level of female participation in the technical fields. While noting the absence of courses on technical drawing, woodwork, and metal work in girls' schools, interviews with educators suggest that lack of information about these courses is a key factor in explaining low female participation in these areas.

In commercial subjects and Home Economics women are in the majority, while men are highly represented in the scientific and technical fields. This suggests that while women are not barred from enrolling in technical subjects, the pattern of course selection does reflect a strong influence of gender stereotypes in education. This pattern is also evident at the level of higher education where the enrollment of female students in the field of engineering at the University has been much lower than that of male students.

Shortage of educational facilities is a serious constraint to the education of boys as well as girls. The problem is made worse by the heavy concentration of educational facilities in Freetown resulting in large-scale rural to urban migration of young people. Consequently, Freetown schools are overcrowded and chronic shortage of space, facilities, and teachers is the norm. The problem is further exacerbated by the strict educational requirement for entrance to secondary schools. This creates a bottleneck at the entry level of secondary schools and severely restricts educational mobility. As a result, the establishment of additional schools by individual women as well as by women's associations has become necessary.

The Educational Role of Women's Associations

The contribution of women's associations to education particularly to female education, and to human resource development, has been significant at several levels. Some notable examples will be discussed later. In the meantime, their contributions can be summarized as follows:

In the first place, most women's associations, regardless of their objectives, raise funds and operate a scholarship fund for the education of girls. Second, associations such as the ex-pupils associations support their Alma Maters directly and also provide scholarships to students. Third, guilds and craft associations offer training courses and apprenticeships in addition to promoting universal standards and providing professional support. Fourth, many programs designed to promote adult literacy have been run by women's associations.

Fifth, vocational education has always been given a lower priority compared to education in academic subjects. Women's associations, to a large extent, have helped to fulfill some of the gaps. Sixth, due to women's multiple roles in society, women's associations have made possible, a more holistic approach that include socialization, personal growth, and education for a culture of peace. Finally, the government does not allocate enough resources to education, due to other priorities, such as the military, protracted recession, and SAPs, leaving a gap that is often filled by women's associations.

Vocational Education

The pattern of the system of education followed was the British grammar school model, with strong emphasis on academic subjects primarily for male students. The development of vocational and technical education was largely ignored until after independence. One of the earliest attempts at reforming this model was provided by the first vocational school for girls, the African Methodist Episcopal Girls' Industrial School. It was founded in 1924 by the African Methodist Episcopal (AME), an African-American church, headed in Freetown by the Reverend Henry Metcalfe Steady.

Mrs. Constance Cummings-John became its principal, shortly after her return to Freetown from England in 1937. In her autobiography, she remarked on the financial and other difficulties experienced by the school including the use of outdated equipment, lack of proper instructional material, and few teachers with the proper teaching credentials.

Some of the challenges were met through the support of individuals. One was Mr. Babington Johnson, a local businessman who provided support to the school as landlord, by charging negligible rent. Under the leadership of Mrs. Cummings-John, it expanded its curriculum from home economics and literacy to industrial courses. Students were taught how to make cooking stoves and furniture, and also learnt weaving techniques.

The emphasis was on self-reliance. A number of fund-raising activities were launched with student participation, resulting in the construction of a new domestic science building with modern equipment.[12] In 1952, the Roosevelt Preparatory School for Girls, named after Eleanor Roosevelt, expanded on this pattern of providing vocational education, combined with academic subjects, on a larger scale. It also emphasized technical work, commercial subjects, and adult education. Once again, Mrs. Cummings-John's role was significant, as she was the founder of this school.

The school's motto, "There is dignity in labour" was expressed in the school song and illustrates the priority given to development, self-reliance, and citizenry as follows:

> Why do we come to school today?
> To make our lives worth living.
> Why all these books, why all these rules?
> They're just to keep us going.
> To make our nation strong and great,
> To draw out what is best in us,
> To work for God and country.
>
> We see in labour dignity,
> We hold our heads above board,
> We struggle on for unity
> And happiness in this world.
> With hearts and hands and heads at work,
> We pledge ourselves to help our race,
> An endless moving team work.
>
> And when our school days are over,
> We pass the baton over
> To others who will take our place
> With dignity and pleasure
> To those who will continue the race.
> It's ours today, it's theirs to be
> A never-ending teamwork.[13]

The national educational policy was subsequently revised in response to criticism of the grammar school model, which placed too much emphasis on academic subjects. Some schools now include agricultural, technical, and commercial subjects in their curriculum. The Sierra Leone Technical Institute has been playing a leading role in promoting vocational education on a coeducational basis for many years. Two associations have been exemplary in promoting female vocational education, namely the Young Women's Christian Association and the Federation of African Women Educationalists (FAWE).

The Young Women's Christian Association (YWCA)

The Young Women's Christian Association (YWCA) is a worldwide association. Its aims are "to bring women and girls of different traditions into a world-wide fellowship through which they grow as Christians, participating in the life and worship of their church and expressing their faith by word and deed," while maintaining the structural continuity of these aims.

Over the years, the association has demonstrated great flexibility in its program, enabling it to be responsive to changing social conditions and the needs of the modern woman in each generation. It has become an association whose philosophy and aims are centered on promoting education, particularly vocational education, good citizenship, self-improvement, and the advancement of women.

It seeks to improve the status of women through the elimination of stereotypes and the rejection of rules that lead to gender-based discrimination. These stereotypes also present obstacles to the formulation of policies for a new role for women through education. Consequently, as much as 80 percent of its funds are channeled into projects designed for promoting education.

A major recipient of funding for education has been its Vocational Institute, which was founded on September 20, 1961. It is located at the headquarters building in Brookfields. In accordance with its constitution, it is governed by a board of directors made up of representatives from the fields of education, industry, commerce, and the government.

The Institute was established as a self-help project by a few members who, as teachers and educators, felt that the need for vocational education in Sierra Leone was not being met. The Institute was funded mainly through donations and fees. Later, small donations were also received from church groups in various countries including Switzerland, Australia, New Zealand, and the United States. The World YWCA also made some financial contributions, as did local businesses and industries.

In 1962, the Swedish politician and Human Rights activist Inga Thorssen visited Sierra Leone and was impressed by the work of the YWCA women. Realizing their financial constraints she made a request for aid from her government in her capacity then as a Member of Parliament. Under the Swedish Aid to Developing Countries Program a grant was made to the Institute. The rest of the money came from local sources, other countries, and the World YWCA.

The Sierra Leone government provided all the technical experts for the building and agreed to pay 95 percent of the teachers' salaries. The International Labor Organization provided a consultant on vocational education and the World YWCA took on the responsibility of training the staff for the Institute.

A modern building that accommodates about 500 students was constructed. In 1971, the total number of students was 570 and staff members numbered 14. By 1974, the number of students increased to 640 and an additional part-time staff increased by four. In 1982, the students numbered 800 and staff members increased by eight. In 1996, the enrollment was estimated at 950 and continued to grow. In the year 2000, full-time enrollment was estimated at 1500.

Although training is given primarily in vocational education, and earlier emphasis has traditionally been on Home Economics, increasing attention is being given to maintaining a high standard in academic subjects as well, particularly English and Arithmetic. Social Studies is also offered in the majority of classes. Girls are generally trained according to their aptitude and interests in one or two of the following subjects: Business Education, sales assistant studies, clothing construction, Home Economics, and arts and crafts.

The first two years are devoted to pre-vocational courses that give a basic general vocational education. Subjects include Home-Making, First Aid, Home Nursing, Arts and Crafts. During the third year, opportunities are provided for exploring training in some specific vocational subjects as well as providing general secondary education. In the fourth and fifth years, students are prepared for national examinations and similar mechanisms of certification.

In addition to these formal classes, special attention is given to extra-curricular activities, which include physical education, music, and discussions on current affairs. Evening classes are offered for working women to upgrade their skills.

Day-release courses are given in business studies, clothing construction, catering, home management, and retail salesmanship.

Other educational activities include the development classes formerly called "Improvement Classes" for girls who have dropped out of school. The objectives are to improve their basic education, teach skills for employment, develop good working habits, improve personal health, and build good moral character.

Financial constraints, poor academic performance, and pregnancy are among the reasons for school dropout which tends to occur in the middle of the student's school career. As a result, sex education was introduced into the curriculum. It is hoped that by attending these classes for two years and receiving a certificate of participation these girls would be able to earn an income or enroll in one of the continuing education programs in the country.

The objective of the YWCA's education program is to develop and train the whole person, not only in terms of acquiring skills, but also in relation to the development of good character and the formation of a positive attitude toward work. Grooming, personal hygiene, poise, posture, and physical education are also essential aspects of the curriculum as is work etiquette and job performance. It is hoped that this would ameliorate the problem of idleness characteristic of girls who drop out of school, as well as unemployed girls.

Due to resource constraints, it has been difficult to maintain such a comprehensive curriculum. A narrower focus has been more appropriate. This consists of nutritional subjects and catering, family life and sex education, needlework and sewing, hairdressing, and handicrafts. In order to discourage idleness and dissipation of youthful energy, the association has also incorporated recreational activities for the students.

It has also worked with other associations on combating social problems among youth, such as alcohol abuse, drug abuse, and pregnancy. Other educational activities have included residential camps during the long holidays, a "summer school" for children under ten, nursery schools, and several recreational programs designed to meet the needs of various age groups.

Adult literacy is another important activity. This includes nutritional projects to demonstrate the benefits of proper nourishment and hygiene. Most village branches have adult literacy classes and a few operate play centers for preschool children. Some village branches tend to concentrate on developing agricultural and technical skills. Instruction is sometimes provided in post-harvest conservation, food preservation methods, small-scale manufacturing, small business management, and bookkeeping.

In the village of Gloucester, a successful vegetable garden project was in operation during the prewar years in conjunction with the Gloucester Village Development Association.[14] Two-thirds of the food grown is marketed, and association members and their families consume a third of the food. Resource constraints often affect the buying of fertilizers and seeds and hiring of male gardeners for the heavier work of clearing and preparing the beds for planting. As a result, the women had to do most of the work themselves. The village branch has other projects, which include a water project for providing taps for schools and a woodwork project for constructing large tables to be used as stalls in the central market in Freetown.

In order to promote most of its activities, the national YWCA has had to develop a nationwide consciousness. This has often involved the public in some of its projects through the use of the mass media for the dissemination of information, in the hope that an enlightened public opinion would stimulate voluntary social action on development-related projects. Financial constraints continue to plague the association, made worse by the decade-long war. It is also facing difficulties in recruiting young members. Despite this fact, it forges on and continues to hold an important position among women's associations in Freetown.

Kankalay

Kankalay, meaning "unity," is a Muslim association which started with the objective of providing mutual aid and promoting the ideals of Islam. Although it will be discussed later under mutual-aid associations, its objectives included promoting education. This began in response to the shortage of schools and the growing concern over the high dropout rate of Muslim girls. The Kissy branch embarked on a plan to build an educational institute as a self-help project in order to promote the education of these girls.

Through the lobbying efforts of Muslim women leaders, especially "Hajas," the government, in 1973, donated a piece of land at Kissy to Kankalay for the construction of a secondary school for girls. The cost of the construction of the school was met primarily through fund-raising activities such as luncheon sales, bazaars, and concerts. Contributions were solicited from government ministers, merchants, and private citizens. Fund raising was also extended to Muslim associations overseas, which resulted in a generous donation from Saudi Arabia.

Admission to secondary schools, as a rule, is limited to those who pass the national entrance examinations to secondary schools. Girls from the Kissy area are generally at a disadvantage because of their generally low socioeconomic position that tends to limit their academic performance. Even those who make it through the first hurdle of gaining entrance and do start their studies may not complete their secondary school education because of lack of funds or other reasons.

The Kankalay school is primarily for Muslim girls who have not passed the entrance examinations or who have dropped out of school. The school provides education for about five hundred students in both Arabic and English at the secondary school level. The emphasis is on vocational education, although academic subjects are regularly and extensively taught.

Kankalay leaders believe that the best foundation for the future of children lies in a sound education. This need was considered to be even stronger for Muslim women who by convention and custom had a later start in receiving formal education.

Forum for African Women Educationalists (FAWE)—Sierra Leone Chapter

One of the most effective associations promoting women's education is FAWE. It is a pan-African association with a chapter in Sierra Leone, established in 1995 and headed by former cabinet minister for education, Christiana Thorpe. Its overall objective is to provide mutual assistance and collaboration in developing national capacity to accelerate the participation of girls and women in education at all levels. This is in support of the universal goal of education for all agreed to by the international community under the auspices of the United Nations Educational, Cultural and Scientific Organization (UNESCO).

FAWE also aims to build public awareness through the media about the social and economic advantages of female education. It hopes to support women and girls in acquiring education for socioeconomic, human resource development as well as for personal growth. In this regard, it has played a leadership role in advocating for an end to the destructive effects of the ten-year rebel war and for peace.

Among its priorities is providing rehabilitation for women and girls who have been victims of the war. Its activities also include the establishment of endowment funds and camp schools for children displaced by the war, and management of the FAWE School for Girls, which consists of 350 primary school students, most of whom are displaced girls. It offers training and helps find employment for girls and young women. According to one of the members:

> The education we are giving to women and girls will contribute, in no small measure, to the rehabilitation of their minds. We do help out with their fees, uniforms, books, and other useful gifts that go toward improving their standard of life. This puts them in a mood of readiness for their final repatriation to their various homes and communities.[15]

Education for a Culture of Peace
FAWE's activities to provide education for a culture of peace through workshops and the media for all levels is a way of ensuring sustainable peace and upholding justice, human rights, and tolerance. This is essential since the devastating effects of war are long lasting and do not cease with the end of hostilities or the signing of peace treaties. Many people are left bereaved, wounded, orphaned, traumatized, or made refugees. FAWE ensures that this type of education includes elements of conflict resolution, reduction of prejudice, re-evaluation of value systems, attitudes, institutional practices, and respect for diversity.

According to the president of FAWE, Christiana Thorpe, the whole population must be involved in moving from a culture of violence to a culture of peace:

> Although conflict is gender neutral, the most vulnerable groups in conflict situations are women and children. The women are raped, widowed, displaced, made refugees, while children are maimed, abused, orphaned and unaccompanied. These two groups tend to bear the brunt of the conflicts. During times of armed conflict and the collapse of communities, the role of women is crucial. They often work to preserve social order. Unfortunately, their concerns are hardly ever focused on the post-conflict rehabilitation initiatives.[16]

A case study of women's traditional mediation and conflict resolution practices carried out in Sierra Leone in 2000, by FAWE, proved this to be true. It revealed that only 19 percent of women responding felt that women form a part of the structure for mediating settlement of disputes; 81 percent stated that women are excluded from the structures. The study also found a number of obstacles to female participation in mediating the resolution of conflicts, which included, time constraints, gender bias, fear, indifference, and professional and social limitations. The obstacles to women's participation were explained as being due to women's multiple gender roles, which create pressures and make it impossible for their participation in mediation exercises. The study made the following conclusions:

> Women make important, but often unrecognized, contributions as peace educators both in their families and in their societies. Thus in post-conflict rehabilitation efforts, an active and visible policy of mainstreaming a gender perspective into all policies and programs should be promoted so that, before decisions are taken, an analysis is made of the effects on women and men, boys and girls respectively. Women must be part and parcel of the peace building process. They are not only stakeholders, but their nature and disposition generally make them natural peacemakers. They are the mothers of both victims and perpetrators, and in times of crisis they often offer solace. Thus the women's participation in peace building or conflict management is essential, crucial and imperative. But to do so women's capabilities and capacities must be strengthened.[17]

Reconciliation is another important objective of FAWE that is built into its educational program. Advocating for forgiveness by both parties is considered necessary, as it is only by forgiving the perpetrators that the victims will be able to start rebuilding their lives and move on. According to FAWE, this is because forgiving is a private and personal affair. It does not need any input from outside. The individual who is hurt needs to unburden himself/herself and be healed of the hurt, so that it does not inhibit their forging ahead or living a meaningful life once more. Thus, one can forgive without reconciling. On the other hand, reconciliation has to be mutual. The victim and the perpetrator decide to forgive each other and continue as friends or good neighbors. So forgiveness is a prerequisite for reconciliation.[18]

FAWE insists that education for a culture of peace would also include a review of the value system. The essential elements for peace includes respect for human rights, freedom, and responsibilities; respect for human life and dignity, public and private property; the democratic and religious aspirations of others; acceptance of the need for equity in economic, social, and political development and appreciation of the value of human solidarity, good governance, and justice.

The association insists that the curriculum of both formal and nonformal education programs must comprise these values, so that they are taught to and imbibed by students, from preschool to tertiary levels. According to FAWE, any initiative for education for a culture of peace should ensure the promotion of good governance as a climate for preventing conflict and maintaining peace and the building of strong foundations of democracy, where universal respect for human rights and freedoms would be observed and guaranteed. Also of importance is tolerance. Above all, sustainable and equitable development, which stresses economic, political, social, and cultural development should be emphasized. Concern for the disadvantaged and marginalized who are often among the most affected victims of war and oppression should be central to any strategy for peace.[19]

FAWE's relationship with the government has been collaborative, especially in the implementation of policies for the education of girls and women. It regards its role in helping to develop the country as particularly important, especially in the postwar rehabilitation era. It considers its educational role with displaced girls as essential in reducing the number of refugees and displaced people roaming the streets as beggars.

The major problem faced by the association is finance, the shortage of which often prevents it from carrying out many of its activities.

Guilds, Professional Associations, and Alumnae Associations

A number of professional associations have membership of both genders, but there are some that are exclusive to women. Some of these are affiliated to international associations, such as the International Federation of Business and Professional Women, and reflect the philosophy and objectives of these associations as well as their own. The objectives usually include raising professional standards, protecting the economic and general welfare of their members, improving working conditions, and protecting the rights of workers. Associations in this category also include the Sierra Leone Nurses Association, the Sierra Leone Home Economics Association, Zonta International, and the Sierra Leone Association of University Women.

Other activities have included services to the community; fund raising for a particular project or scholarship; running workshops as part of extension programs; setting up nursery schools and vocational institutes; helping rural women with rural projects; participating in literacy campaigns; and helping with the rehabilitation of female victims of war.

In addition, the major secondary schools for girls have active alumnae associations, both at home and abroad, that raise funds for the schools. They are usually referred to as "Old Girls Associations." For Freetown, these include the Annie Walsh Memorial School, the Freetown Secondary School for Girls, the Methodist Girls High School, and the St. Joseph's Convent Secondary School for Girls.

Female Education and Socioeconomic Development

Women's associations serve as indispensable educational institutions and resources promoting female education, which is an essential aspect of socioeconomic development. In so doing, they help to rectify some of the serious deficiencies in the educational system and in the shortage of educational facilities. By providing both formal and nonformal education, they expand educational opportunities in holistic ways to include vocational education, skills training, and personal education and cater for the more varied demands of both the formal and informal labor market. Non-formal education has also been a significant, but often under-reported aspect of women's education in Sierra Leone.

The relationship between formal education and religion is complex, when one considers its possible gender implications. Because of the close relationship between education and religion in Sierra Leone education has tended to reinforce the patriarchal values inherent in most religions. On the other hand, it provided women with an opportunity to acquire knowledge and skills that were essential for participating in the Western-oriented colonial and postcolonial countries. This was evident in the early curriculum, which tended to confine women to domestic-oriented subjects and to reinforce the public/private dichotomy extant in Western-type nuclear family structures that were being reinforced by Christianity.

The last twenty years have witnessed a major paradigm shift in development thinking. The overemphasis on economic growth has given way to a more human-centered approach, encompassing concerns about equity, social development, and ecological sustainability. Quality of life indicators such as life expectancy, literacy, infant mortality, income, political stability, the status of women, and human rights have now assumed center stage. Equally important is the increasing emphasis being given to gender equality and the advancement of women in human resource development.

Human Resource Development

Human resource development is not new to Africa. It was an intrinsic aspect of indigenous African cultures, providing knowledge, skills, a normative design for living, and a supportive social infrastructure. These were necessary to ensure economic production, social reproduction, and human well-being. In the policy realm also, human resource development has been part of the development strategy of all African countries for more than twenty years.

In 1980, the Lagos Plan of Action, a policy document developed for the African region, identified human resource development as a priority for Africa. In this millennium, the New Partnership for African Development (NEPAD) ranked human resource development highly in its strategy to eradicate poverty and reclaim the policy initiative for African development.

The gender dimension of human resource development gained prominence in the 1985 Arusha Strategies for the Advancement of Women. It stated that "policy makers and planners must recognize that talent is not distributed by sex and that any policy that closes off from development and fails to use the largest part of the national pool of talent is suicidal."[20] In 1990, the African Development Bank (ADB), warned against neglecting women's potential:

> The continent cannot begin to solve its development problems until policies and mechanisms are established to remove barriers inhibiting the maximization of women's development efforts and the realization by women of adequate returns from their labor.[21]

Most countries in Africa accept, in principle, the idea of universal primary education but are far from implementing it for both women and men. Some like Seychelles, Botswana, and Zimbabwe have gone very far in achieving it. Yet, despite increasing improvements in the enrollment of girls in primary schools, boys continue to fare better at the secondary and tertiary levels. University enrollment for women in Africa stands at 25 percent that of men. Sociocultural constraints (which include domestic responsibilities, early marriage, and preference for boys' education) have resulted in the disinvestments in girls.

Furthermore, policies related to globalization, such as privatization and liberalization, are taking their toll on education in general and on female education in particular. Among the many challenges to female education in Africa are SAPs and the debt burden. SAPs, which mandate the removal of subsidies for education and health, have struck at the heart of the social sector, a vital sector for women.

Scholars have repeatedly pointed to the need to identify and eliminate barriers and bottlenecks that prevent women from equal educational opportunities, especially at the higher level of education and at universities.[22] A study by Beoku-Betts concludes that:

> Educational institutions are not structured to be gender neutral, but are, in fact, designed to reproduce conventional gender identities of masculinity and femininity.[23]

Change in this direction is important because women provide the main building blocks for human resource development, which begins with the physical well-being and security of the human being. There is much truth to the saying that "when you educate a woman, you educate a nation." It is clear from the foregoing that, in the building of human resources and the human capital of women in Sierra Leone, women's associations have played and continue to play a pivotal role.

In summary, women's associations serve as indispensable educational institutions and resources promoting female education, which is an essential aspect of socioeconomic development. In so doing, they help to rectify some of the serious deficiencies in the educational system and in the shortage of educational facilities. By providing both formal and nonformal education, they expand educational opportunities in holistic ways to include vocational education, skills training, and personal education.

They also cater to the more varied demands of both the formal and informal labor market. Nonformal education has been a significant but often underreported aspect of women's education in Sierra Leone and has been promoted primarily by women's associations.

The relationship between formal education and religion is complex, when one considers the possible gender implications. On the one hand, both Christianity and Islam reinforce patriarchal values. Some of the gender biases were reflected in the curriculum that tended to confine women to domestically oriented subjects, and to reinforce the public/private dichotomy, extant in Western-type nuclear family structures and in the Sharia. On the other hand, both religions incorporate democratic principles and values of equality and justice that, in theory, should promote gender equality.

The role played by traditional women's associations, especially secret societies, in promoting nonformal education has been exemplary and will be discussed next.

Chapter Six
"Traditional" Associations

Secret Societies

This chapter examines what is being termed "traditional" associations, because historically, they operated in indigenous African cultural contexts, and have their origins in essentially self-contained ethnic communities. The focus is primarily on secret societies which, though operating in Freetown, have their origins and most intensive manifestations outside of Freetown. Secret societies are important because they can serve as mechanisms for the mobilization of women for development and democratization. They also represent female power bases that have come into play in resisting underdevelopment and authoritarianism.

Secret societies are among the best examples of women's associations serving as cultural mechanisms for socialization, nonformal education, and mutual support. As regulatory religious institutions, they cut across kinship ties. Membership is obligatory and confirmed by an oath of secrecy. They are significant enough to warrant some general background discussion.[1]

In the majority of cases, traditional secret societies restrict their membership to one gender but may make exceptions for special honorary and hereditary positions. Among groups with a bilateral kinship system and in secret societies that are more specialized in function, membership can cut across gender lines. For example, the Sherbro, who technically speaking are a bilateral group, have had societies whose membership is made up of both men and women. Examples of these include the Thoma, Tuntu, and Yassi that are also specialized as medical societies. For societies that guard the secrets to traditional medicine, membership is often hereditary or determined by a long period of training rather than on gender.

Another example of a society with both male and female membership is the Humoi society among the Mende. This is a specialized society that regulates sexual relations through a series of taboos and prohibitions, the infringement of which is believed to lead to illness. Among the bilateral Krios based in Freetown, both Christians and Muslims belong to Hunting secret societies imported from the Yoruba of Nigeria. These societies have a predominantly male membership but women are increasingly becoming members.

Other societies that restrict membership to one gender include Masonic Lodges who are imported from Europe and popular among the professional groups in

Freetown, especially the Krios.[2] Examples of other societies with exclusively male membership tend to have a political thrust and may incorporate highly specialized military functions. Examples of these are the Wunde among the Mende and Temne, which is largely responsible for military affairs, and the Ragbenle, found among the Temne.

Poro and Sande

The oldest and best-known secret societies are the Poro (male) and the Sande/Bondo (female) initiation societies widespread among several ethnic groups in Sierra Leone and Liberia. These are religious associations with important educational and development-oriented functions. They are also regulatory mechanisms of social control and judicial mediation.

These societies are revered because of the "secret" nature of their traditional "lore" that enhances their function of maintaining law and order meting out punishment, and promoting the socioeconomic development of their communities. They are also mystified by patron saints and by the spirits of dead leaders. These spiritual entities are believed to have the power to influence the life of the living and are often personified as masked figures from the supernatural world.

These masked figures appear during ceremonial parades, command great respect and awe, and appear on ritual occasions to reinforce norms and values as well as social cohesion. The corporate identity of the group is regenerated at each initiation ceremony. This facilitates the incorporation of new members as well as the development of group solidarity and continuity.

Both Sande and Poro emphasize male and female socioeconomic spheres of activity and impose regulations to ensure that each is maintained with a certain degree of autonomy. Some of these regulations determine procedures and obligations in the gender division of labor and regulate behavior between men and women. As a result, Sande has functioned as an association that protects and defends women's rights as defined by the traditional lore and custom. For example, Sande leaders can reprimand and impose punishment on men who disrespect use violence against women or mistreat women. As a corollary, Poro leaders can reprimand socially unacceptable behavior by women toward men. An institutional mechanism is thus created which promotes mutual respect, interdependence and complementary rights and obligations in male/female relationships.

Sande/Bondo Female Society

Sande is probably the most widespread women's association in Sierra Leone. Like its male counterpart, the Poro, it embodies a repertoire of cultural norms, indigenous knowledge, beliefs, ethics, arts, and crafts that have ensured the economic, social, cultural, biological survival of generations. Its ritual is a necessary precondition for biological and social transformation from childhood to womanhood and marriage.

Marriage, a supremely important institution is an inevitable stage in the life cycle guaranteeing regeneration and continuity of the group. Although in theory Sande membership is not compulsory, it is, like marriage, desirable and inevitable since it is promoted as the cornerstone of the ideology of womanhood, fertility, and the survival of the group. As a result it has continued in full force in spite of opposition stemming from its encounter with Christian missionaries and from the pressures from social change and urbanization.

Origin
The origin of Sande is not known for certain, but there are records of associations bearing close similarities to Sande and Poro which existed in the sixteenth century. The name "Sande" was first used in a literary sense by Olfert Dapper, who also suggested a possible Gola origin, because the High Priestess of Sande then resided in Gola territory.

This claim has been substantiated by modern scholars who have noted the fundamentalist tenets of Gola's Sande societies as well as their conservative and parochial proclivities.[3] Equally significant are the close cultural ties and linguistic similarities that exist between the Gola of Liberia and the Mende and Gola of Sierra Leone.[4] Whatever may be the specific origin, it is clear that both Poro and Sande are important and somewhat unique social institutions in this central West African cultural area.

Like Poro, Sande reveals much local variation being organized according to local chapters or lodges, each headed by a Sowei. The Sande initiation camp, known among the Mende as Kpangu, is usually held in a secluded area in the forest, during the dry season from December to April. In the past, the initiate would reside in these camps for a period of one year or longer, but this period has been shortened to an average of four weeks or less. Social change, schooling, and migration have brought new demands on the time available to the young initiates.

Traditionally, the camp would have a sacred grove where departed leaders are buried and where sacrifices and initiations are held. The interaction with the supernatural is a necessary component of the transformation of status from childhood to adulthood. It also facilitates the revelation of the special "secret" knowledge about the society and about the virtues of womanhood that ensure fertility.

The cultural management of sexuality, fertility, pregnancy, and childbirth is the most important function of the Sande and the leader is usually a midwife. Sande, therefore, sustains the most vital elements of the culture that are decidedly female. The traditional midwife or Birth Attendant has a high status in the community.[5] This is derived from her leadership position in the Sande society and recognition as the authority on fertility, female sexuality, as well as on gynecological ailments, children's diseases, and child care. Sande leaders are knowledgeable about the epidemiology and history and the medicinal resources of their communities.

The Organization of Sande
Sande is organized hierarchically, and on the basis of kinship affiliation and cuts across kinship ties. The highest ranking officers are usually related to high-ranking families in the community. The hierarchical structure serves to enhance its secrecy. At each level of the hierarchy, new and more exclusive secrets are revealed and additional instruction and fees are usually required.[6]

Although Sande does not have a politically explicit function, female chiefs and paramount chiefs are important members and can strengthen their position by tapping on the power base of Sande. Some chiefs, like Madam Yako, were reputed to have a strong influence in Sande, which enabled them to make alliances through the marriage of Sande women, who are often their relatives or protégés, to prominent chiefs. This often consolidated their position and expanded their influence.[7]

In relations with men, strict sanctions and prohibitions are observed. Men have no access to the secret knowledge and are forbidden to enter the camp while it is in session. In addition, men are forbidden to speak about Sande at the risk of contacting a serious illness.

Among the Mende, Sande leadership is organized into three grades. The highest grade is occupied by the Majo, the supreme head of one or several camps, and the Sowei (plural Soweisia), who usually heads only one camp. In the next grade is the Ligba (plural Ligbanga) responsible for performing the operation of circumcision, for which a fee is paid. Below these are the Klawa officers, who usually act as matrons, responsible for the welfare of the initiates. Among the Temne, the leader is known as Ddigba, which is the term also commonly used in Freetown.

Upon their death, Sande leaders become ancestors and protectors of Sande members and provide a link to the high God Ngewo. As a result their death is observed by elaborate ritual and special taboos. The masks (Sowei) worn during the Sande rituals personify the spirits of the dead leaders and ancestors.

Initiation
A girl is usually initiated at puberty between the ages of 13 and 16 and must be sponsored by an adult member, usually her mother. Initiation marks the biological transmission from childhood to adulthood and the social transformation from girlhood to womanhood. It also confers full membership into the ethnic groups and is closely linked to one's social identity. The dramatic change of status is marked by elaborate rituals and rites of passage that conform to Van Gennep's classical stages of separation, transition, and incorporation.[8]

In between these "secret" stages, which are marked by intense ritual ceremonies, are also the "open" stages during which the initiate can interact with relatives and congregate in more public places. The "open" stages have provided opportunities for innovation and modernization of the Sande society. It was during one such period that the introduction of new courses to the Sande curriculum such as anatomy, physiology, sanitation, first aid, and domestic science was first made.[9]

The first stage of initiation commences soon after admission of the initiate, with clitoridectomy, regarded as necessary for achieving full womanhood and for ensuring fertility as well as sexual gratification. Many initiates are usually already engaged to be married or become engaged soon after initiation.

Other explanations given for this operation emphasize the promotion of modesty in the sexual behavior of women, an important aspect of the social construction of the ideology of womanhood and a system of morality for women. It is also regarded as proof of having endured a physical ordeal as a woman in preparation for the ordeal of childbirth.

Significantly, the operation is a vital aspect of the change of status and is intrinsically linked with identity, solidarity with the group, fertility, and marriage. This explains why many initiated members of Sande vigorously and intensely defend the practice when faced with criticisms from outsiders.

At a second stage of initiation, known as the transitional stage, the initiates are usually allowed to move about and visit their parents escorted by the Ligba. They are allowed to perform light work such as collecting twigs, making fire, sweeping, and so forth. At this stage instruction in childcare, sanitation, arts and crafts, ethics, character development, and respect for elders are usually given. The girls are also taught the songs and dances of the society many of which have deep moral and ethical meaning.

At this stage also, the initiates have an ambiguous status, which makes them vulnerable and are considered to be in potential danger. As a result they require protection. Their bodies are usually smeared with white chalk as a protective cover against evil spirits and they are clad in distinctive attire composed of a short skirt, beads, and a special dress. At the end of the transitional stage there is a purification ritual performed after the wounds have completely healed.

The initiate then undergoes ritual cleansing through ceremonial washing in the river. Symbolically, this removes all vestiges and ambiguities of the old status of childhood and girlhood and marks the beginning of her status as an adult. Regular clothes can now be worn and visits are usually made to the chief of the village and to relatives, accompanied by another adult. At this time also the curriculum is extended to include instruction in basketry, spinning, and the art of healing.[10]

The final stage formally marks the complete change of status to womanhood through a coming-out ceremony. A public parade is held at which there is jubilation, dancing, and singing. The new adults are adorned in special clothing and accessories. A masked dancer (Sowei Ndolimo) symbolizing the ancestral spirit of the Sande performs special dances to music provided by the Shegure and drums played by women.[11] The girls are finally escorted to their parents' home, where the "new woman" assumes a prominent position of display, amidst much celebration, rejoicing, and admiration by potential suitors.

While recognizing local and regional variations and noting urban changes, a number of studies of Sande have tended to emphasize its conservative elements and to present an idyllic picture of ritual transformation amidst the celebration of cultural values. One exception is a study of the Sande society among the Kpelle ethnic group in neighboring Liberia by Bledsoe.[12] Although negative in tone and simplistic in terms of reducing actions of Kpelle women to greed and the desire to exploit the labor of children, the study examines stratification among the Sande in a new light. It suggests that the monopolization of power by the leaders of the society creates tensions that have been overlooked by other studies.

Poro—the Male Counterpart
A brief discussion of the Poro male society, based on published material, is in order to illustrate the relations between Poro and Sande and elements of their complementarity. Poro exists primarily among the Mende, Temne, Lokko, Kono, and Vai although there has been considerable diffusion to other ethnic groups. It is a decentralized organization made up of local Lodges each with its own leadership.[13] According to the literature, the core leadership comprises the Sowa, the founder of the Lodge, the Gbeni, the spiritual leader who personifies the guardian spirit of the society, and the Mabole—the matron (a female member) who usually looks after the young initiates.

In some cases, these positions are hereditary based on kinship ties to the founder of the Lodge. Female political leaders in high office can become members. In special cases, membership becomes mandatory for women who have acquired the Poro secret accidentally or by design. These women have restricted social mobility within the ranks of the organization and can never assume leadership roles.

Initiation is a necessary stage for biological and social maturity and is a rite of passage for the transition from childhood to adulthood. It is compulsory for male adolescents and involves circumcision. Poro promotes the ideology of manhood, which encompasses the values of a patriarchal society. The rites are usually held in seclusion and not concurrently with the Sande female initiation sessions. This emphasizes the uniqueness of each gender in their contribution to adult life, marriage, fertility, and ultimately to the functioning and development of society and the continuity of the group.

Formal instruction in politics, law, gymnastics, and oral tradition are combined with the learning of arts and crafts, singing and dancing. Moral and ethical subjects, as well as the values of self-discipline, cooperation, obedience, and respect for elders are all included in the curriculum. These helped to produce a well-rounded and mature personality compatible with the requirements of responsible citizenry. In addition to circumcision, the initiate may also undergo a ceremonial ordeal (similar to hazing in fraternities in the United States) that involves enduring physical suffering with demonstrated acts of courage.

Poro is recognizably a para-political association in its function as the religious arm of the chieftaincy. The core of Poro leadership has executive, administrative, religious, and legal functions; sets the code of conduct and ethics and makes decisions on important economic and political matters.[14]

The power of Poro officials is mainly religious since they cannot assume the role of chief. However, their religious sanction is a vital ingredient that determines the efficacy of the secular power of chiefs. The Poro core leadership also arranges treaties and alliances among chiefdoms and ethnic groups.

In this regard, Poro has served to develop a number of confederacies and is often regarded as a direct or indirect player in modern politics. The Kamajors, a citizen's militia group that was influential in the pro-government combat with the rebels during the ten-year war, built their solidarity, in part, on ethnic affiliation and alliances to the Poro society.

Poro influence is widespread throughout Sierra Leone. It is generally believed that many political and economic arrangements and negotiations occurring in the modern context have been influenced to some extent by Poro networks. Poro has also been known to play an economic role in rural communities, regulating economic activities, particularly with regard to the deployment of male labor during busy farming seasons. Initiates are taught economic skills to strengthen their role as producers of food, goods, and services.

Sande and Social Change

Returning to Sande, it is clear that in the context of rapid social change a number of contradictions become apparent. On one level, Sande is a conservative institution functioning to maintain traditional norms and values and regulating social conduct. On another level, Sande is an institution, which is intrinsically committed to change. It registers the change of status from childhood to adulthood in a profoundly religious and dramatic way. In addition, the symbolic change of location from the village to the forest and back to the village through initiation can be viewed as an aspect of Sande's own transition and inherent flexibility.

Sande's accommodating structure and dynamic constitution have facilitated the introduction of several reform measures in the past and are contributing to changes taking place in more modern times. Before becoming the first Prime Minister of Sierra Leone, the late Sir Milton Margai introduced several modern ideas of childcare, hygiene, sterilization techniques, anatomy, physiology, obstetrics, and home economics into the curriculum of Sande when he was a senior medical officer in the Provinces.

He did this partly to upgrade the standards and ensure safety and partly in recognition of the valuable instructional role of Sande.[15] The association played an important cultural role in the education and socialization of young girls and in promoting values of solidarity, group identity, and cooperation.

The Urbanized Secret

By virtue of their structure and characteristics, secret societies are not easily amenable to research or the application of survey techniques. They operate on an oath of secrecy and possess no written records or documents. Researchers have had to maintain confidentiality as far as individual informants are concerned.

Much of Sande's secret, however, is an open secret. Within Sande itself, conflict and change are best exemplified in an urban context. The context of ritual and its hierarchical structure is diluted when transferred to an urban area. It loses its secrecy and becomes subject to urban norms and greater population density, size, and heterogeneity. In this regard, the secret becomes an "urbanized secret."

Sande was introduced to the neighboring villages around Freetown in the nineteenth century and later to the city.[16] In Freetown, it is commonly known as "Bondo" and the leaders are called "Digbas." As a result of rapid rural to urban migration over the last 80 years, various ethnic groups have been able to establish themselves as stable urban communities in Freetown and to preserve many of their institutions in the city.

Before secret societies become well established in the cities, women migrants would become affiliated with Sande chapters in the neighboring villages. Traditional and cultural associations known as "Dancing Kompins" provided recreational and mutual-aid functions that enabled the new migrant to make initial adjustments to the urban environment.[17]

In time, secret societies took over many of the functions of mutual aid "Dancing Kompins." They serve as recreational dancing groups and perform the most important functions of initiation into womanhood. They have also inspired the founding of newer types of mutual-aid societies, which will be discussed later.

Significant differences exist between the secret societies in Freetown and those in the Provinces. The urban associations have assumed a more secular characteristic in the city. They do not have camps with sacred groves in the forest where ancestral leaders are buried.

Ecological and demographic factors have resulted in high-density living and physical proximity making isolation and privacy impossible. The period of seclusion is shorter and the ceremonies more simplified. A Digba could perform ceremonies in an enclosure in the backyard of a small house or ajoini (small house in the same yard) where she resides often in multiple-family settings.

Initiates receive their instruction and learn sacred songs within earshot of neighbors and other tenants and dwellers. It is not unusual for small children in the neighborhood, unrelated to the events of Sande, to become conversant with these songs and to chant them freely.

In the urban setting also, the organization of Sande is less hierarchical and amorphous in terms of time and space. There are fewer functions and the Digba is often the only identifiable leader as opposed to hierarchy of leaders in a specific secluded and sacred setting. Initiation ceremonies become temporary "happenings," which are convened for the specific purpose of initiation and then dispersed.

They are small and only take on the characteristics of an association or a society in parades and dances at coming-out (pul bondo) ceremonies. A number of informants claimed that with the exception of a few organized Sande societies such as those found among the Mende at Ginger Hall in Kissy, many secret societies in Freetown were not organized.

Membership has also undergone changes. Social pressures promoting initiation still exist in urban areas, but it is less obligatory and there is greater flexibility and an element of choice. Women sponsoring their daughters for membership tend to do so on a voluntary basis, and some young girls have been known to refuse to be initiated.

Most of the urban initiates are exposed to new and different ideas about fertility and womanhood and may become skeptical about the claim that these qualities can only be conferred through Sande. They are also likely to be attending secular schools that expose them to urban and Western values. School attendance has necessitated the shortening of the period of seclusion to coincide with school holidays. It may also have contributed to the reduction of some of the institutional functions of these associations, especially those related to educational activities.

The impact of urbanization is most marked during the coming-out ceremonies, revealing the influence of urban norms, values, and aesthetics. The typical traditional outfit consisted of short skirts, head tie wraps, chest beads, and anklets made of tiny bells and adorned with white chalk marks covering a large portion of the face and parts of the body. The newly initiated would walk through the streets in single file and dance in a restrained manner, accompanied by Shegure music. Usually there were no more than six new initiates and about ten followers.

More recently, the new "graduates" are more likely to be dressed in the latest styles, reflecting a combination of western and African fashions. Cosmetics, perfumes, and Western style jewelry are used profusely and hairstyles are elaborately fashioned. Usually, each set of new graduates endeavors to dress in the latest or most exotic styles. There is a strong competitive element as one graduate tries to outdo the other in terms of adornment. Parades can also become quite elaborate, depending on the social ranking of the graduates. Processions can number as many as 500, comprising initiates, society members, relatives, and friends.

During the school holidays, there is at least one parade a day. The music has also become more varied and is no longer confined to the traditional Shegure. It can include mailo jazz and other bands and the music has also become much louder. The dancing is less restrained and the procession less orderly. After the street parade, the subsequent celebrations at the home of the initiates have also become more elaborate and expensive. Fees for initiation have increased as a result of inflation. In addition, Digbas receive gifts of food, clothing, and livestock from the parents and relatives of the girls.

The "pul bondo" ceremonies have also become highly commercialized as families try to compete with each other at the elaborate celebrations. These ceremonies can be rightfully regarded as manifestations of an embryonic form of class consciousness through the conspicuous consumption by the more affluent families that would not be as marked in the traditional rural setting.

Any visitor to Freetown in the 1980s would have been struck by the frequency of "pul bondo" ceremonies, but they were less frequent in the 1990s and in the new millennium. This is due to several factors, the most important being protracted economic recession and the ten-year rebel war. Another could be related to the criticisms that have been meted against some aspects of Sande, notably female circumcision.

Sande and the Health Care System

The skills and knowledge of the Traditional Birth Attendant continue to be needed by the majority of women in rural communities. The medical profession has long recognized this and initiated a program, in conjunction with the World Health Organization, to utilize and incorporate the traditional midwife into the modern health care system through training and cooperation.

In 1974, a program was inaugurated by the Ministry of Health, under the direction of Dr. Belmont Williams, an obstetrician and gynecologist, who also served as the chief medical officer. The objective was to bring the Traditional Birth Attendants, who deliver 70 percent of the country's babies within the centrally organized health service system. In this program, recognition was given to the valuable skills and experiences of the Traditional Birth Attendant. A systematic attempt was also made to learn about the medicinal and nutritional properties and values of locally-grown medicinal herbs used by these practitioners.[18]

Although the medical profession recognizes and incorporates many of the valuable aspects of the Traditional Birth Attendant, not all of their work is valued. Female circumcision performed by Sande is not condoned by them, because of its health consequences and the complications caused during pregnancy and delivery.[19] Some medical practitioners, such as Dr. Koso-Thomas, who conducted a study on female circumcision in Sierra Leone, have become well-known advocates for an end to this practice.[20]

Female Circumcision: A Global Practice

Though hardly known and acknowledged, female circumcision has been practiced in all regions of the world. In Europe, this operation was performed on women to control their behavior. It sometimes included the removal of women's ovaries that were viewed as the seat of women's disobedient and disruptive behavior. Sexual surgery was not uncommon in the late nineteenth and early twentieth centuries.

According to Barker-Benfield, "clitoridectomy was the first operation performed to check women's disorder."[21] In the United States, female circumcision was performed by physicians for the same reason and the last operation was conducted in 1925.[22] Babies born with ambiguous genitalia, often manifested in both sexual organs being present and a large clitoris, could be subject to having them surgically removed.

In the West also, numerous mutilating surgeries are performed for cosmetic reasons, which often involves the drastic alteration or removal of body parts. Reconstructions to narrow the vaginal canal are also performed to increase the pleasure of men during intercourse. Seldom does one reads or hears criticism about these types of surgery for what has been termed "designer vaginas," that cannot be justified on medical grounds. Also rare is information about the complications resulting from these forms of plastic surgery, including death. With few exceptions, criticism of these practices is muted.[23]

It is generally believed that Sande officials in Freetown are equally concerned about complications from these operations, but that ending the practice is not an option for them because of cultural and economic reasons. As a result, some are taking greater precautions. On several occasions, I was told by informants that, as a preventive measure, some Digbas and Sande officials in Freetown arrange to have tetanus shots given to initiates. Others keep a good supply of antibiotics, notably penicillin, for use during the operation.

The more serious complications include scarring which can later result in severe lacerations and excessive bleeding during delivery. In addition, the difficulties posed by improper vaginal examination may aggravate cases of cephalo-pelvic disproportion, a major obstetrical problem in Sierra Leone.[24]

Views about this practice from informants were mixed. According to my younger informants of both genders, a substantial number of young men are increasingly opposed to clitoridectomy and try to influence their girlfriends against it. Some regard it as an anachronistic and archaic practice. Older women, especially grandmothers, are the greatest supporters of the practice. Among the professional groups some men expressed strong opposition, especially when it involves the possibility of their own daughters undergoing this operation.

At one medical association meeting, which I attended in 1996, a male physician stated publicly, that although he belonged to one of the ethnic groups that adhered to this practice, he would never allow his daughters to go through it. Older adolescent schoolgirls have also been known to independently refuse initiation, finding it difficult to justify the physical ordeal. To counteract possible adolescent defiance, some families initiate their daughters at a younger age, when they can offer minimal resistance.

Conflicts produced through social change have also been reflected in Sande. The influence of Christianity, Islam, Western education, political change, and the International Women's Movement have also affected Sande. Christian and Islamic religious leaders tried unsuccessfully to discourage the activities in the past. For example, Christian missionaries from Europe disparagingly referred to the masked figure of

Sande and Poro as "devils" and were strongly opposed to female circumcision and polygamy.

Although Islam's condemnation was less severe, whenever Islam became firmly established such as in the parts of the northern areas, secret societies reputedly tended to decline in influence. However, since Muslims also practice female circumcision and polygamy, Islamic leaders have not openly condemned these practices. In general, female circumcision is practiced by almost all ethnic groups in Freetown, with the exception of the Krios, who do not have a tradition of female secret society initiation.

Social Change, Female Circumcision, and Controversies

Social change, increasing secularization and urbanization, as well as Western education and overseas travel, are already eroding traditional customs and religions in many respects. Sensitive campaigns, as well as the work of the United Nations, particularly the World Health Organization (WHO), have been launched against clitoridectomy. African women themselves, including those who have firsthand experience of this practice, have condemned this practice, both nationally and internationally.

One noteworthy example from Nigeria is Olowe Jo, an anti-female circumcision activist who has appeared at the National Assembly in Nigeria, requesting that this practice be abolished.[25] Equally significant is the recommendation of the African Platform for Action for abolishing this practice adopted at the preparatory meeting for the Fourth World Conference on Women held in Dakar in 1994.

For female circumcision to cease completely, it will be necessary to dispel some of the myths and sensationalism surrounding it. While it can be argued that it is an aspect of patriarchal ideology to control and dominate women and their sexuality, it is important to understand its continuing appeal to women who have experienced it. First, it ranks high in the system of merit for women and defines their identity. Second, it is exclusively a female practice conducted by female practitioners and has systems of reward and status acquisition attached to it. Third, it builds female solidarity and provides a power base for women.

In addition, the context in which it is performed in Freetown is linked to the cultural practice of initiation from girlhood to womanhood and has strong incentives for developing female identity and female bonding. A friend of mine, a university graduate, and who still believes in clitoridectomy as part of Sande initiation, expressed to me her fear that her daughter (who was growing up in the United States) would be excommunicated by her ethnic group if she does not undergo initiation and the operation. Ironically, her husband, who is from an ethnic group that also adheres to this practice, was vehemently against his daughter being initiated. In the end, the initiation did not take place as the daughter was also opposed to it.

The solidarity created by bonds of sorority can function in many respects like a support group or a sorority. Women assist each other in child care, trade, and in dealing with situations of crisis. The social significance in the urban area has far-reaching implications than the initial ritual itself, for it provides women with a type of social insurance policy.

Feminist-inspired conferences organized on the subject of female circumcision and mutilation often become highly controversial. In 1996, a Freetown newspaper,

"For Di People," published a series of articles against female circumcision (the term used locally or its linguistic equivalent) that sparked a hostile protest march of thousands of women, determined to defend the practice against the newspaper. The protest was led by Haja Sally Sasso, President of the National Council of Muslim Women, who vowed that attempts to ban the practice would lead to war.

The controversy led to mobilization of the Soweis to protect their rights as leaders of an association perceived as under siege. An association of Sowei was later formed. According to Haja Sasso:

> I am only doing this to protect our culture. I do not want to see this ceremony eradicated, because it binds us, we the women, together. We respect each other in this way and we feel free together because of it.[26]

In a follow-up interview with Haja Sasso, she insisted that the international community and those that follow them were misguided and ignorant about the custom and had no right to interfere with it. She insisted that this was an important sorority that gave women a sense of security and solidarity and helped them to advance their views and wishes. She could not accept the sudden upsurge of campaigns against the practice after several years of no resistance. She was convinced that the "frustrated Western feminists" were behind it. She concluded the interview with the following statement:

> Why don't they campaign to end racism and violation of human rights against Black people instead of interfering in our culture?[27]

Later, a major Sande ceremony, marking the end of war, was held for displaced girls at the refugee camp in Grafton, just outside Freetown, as a special celebration for their imminent return to their homes. Several of the girls were reported to have had medical complications, including death, resulting from the operation. This led to large-scale publicity against the practice. Despite this, support for the custom is still strong in some circles. One of the female leaders of the refugee camps reputedly said:

> I have grown to the age of 50 years, and this is the first time anyone has come forward to ask me why we do those ceremonies. It does not matter what other people think because we are happy with our customs. We will carry on with our lives.[28]

There is no question about the value placed on bonding and identification. At the same time, a measure of ambiguity exists, and becomes apparent as an initiate tries to come to terms with the pain of the operation. The following personal account best illustrates this dilemma as a girl contemplates the meaning of becoming a woman.

> What an experience, Mayepe thought. She remembered that all the girls would soon go through the same thing and she prayed for them. No wonder there was so much bonding between women in the village, Mayepe thought, especially during initiation. She now realized that she had become part of the bond, but the most important bond of all was going to be between Mayepe and the other fifteen women. "Mayepe" she heard the soft voice of her mother calling her. She opened her eyes and saw her mother's gentle face full of tears. "Congratulations, my daughter, you are now a Sande Nyahin. You and I can now talk as equals." The passage into womanhood was more painful and less fun

than she had been made to believe. As Mayepe lay bleeding and weak, she began to feel betrayed by everyone she loved. But if it brought her closer to her mother, and other women, who had gone through the same experience, then it was worth it. Suddenly, another surge of pain shocked her body. Then again, she thought, maybe it wasn't.[29]

Sierra Leonean women, who adhere to this practice, have sometimes been targets of scorn and hostility in the United States. Immigrants in the Washington area have reported that co-workers often bluntly ask them if they had undergone female circumcision. According to D'Alisera, an anthropologist who has conducted fieldwork among Sierra Leonean immigrants in Washington, D.C.:

> One woman felt people were looking at her and talking to her as if all she was a big genital that had been mutilated[30]

As noted earlier, the Bondo society is a religious association with educational functions related to rites of passage, ideologies of fertility, womanhood, and female solidarity. It also represents an important power base for women since it creates female bonds across ethnic and class lines. It has important economic incentives related to preparing a woman for marriage. Financial security, wealth, and privilege can result for female leaders of the association and female chiefs.

It has also served an important function of exclusion and inclusion in modern politics. It grants legitimacy to female political officials who have undergone this experience and can seek to marginalize those that have not. For example, as stated earlier, it was arguably used after the elections of 1996 in a failed attempt to delay the nomination of a Minister for Gender and Youth, who had not undergone initiation. Since initiation was not a requirement of the proposed Minister's own ethnic group, the argument used was that she would be unfamiliar "with our adored customs."[31]

Some female political leaders have adopted a more moderate position. One of them is Zainab Bangura, who later became the head of the National Commission for Democracy and a Presidential candidate in the 2002 elections. She insists that understanding the rituals that surround the practice and trying to modify them would be a better approach. She gives the following reasons for this:

> First, the rituals are an important part of a whole set of procedures that educated young girls and passed knowledge of womanhood from one generation to the next. Second, much of this had been eclipsed by the ritual of female circumcision and need to take a more important place again. Third, women should be reassured that something of value will not be taken away from them because of one ritual.
> For me, you cannot bring a Western approach, lecturing people about their customs. The more you decide you are going to take something like this on, the more you are going to face resistance. Instead, a dialogue has to be established, and women have to understand that Sierra Leone is part of a global community and should not be left out . . . . We could begin by telling women that Bondo has been trivialized by reducing it to a circumcision ceremony. Instead, the institution could be modernized by teaching abstinence or sexual education to young girls.[32]

A male anthropologist from Nigeria throws some light on the cultural logic behind female circumcision. This was intended to diffuse what he considers the ignorance behind the sensationalizing and mercenary proclivities of some Western feminists. While not condoning female circumcision, he offers solutions that are vastly different from the simplistic feminist explanations of male dominance. According to Babatunde:

> The logic of the practice is couched in the anthropological term of prestation, a gift that you give under pain of sanction for which you receive a gift in return. The logic of clitoridectomy is that by taking a tiny bit of the sacred instrument of fertility as an offering, the god of fertility will bless you with more children.[33]

He proposes an effective strategy for an alternative that will include efforts to improve health care delivery in countries where infant mortality and maternal morbidity are high. This will reduce the value placed on bearing many children to ensure the survival of a few. He noted that in 1980, when Nigeria's earlier investments of oil resources began to pay off in improved health care, the social emphasis on procreation quickly changed.

He claims that this approach will end clitoridectomy faster than the over-sensationalizing by Western feminists and authors, such as Alice Walker, who have made huge profits from the erroneous feminist propaganda against female circumcision.[34]

The Enduring Features of Sande

Sande is an important power base for women. It is also an effective mechanism for lifelong bonding and female solidarity. It utilizes its membership to create important networking opportunities with women who have political access or are in a political position themselves. It can also act as a pressure group with important political functions. For example, its membership can be a critical factor in the selection of female candidates for political office and in deciding the strength of the ballot.

Women who have not been initiated into Sande are usually excluded from some acts of bonding such as eating from the same bowl. They are sometimes disparagingly referred to as "children" by Sande members. In Freetown, tensions between Sande and non-Sande members are particularly marked when issues of female circumcision are discussed. In addition, opposition to Sande drumming on Sunday morning during Christian worship has been a point of contention in forging unity among women.[35]

There are numerous campaigns against female circumcision, both internationally and locally and it seems clear that female circumcision will soon be a thing of the past. This would be most desirable. Nonetheless, the issue should be viewed in a balanced perspective. As a highly emotional topic it can be blown out of proportion. As a sexually charged issue, it has the potential for controversy since sexuality, including one's own sexuality, is a topic about which very few women are comfortable discussing.

Although there are serious health consequences for some women who undergo this operation, health problems are also caused by many other factors such as underdevelopment, poverty, inadequate living conditions, lack of water supply and sanitation facilities, lack of food, and economic exploitation. Western critiques of female circumcision do not often show the same concern for human devastation, suffering, and the health consequences of inhumane global economic policies emanating from their countries.

Sande is a valuable institution to its members. Regardless of the controversies surrounding female circumcision, it is closely linked to concepts and ideologies about womanhood, identity, and solidarity. Its members consider it an essential part of the culture. For this reason, it has received the public support and endorsement of leaders of the Sierra Leone People's Party (SLPP) and the Armed Forces Ruling Council (AFRC) regimes. It represents a critical power base for women that can translate into political capital.

Sande is a dynamic institution receptive to change and reform that are often inspired from within the country. It is very likely that these factors will influence the association toward re-evaluating the practice and searching for alternatives that do not have undesirable medical and physiological consequences.

National and international pressures are increasingly brought to bear on this practice. The World Assembly of the World Health Organization passed a resolution calling for an end to the practice. The organization has also implemented a major worldwide campaign aimed at its cessation for health reasons. In June 2003, the European Commission, in conjunction with an association of the Egyptian government and led by the first lady, Mrs. Susan Mubarak, met in Cairo to pass a resolution outlawing female circumcision.

The commercial aspect that Sande has assumed in the city and its relation to the development of urban class-consciousness, tend to emphasize its economic aspects rather than its cultural and educational aspects. These factors, combined with the relative lack of privacy in the urban environment have diluted some of its original functions.

The erosion of traditional sanctions in the urban context has, with few exceptions, transformed Sande from a highly ritualistic and religious traditional institution with educational functions, to a secular social agency performing the functions of initiation into womanhood in a bureaucratic and public fashion. Still, the educational functions could be revived to include new curricula that would expand women's knowledge and further strengthen their power base.

Another significant urban effect is the development of female solidarity across ethnic lines, thereby making Sande an integrative agent for women from various ethnic groups. This aspect is sometimes extended to voluntary associations based on a mutual-aid model, since Sande provides support for women in a difficult urban environment.

As an association dealing with human fertility and motherhood, it has the potential of helping to strengthen health care delivery systems through continuing collaboration with the medical establishment, in conjunction with the World Health Organization. The mutual-aid associations that it has inspired, and which can also promote development and take on a political dimension, will be discussed next.

Chapter Seven
Mutual-Aid Associations

Introduction

Mutual-aid associations or "friendly societies," have as their main aim the rendering of material, social, and moral support to members in times of bereavement, emergencies, and destitution. In this regard, they function like social welfare agencies and help to compensate for the limitations and resource constraints of the social welfare department of the government. For example, the expenditure for social security benefits as a percentage of the Gross Domestic Product (GDP) for 1996 to 1997 was 0.6 percent.[1] Freetown's history is characterized by the formation of mutual-aid associations among successive groups of rural to urban migrants and settled populations, regardless of origin.[2]

Most of the systematic studies of mutual-aid associations, conducted by men, have focused on those formed by male migrants, due in part to the higher degree of male migration versus female migration at the time of the studies.[3] In more recent times, women have constituted a significant proportion of migrants, a factor, which has been influential in the formation of mutual-aid associations among women.

In most cases, among the associations formed by women, mutual-aid associations have been inspired by secret society membership and can be regarded as spin-offs from the Sande society. The mutual-aid associations presented here are characterized by their functions. A case study of one of them showed that the Sabanoh Women's Association, performed the following functions: strengthening the values of secret society affiliation; alleviating tensions between statutory and customary marriage; ensuring the imposition of moral sanctions; promoting multiple mothering; providing recreational functions; and enhancing solidarity and political expression.

Secret Society Affiliation

The influence of Sande in Freetown has made possible the development of newer types of mutual-aid associations among urban female migrants from rural areas. Many of the rules and regulations of Sande are informally applied as well as the sanctions on behavior and on interpersonal relationships, particularly with men.

Some of the members were usually initiated into the same secret society. In addition, some of their ancestors may have come from the same village or shared affinal ties through marriage. On several levels, primary groups ties would be strongly reflected in membership composition so that the association serves as a substitute kinship group providing mutual-aid and support.

Secret society rituals provide opportunities to strengthen sororal bonds and guarantee a type of social insurance against hardship and destitution. Thus, although these are voluntary associations, prior initiation into Sande can become a de facto requirement for membership.

The President of the Sabanoh Women's Association, explained in an interview that prior initiation in Sande was a requirement. This is because participation at Sande coming-out ceremonies ("pul Bondo") is one of the association's most important social activities. Events such as this involve communal eating rituals that reinforce the Sande experience of sororal bonding. According to her:

> We have to eat together from the same dish and society rules forbid eating together in this fashion at such gatherings unless one has been initiated.

All new members have to be approved by the President who is also the founder, and each is given a badge. In addition, a verbal pledge is made to abide by the rules and regulations and to attend its weekly meetings punctually. The meetings are chaired by the President assisted by an executive committee of five persons. Meetings are conducted in the Temne and Krio languages. The President has wide-ranging powers and can replace any member of the executive for incompetence. Rank and file members who fail to abide by the rules and regulations and to comport themselves in a proper manner are expelled.

There is no term limitation to the presidency, so in theory, the office can be held for life. However, elections are held annually for the other offices in the executive committee. Eighty percent of the membership is illiterate, although literacy is a requirement for the offices of treasurer and secretary. The signature or thumbprint of both the President and the Vice-President are needed for withdrawals from the association's bank account. Funds are usually allocated for funerals, medical expenses for grave illnesses, lawsuits, childbirth ceremonies, cost for divorce, "pul bondo" ceremonies of members' daughters, and general hardship.

Similar associations in Freetown have included the Tasutekeh Women's Society and the Kissy Road Women's Union. Most of the members of these associations are women who have lived in Freetown for a period of less than ten years. They often face problems such as unemployment, poor housing, inadequate facilities and services, and lack of access to resources.

For many of these women, these associations serve as coping mechanisms against poverty and hardship. Much of the economic support comes in the form of mobilizing resources through the osusu mechanism. Because many of these associations are involved in subsistence trading, resources are pulled together in special circumstances, such as providing start-up funds for trading.

Economic support can also come in the form of information about market access, buying in bulk, producing collectively, lowering transportation costs, and using discounted wholesale opportunities. In addition, they can provide the rudimentary requirements for participation in an urban environment through socialization to urban norms. A number of them provide opportunities for the acquisition of basic skills for participating in the informal labor market, such as participating in functional literacy classes and bookkeeping.

Tensions Between Statutory and Customary Marriages

Women, who are married by customary law, or live in a relationship as a concubine, often feel unprotected against destitution due to the difficulties of upholding customary marriage in the city. Being married by customary law or living as a concubine does not offer women legal protection in the city. Joining associations with other women in similar circumstances for the purpose of mutual-aid is viewed as a partial solution to the insecurity of their customary marital status.

Statutory marriage was instituted during the colonial era and reflected Victorian patriarchal traditions and Christian morality with strong biases in favor of men. This is particularly marked in rules of inheritance and in domestic authority. Despite its shortcomings, statutory marriage is regarded as offering some protection and economic support for a wife and her children and is therefore viewed as an important "resource" among women in customary marriages. In an ironic way these women would rather choose dependency in statutory marriage, as a safety valve against the realities of a precarious urban economy, than live the unprotected lives of marriage by customary law.

According to those interviewed, there is a tendency to feel deprived of the securities offered through statutory marriage when in the urban area. In many instances, this is an unrealistic view since statutory marriages are also prone to instability as a result of economic and other stresses in the urban environment. Even though these marriages may not end in divorce, studies have shown that instability has been a feature of both statutory and customary marriages.[4]

In the absence of traditional sanctions supporting customary marriages in the city, women feel the need to support each other in case their male support is suddenly withdrawn or cannot be guaranteed. Many of these women would like to be self-supporting and plan to achieve a measure of economic security through petty trading. Unfortunately, much of this trading is subsistence trading in the informal sector, which is extremely vulnerable to economic downturns.

The much-needed additional security is provided by mutual-aid associations. Although the rules and regulations of most of these associations do not actually make provision for destitution in the event of a marriage breaking up, there is a "general hardship" category, which covers such crises. From interviews with the members, it would appear that this interpretation is used most effectively during situations of a sudden breakdown of marriage or similar bonds.

Some women idealized customary marriage in the traditional rural setting since their marital status was upheld by custom and cultural sanctions. In the urban environment, sudden destitution could be a real threat.

Some scholars have argued that in urban Africa some women have exchanged family and other traditional forms of security for the prospect of personal liberty and the desire to be "free." They further argued that the freedom influenced the formation of mutual-aid associations. This may be true in a few cases but in the mutual-aid associations studied in Freetown the women got together for the opposite reason. They felt vulnerable *without* the family and other traditional forms of security and as a result experienced severe hardships.

Unless these women can improve their positions through successful trading, employment, adult literacy, or vocational education, they are trapped in a situation which prescribes two sets of norms—one rural and backed by customary law, and the other urban and contractual. As many are caught between the two, the discussions at some of the meetings I attended were often centered on the anxieties and conflicts surrounding customary marriage in an urban environment that give more weight to statutory marriage. As a theoretical reflection, it can be said that the oppression of women by statutory marriage as viewed by Marxist feminists, will likely not gain much popularity among women who view these marriages as sources of economic security.

This is aggravated by the fact that many of the husbands are reluctant to enter into statutory marriages. Although destitution, abandonment, separation, and divorce can occur in statutory marriages, from the point of view of these women, statutory marriage still offers them better economic security and legal protection through which they could have some recourse.

Imposition of Moral Sanctions

In the absence of effective traditional sanctions in the urban environment, guidelines for a code of conduct for women are usually provided by these associations. These rules ensure that the members are modest in their behavior and that married women do not enter into casual relations with other men. Modesty in behavior is a cultural prescription since it is one of the virtues inculcated during initiation by the Sande society. It is therefore an important aspect of the system of female morality.

In the city, some of this morality is threatened by the breakdown of traditional mores. A number of these associations seek to control morality to some extent by imposing strict rules of conduct on their members. The following Rules and Regulations of the Sabanoh Women's Association best illustrate the point:

> Any member who tries to bring the club into disgrace through bad suspicions by their husbands, parents and guardians will be expelled from the club. For example, if any member leaves her house to attend a function of the society she must not go to any other place. The President shall not be afraid to expose any such member if she is asked about her whereabouts, or if she notices the member left home for the meeting or function of the society but did not attend.

While modesty in behavior is seen as important for women, the association recognizes its gender implications and considers it of equal importance for men if the integrity of the system of morality for women. As a result, the association may have members who offer informal counseling to couples having marital difficulties, and try to influence men who are negligent in their family responsibilities.

Multiple Mothering

To some extent these associations serve as extended families in the city by offering assistance with childcare, housework, trading activities, and health care. They also provide an informal network for the collective socialization of children. Older women help with childcare and at the birth of a baby, at least three members will be assigned to assist the new mother. A form of multiple mothering is evident, which is characteristic of the kin-based support group, more prevalent in rural areas.

Since members live in the same neighborhood, collective childcare and multiple mothering are often feasible, expedient, and desirable. This form of mothering allows for the consolidation of various resources and talents. Discipline is a problem for children in their teens and it is not unusual for older women to be asked for disciplinary advice and counseling. Male kinfolk and neighbors can also be called upon for disciplinary responsibilities that extend beyond the kin group. This often benefits other members of the association without male kin who can serve as disciplinarians and role models.

Recreational Activities

A number of mutual-aid societies have functions that provide important recreational avenues for socializing. Although personality conflicts and grievances surface from time to time, the atmosphere at most meetings is one of levity and humor, and the spirit of friendship and conviviality is fostered as a principle. Some fund-raising activities have a social component and are often accompanied by singing and dancing. They are designed to augment subscription by members. Other recreational activities are directly related to Sande ceremonies, which often involve the initiation of children and close relatives of members.

Members usually make special contributions to the new graduates and their families for "pul bondo" ceremonies. As a symbol of the solidarity fostered by Sande, members wear the same dress material, according to a custom known as "ashuobi" during these celebrations.[5] Feasting, singing, and dancing usually take place in the home of the member whose family or relative is being honored.

It is also customary for other occasions such as marriages, childbirth, and after-death ceremonies and so forth to be celebrated collectively. These activities renew friendships and help cement the bonds of sorority and solidarity, derived from initiation into Sande society.

Promoting Solidarity and Political Expression

The motto one of these associations, the Sabanoh Women's Society ("wan word," i.e., "of one mind") underscores the solidarity and sense of unity fostered by these associations. In addition to the solidarity fostered by Sande membership, most of the members belong to the Temne ethnic group and are Muslims by faith. Many claim to have come originally from the same geographical area.

Most of the women are petty traders and some are involved in the gara-dyeing industry. Only 15 women out of a total of 124 in three mutual-aid societies are wage earners. Of these, ten are sales workers, one is a hairdresser, another a clerk, and three are maids.

Women of similar backgrounds, socioeconomic status, and common experiences can be said to share primary group identification in the city. The common experience of secret society initiation already constitutes a group with strong unifying bonds. Mutual-aid societies such as the Sabanoh Women's Society, merely formalize these bonds and provide opportunities for the further expression of solidarity. In this regard, they are already mobilized to take advantage of political opportunities and to lobby directly or indirectly, for issues of interest to them.

Children playing at Lumley Beach. Courtesy: Azania Steady.

The solidarity developed in Sande and fostered through these secular associations has found political expression on a number of occasions. It has been particularly instrumental in political participation at the national level. Secret societies traditionally had political functions, especially among the Mende and Sherbro, where female chiefs and paramount chiefs with executive political powers have been commonplace. Many of these chiefs also have leadership roles in Sande.

Secret societies and mutual-aid associations that are already mobilized are important assets to political parties and seek their affiliation. This often takes the form of membership in the Women's Wings of political parties through which they can become active in campaigning for candidates in local and national elections. Although the majority of candidates for political office tend to be men, in the event that there is a woman candidate that can promote their interests, these associations have campaigned for women candidates in local and national elections.

Politics in Freetown often takes on an ethnic dimension and political parties have traditionally developed along the lines of one dominant ethnic group serving as the base. As a result, women's associations that have a strong ethnic base have become important avenues for political expression. But there is a kind of paradox. On the one hand, they tend to reinforce ethnic cleavages by invoking consciousness of ethnic affiliation and common experiences. On the other hand, by linking up with mass political parties, which albeit have strong ethnic flavors, they are contributing to the promotion of a single nationalism and democratization of the political process. This often involves promoting development objectives aimed at improving living standards and improving economic opportunities for women.

CHAPTER EIGHT
ISLAM AND WOMEN'S ASSOCIATIONS

Introduction

The process of Islamization, which presumably began in sub-Saharan Africa around the eleventh century, became intensified in the eighteenth and nineteenth centuries. Islam was introduced to Freetown primarily through trade and individual contacts, although the Futa Jallon Jihad of 1727 had the most widespread and lasting effect.[1] Records show that an Islamic institution, the Islamic University of Foday Tarawaly, thrived at Gbile around 1870. One of the earliest Muslim communities in Freetown was Foulah Town, founded in 1819 and composed mostly of Madingo and Foulah migrants from further north who practiced a form of orthodox Islam characteristic of the western Sudanese variety.[2]

The Oku, a predominantly Muslim group of Yoruba origin, formed a distinctive community within the larger group of Liberated Africans. These were people freed in Freetown, en route to being enslaved through the trans-Atlantic Slave Trade process. Many became Christians but some, like the Oku, resisted conversion to Christianity in the nineteenth century and have successfully practiced Islam mainly at Fourah Bay, Foulah Town, and Aberdeen for over two centuries. Their Yoruba connection, coupled with their role as traders resulted in relatively greater economic and individual independence for Oku Women.

In 1932, an attempt was made to unite the various Muslim communities through the creation of the Sierra Leone Congress, which set up Islamic schools and colleges. Most Sierra Leonean Muslims belong to the Sunni sect but in 1939 missionaries from Pakistan introduced the Ahmadiyya sect as a more "progressive" form of Islam and tried to win converts through their greater emphasis on education. Partly in response to this, and as a result of the growing importance of Islam in Freetown, the Muslim Association was founded in 1942 with the main objective of setting up Islamic educational institutions.

Both Western and Islamic education are sought by Muslims as Western education is given a high value as one of the main avenues for social mobility. Muslim women receive formal education in large numbers due primarily to the role of the Ahmadiyya sect.

A large number of middle-class and affluent Sierra Leoneans practice the Islamic religion. At the same time, it also has strong grassroots appeal to low-income groups, marginal groups, and newcomers to the city. Islam appeals to all socioeconomic

Federation of Muslim Women marching for peace on International Women's Day in Freetown. Courtesy: Christopher Greene.

groups. In Freetown, and in Sierra Leone as a whole, it is a religion of social and political significance. The religious leaders who constitute the board of Imams are among the most highly respected and influential leaders of the country. Muslims have held and continue to hold high office in all spheres of society including the mayoralty, cabinet ministers, and so forth. The current Head of State Alhaji Dr. Ahmad Tejan Kabbah is a Muslim.

In the international context, the growing wave of Islamic nationalism has inspired similar sentiments among Muslims in Sierra Leone. Diplomatic relations and communication channels have increased with Muslims in North Africa. Sierra Leone formally became a member of the Islamic Conference in 1982. Such a membership entitles the country to receive development funds from the Arab Fund for Development. Some Muslim organizations, including women's associations, have received financial assistance from Arab countries and maintained links with individuals in these countries. Other activities reinforcing this international link include the popular Pilgrimage to Mecca, the annual observance of Ramadan, and the feast of Eid-ul-Fitri.

Women and Islam

The Islamic tradition in Freetown offers great flexibility and represents the amalgamation of various cultural and religious elements. This characteristic has influenced women's role and status. The perception of Islam as an African religion, or at least as a religion closely related to African belief systems, has also been influential.[3]

In Sierra Leone as well as in many African countries outside the Mediterranean Islamic cultural area, Islam represents a synthesis of Muslim and African cultural elements—a factor that is important in understanding its specific form, its secular proclivities, and its influence on women in Freetown.

Among the Oku, one of the more established Muslim groups in Freetown, some women who were interviewed regarded the restrictions placed on women by Islam as things of the past. One of them stated:

> The role which the average Muslim girl was expected to play in life had to be inferior to that of her menfolk. She was expected to be a good wife and housekeeper and to bring up her children—not to take up a career. She was to be under the direct control of her husband and always submissive to his will. She was never regarded as equal to her husband whether in status or in intellectual ability.

This was confirmed by an Oku scholar, who noted the following:

> Only the prosperous Oku gave their daughters equal educational opportunities with their sons. Women received non-formal education for extended periods in institutions such as the Sunna which prepared them for their domestic roles. Instruction in Arabic or opportunities for advanced education were limited.[4]

According to informants, a certain degree of female emancipation was possible through trade, the traditional occupation of Oku women. Girls were taught from a very early age how to trade so that by the time they were married they were experts and could become economically self-sufficient. This was sometimes necessary because, according to Islamic law, a husband could have up to four wives and may be unable to support all of them. The only other traditional occupation for Oku women was dressmaking, the skills of which were acquired through apprenticeship with relatives. During these periods, the girls were expected to help with household chores.

Within the last three decades, considerable change has taken place in the general position of Muslim women. Both Muslim and Christian girls routinely receive formal Western education and the barrier to education is likely to be economic than religious. There are several Muslim women graduates from universities, in both Sierra Leone and abroad, and Muslim women hold positions in all professions and in political life.

Due to the changes in the position of Oku women, a number of their institutions are being affected. Girls are no longer secluded for long periods during the Sunna ceremony, equivalent to the initiation of Sande. In addition, the long period of seclusion for widows, which could last for two or more months, has been considerably reduced. Similarly, rules pertaining to women's attire, as stipulated by Islamic law, are no longer strictly followed.

As a religion in general, Islam is noted for its strict ritual separation of the sexes, the most extreme being female seclusion in the Purdah. Men are more fully involved with the formal aspects of religion and women's religious participation is considered marginal. This gendered dichotomy has been referred to as the "sexual division of religion."[5]

Mernissi is even more explicit about gender inequalities in Islam in her study of Muslim women in Morocco. She states that the UMMA, the community of Muslim believers, is primarily male and that women's position in the UMMA universe is ambiguous:

> Allah does not talk to them, we can therefore assume that UMMA is primarily male believers.[6]

In Morocco, the ritual separation of the sexes has resulted in the development of secondary cults and practices that are predominantly female. These include ritual performances for the dead; offerings at the tombs of Saints; non-Islamic cults involving sacrifices to nature spirits and participation in possession cults such as Zar, Bori, and Pepe.

Throughout Muslim Northeast Africa and the Middle East, spirit possession cults (Zar) exist with membership that is predominantly female. These cults have been viewed as important for promoting female solidarity and facilitating women's adaptation and integration into the heterogeneous urban society. One study has shown how dancing groups of Muslim women perform similar functions in urban Kenya.[7]

In Freetown, these cults and ritual groups do not appear to exist among Muslim women in any significant way or in large numbers. Muslim women in Freetown, with the exception of a few orthodox groups, practice a more flexible, secular, and individualistic form of Islam which gives them a certain degree of freedom and independence.

On the whole, they are less subject to the strict interpretation of Islamic doctrines concerning the place of women. This is possible partly because Islamic doctrines have not been rigidly adopted. For many groups in Freetown, conversion to Islam does not require total religious commitment. It also does not preclude continuing beliefs and practices of other religions, including African traditional religions.

In keeping with Islamic tradition, men and women are usually segregated during ritual occasions but women frequently use the Mosque as meeting places for their associations. Among the more orthodox Muslim groups, such as the Fullah and Hausa, women tend to be confined to the home. Their participation in religious rituals in the Mosque is often restricted. The strict observance of seclusion (Purdah) is rare, even for these orthodox Muslims.

There are examples of women who grew up in an environment that encouraged female religious education for women and who have become Islamic scholars. These women participate in various religious festivities by leading the singing in Mosques and by taking part in public recitals in Arabic. One such example was an informant, a descendant of a religious leader, who was one of the first Sierra Leonean Muslim leaders to make a pilgrimage to Mecca. According to one informant, Islam is a good religion for women.

> For me Islam is the best religion. It respects women. I have never contemplated changing it for another. Today, an increasing number of Muslim women are attending school and several have obtained university degrees. Whereas in the past marriage was seen as the most important goal for a Muslim girl, mothers now desire a good education and a good job for their daughters in addition to marriage.

Women in Freetown are developing their own parallel spheres of influence within the religion through their associations and through participation in the pilgrimage to Mecca—the Hajj. They are reexamining Islam and providing mutual support for each other. By making the pilgrimage, they earn the title of "Haja." The Hajj provides an avenue for achieving status in the Muslim community, where religion is strictly male dominated.

The Hajj also fulfills a deep spiritual need to be at the most holy place for Muslims and to offer prayers to Allah from the sacred land. As will be seen later, some Hajas hold important pivotal positions in the mobilization of women for mutual support and serve as links between grassroots women and the government.

Hajas automatically receive a high spiritual status and its accompanying respect in the community. They wear a distinctive head-tie and status-conferring dress. This status is even more special since the high cost involved in making the pilgrimage precludes a large number of women of lower income from becoming Hajas.

A number of Muslim women's associations, comprised predominantly of younger Muslim women with advanced education, have been studying the Koran and Sharia with a deconstructionist and revisionist approach. They argue that these religious doctrines guaranteed women's rights and that the erosion of women's rights has been the result of male-biased interpretations of religious texts. These associations act as pressure groups to raise awareness and effect change. They emphasize the need for a reinterpretation of these classical texts to reflect the protection of women's rights that are inherent in them.

Limitations of Islamic Influence on Social Institutions

The influence of Islam on social institutions has not been as extensive as in North Africa and consequently affects the status and position of women less profoundly. In the first place, the various Muslim groups, representing different ethnic groups and neighborhoods, belong to several Islamic communities known as Jamaa. The elders of these communities form a council with legal responsibilities to interpret the Sharia, and to a certain extent regulate the lives of the members of the Jamaa. Each Jamaa can give its own slant to the interpretation of the Sharia.

In the second place, whenever strong patriarchal authority is exercised, there is greater likelihood of it being evident in smaller households where males are the sole breadwinners. This factor tends to reinforce patriarchal control. It is also likely to be more a characteristic of orthodox groups in which women tend to be confined to the domestic sphere. Most Muslim women in Freetown are involved in income-generating activities, primarily trade, making them less dependent on men. Given the difficult economic conditions of Freetown, it is highly unlikely that most men can ever be the sole breadwinners of their households. This would tend to decrease their ability to exercise absolute patriarchal dominance. As women traders often say, "It is better to depend on trading than to depend on a man."

Third, another important difference pertains to marriage. The alliance between two families, which African marriage fosters, is likely to be less affected by Islam. As Trimingham points out, under Islam, marriage is more of a contract between a man and a woman than an alliance between families.[8] According to informants also, this has had several consequences. One is a lesser degree of family supervision of the married couple and greater individualism for women within the marriage.

Fourth, Islam facilitates the divorce of a man by granting him the right to employ the Islamic rule of repudiation. According to this rule, a man only has to repeat the phrase "I divorce you" three times and the divorce is final. In Freetown, among the non-orthodox groups a woman may also initiate divorce by virtue of the individualistic nature of the marriage contract and the degree of her economic independence. Furthermore, many women still adhere to a tradition of customary African marriage even when they have converted to Islam.

Finally, both Islamic and African marriages (unlike in Christianity) are by custom polygamous. However, this practice is more difficult to maintain in a non-agricultural area such as Freetown and can become quite expensive. It is customary for women in polygamous households to generate their own income and be solely or partly responsible for their children's upkeep and education. Polygamy offers economic security in the urban area only if the husband is prosperous, but in such instances, the independence of women in the household is likely to be curtailed.

As trading is the most widespread economic activity among women in Freetown, Muslim women are mobile and can exercise a certain degree of autonomy and control over their lives. The stereotype of the subordinate and restricted Muslim woman does not apply to the majority of Muslim women in Freetown. This factor, in addition to living in a culturally heterogeneous urban area, encourages the development of a worldview that anticipates and responds to change in gender roles and expectations. In addition, the frequency of intermarriage and informal adoptions across ethnic and religious lines have resulted in less rigid adherence to norms and values restricting and regulating women's lives.

Muslim communities are usually closely knit and the neighborhood mosque provides a central focus. Within domestic and kin groups, women operate essentially in their own sphere, and are dynamic forces behind the social celebrations related to life cycle events. During major feasts such as Eid-ul-Fitri or during Ramadan, they demonstrate their identification with their communities by dressing alike in ashuobi fashion and celebrating together.

Kinship ties are also kept alive through gifts and frequent visits with kinfolk and exchange of services. From an early age, women develop strong informal female support systems, through primary group ties derived from kinship, common neighborhood activities, and trading.

Ironically, one of the most important factors that contribute to greater freedom for the majority of Muslim women is economic hardship. The underdeveloped urban environment leads to frequent unemployment of men and chronic unemployment of women. Trading in the informal sector is the only option for the large majority of Muslim women. This tends to enhance their freedom, mobility and independence to some extent.

Examples of Muslim Women's Associations

A study of Muslim women's associations is useful in understanding the dynamic interaction between religion, economics, and gender roles. It is a study that can lead to an appreciation of how Muslim women, particularly women of low income, cope with problems of economic survival in Freetown through associations for mutual support. These associations do not fit neatly into the category of associations developed by Muslim women in other areas in response to the "sexual division of religion."

Although the religious base of these associations is necessary for identification and mobilization, they hold a secondary position to more pragmatic and secular functions. Many of them were developed in response to the exigencies of socioeconomic survival in the urban environment. Those that grant material assistance to their members are among the more popular, and usually have strong grass-roots support. This support, on occasion, has been instrumental in mobilizing whole communities for self-help projects. Though Islam has provided the unifying base for rallying support, members are motivated for economic and social reasons rather than for purely religious reasons.

Philanthropic deeds, which include contributions to charity and almsgiving to the poor, are among other significant aspects. Contributions to charity are often solicited by invoking the Islamic spirit of philanthropy. For example, the following plea is frequently made to members at meetings.

> We know that a good Muslim should be generous. The spirit of generosity is in keeping with Islamic teachings about giving alms to the poor. So give freely for charity and prove that you are a good Muslim.

The following have been among the most active associations of Muslim women in Freetown:

- The All Muslim Women's Association
- The Amalgamated Muslim Women's Movement
- Tarikful Islam
- The Federation of Muslim Women

- The National Council of Muslim Women
- The Committee of Fulah Town Ladies
- The Young Women's Muslim Association.

Their functions are mainly educational, although almost all of them provide mutual aid to their members. Learning the Koran and about other religious texts is one of their most important functions. They also raise funds for scholarship for Muslim girls. Their secondary function is philanthropic. The All Muslim Women's Association raises funds through luncheon sales and dances, and makes donations to organized charities.

The Young Women's Muslim Association functions primarily as a mutual-aid group, but also serves as a thrift and credit society for raising loans. The Sierra Leone Muslim Women's Federation maintains close contacts with other associations of Muslim women, and serves as an umbrella association that coordinates some of their combined activities.

One of the most important Islamic-based grass-roots women's organizations is Kankalay, located about two miles east of central Freetown in a low income, unplanned peri-urban area which can be classified as a shanty town. Although there are a few professional families of middle income and some modern houses in the area, the majority of houses are in very poor condition and services are inadequate.

Most of the inhabitants are manual workers, laborers, and traders of low income representing first- and second-generation migrants to the city from rural areas. The majority of women subsist in the informal sector as petty traders. Unemployment is rife and life for many of these women is precarious.

Women of this area have had to develop strategies for surviving in the urban environment. One such strategy is mobilizing around Islam. Ethnicity has also been an important factor in mobilization, as has been shown for Muslim women's associations in parts of East Africa.[9] Since 50 percent of the members of Kankalay are Temne, the meaning of Kankalay "one word" or "unity" may imply mobilization and solidarity along ethnic lines. This, however, could be misleading, for the solidarity in this case is designed to provide protection against urban poverty that affects all ethnic groups.

In Kankalay, men are not viewed as adversaries but as fellow victims of economic hardship. In fact, the association, which has branches in other parts of the country, once had a man as its national president. Women presidents are predominant in the branch organizations but most of the secretaries and treasurers have been men, due to their advantages in education, accounting acumen, and negotiating skills.

As most of the activities of Kankalay are aimed at improving the community through self-help, both men and women view these efforts as beneficial to the community as a whole. In their everyday struggle for survival, cooperative values are stressed. Providing relief in times of need is one of its most attractive features, and is an important mechanism for recruitment of women of low income.

Mutual aid usually entails rendering assistance during illness, occasions such as the birth of a child, and early infant care. The association assumes full responsibility for the funeral expenses of bereaved members. Funerals are treated as emergencies that grant the president the right to call a meeting at very short notice and to request donations for funeral expenses. A delegation is usually selected to represent the association and to assist with funeral arrangements if necessary. Funerals are also used as occasions to recruit new members and to cement bonds of solidarity.

Owing to their importance as public events, funerals also serve a latent function of mobilizing the whole community. It is not customary for Muslim women to attend funerals. Hence, women would congregate at the home of the bereaved as well as in the surrounding areas for an informal gathering. At this time, the home assumes public significance and the fact that women dominate this space heightens its ideological significance.

The predominance of women in this space, usually considered as the private space, is a recognition of their life-giving and life-sustaining importance, made possible through their roles in social reproduction. The public space of the home where the funeral is being held, also signals the role of women in heralding the deceased to the "new birth" in the world of the ancestors.

Petty trading, the occupation of almost all the Kankalay members, has necessitated the involvement of Kankalay in trading activities and in the incorporation of a savings and credit facility. This had led to the establishment of an Osusu—a rotating credit association—within its structure. The principle of the Osusu discussed earlier, is that each member contributes the same amount each month and once a month, one member is allowed to make a withdrawal of the total amount.

Kankalay also functions as a cooperative and once operated a retail distribution service. The association would buy foodstuffs, especially rice, at wholesale prices and distribute them to members for retail trading. Rice is a "political" commodity as well as a nutritional staple, and its shortage can precipitate political unrest. Most regimes have capitalized on this by offering rice subsidies to political supporters and their wives. Some regimes have required party membership cards to be shown when purchasing rice.

By belonging to Kankalay, women have on occasion been able to buy rice from the association at the same cost, without having the additional expense of joining a political party and securing a membership card for a fee. A number of Kankalay leaders have been members of the political party in power in their individual capacity. Some have assumed leadership roles in the women's wing of the ruling party, enabling them to maintain links with ruling parties.

Political Links Through Hajas

One effective strategy for survival used by low-income groups in Freetown is to establish links with individuals, who are not only community leaders, but also have access to the government. It is primarily based on a patron/client model but is inspired by a "collective" ethos rather than individual relationships. Some leaders of Kankalay have links with the government that are based on their status as religious leaders such as Hajas who have made the Pilgrimage to Mecca (the Hajj).

Hajas now constitute an important group and have often assumed responsibility for acting as liaison between Muslim women and the government. This has occasionally drawn criticisms of promoting self-interest, political aggrandizement, or patronage, but Hajas usually enjoy a high status in the community. Many Hajas have performed a number of important political functions of benefit to women at the grass roots level, and have helped women's associations to become effective pressure groups. Hajas are assuming national stature as leaders and may become a force to be reckoned with as they get more involved in national politics. In the 2002 elections, one of them, Haja Memunah Conteh, was the running mate of Dr. John Karefa-Smart, a Presidential candidate.

The pilgrimage to Mecca is real in one sense and symbolic in others. Hajas continue to make political pilgrimages to the President and Cabinet Ministers to request assistance as well as to lobby for the improvement and development of their communities. It was in response to this form of lobbying that the government provided Kankalay with some land for the construction of a secondary school for girls at Kissy.

A shift in emphasis has occurred in Muslim women's associations. Whereas in the past they functioned primarily as religious institutions for religious instruction of women, their emphasis has become more secular and has taken on economic and political functions. The shift can be viewed as a response to the marginalization of low-income groups in Freetown, many of whom belong to the Islamic faith.

In summary, Muslim women have very pragmatic views toward life and socioeconomic development. Islam is an important religious resource for facilitating the achievement of some of their development goals. For many, an important motivation in joining these associations is to mobilize resources through group action that offers material support as well as moral, and religious solidarity.

Members of these associations also seek assistance for the education of their children with the hope of future employment. The concept of self-help is strongly developed, as is the realization that linkages to political and religious leaders are vital strategies for the development and advancement of the community. Kankalay is a good example of one association that has brought into being adequate mechanisms for achieving these goals, and serves as an illustration of women supporting and empowering themselves for development and for political alliance-building.

Chapter Nine
Christianity and Women's Associations

Introduction

Christianity, like Islam, has played an important role in women's mobilization and also in their collective action. Like Islam it also represents a religion that is based on patriarchal values and that is relatively conservative. Women's associations to some extent have served to modify some of the most blatant forms of male domination, but both religions are run by a clergy and decision-making bodies that are predominantly male.

At the same time, Christianity like Islam arguably promotes a moral system that stresses equality and justice; compassion for others; philanthropy and human dignity. All of these values echo sentiments that are in keeping with development and democratization, the primary focus of women's associations.

Christians can be found among all ethnic groups. However, one group, the Krios have practiced Christianity for over three centuries. As a result, the following focuses on women's associations among Krio women. With few exceptions, much has not changed within the last thirty years and many of the features are still relevant today.[1]

One important change is the number of women lay preachers and ministers from all ethnic groups that now play active roles as leaders in several churches. Following the pioneering example of Jane Bloomer of the Martha Davis Confidential Association, women like Madam Dumbuya now lead large independent churches. Also of importance is the increase in the number of women's associations affiliated to churches that conduct services in one of the local languages.

As a result of the rebel war, Christian women's associations have increasingly been mixing religion and politics by staging public prayer campaigns and advocating for peace. Many believe that the cumulative prayers of the people of Sierra Leone, especially the women, could be credited with the eventual end of the war and the subsequent process of peace building.

Historically, Christianity was the integrative catalyst for the various groups that settled in Freetown in the late eighteenth and early nineteenth centuries. This was a result of the repatriation of enslaved people of the trans-Atlantic Slave Trade and Liberated Africans to Freetown. The descendants of these various groups became known as Krios. Christian missionaries, particularly of the Anglican Church Missionary

Society, were primarily responsible for proselytizing and maintaining Christianity and Western education among the early groups, especially the Liberated Africans, who later became part of the amalgamated Krio group.

Those among the Krios that came from Britain and Nova Scotia were already practicing Christianity as Wesleyans, Baptists, Catholics, and members of the Countess of Huntingdon Connexion.[2] In time, other denominations led by evangelical missionaries from the United States were established by various mission groups. Nonetheless, the Protestant influence remained dominant.[3]

Christianity in Freetown, therefore, has a strong Krio influence, although most ethnic groups, especially the Mende and Sherbro, have long practiced Christianity either in the mainstream churches or in churches that use local languages extensively. The Okus, part of the extended Krio group, are predominantly Muslims and play important roles in advancing Islam in many of the mosques in Freetown and in their community.

Freetown's acceptance of Christianity was not without some important modifications. The local version did not represent a complete break with the church, but developed its own special brand and flavor: Even the so-called Western Services in the Krio churches, the oldest in Freetown, have a distinctive quality of their own, which is not imported from Europe.[4]

Freetown was one of the first communities in tropical Africa to embark on the task of reconciling African life and Christian values, and of infusing a dynamic African element into the religion through prayer meetings and special liturgical experiences. Voluntary associations developed as extra-curricular entities of the church. This promoted the development of new leadership structures, greater individual expression, and greater female participation.

Making Christianity indigenous is credited to the work of some of the earliest African clerics such as "Holy Johnson" who insisted on an African emphasis to Christianity and worked on developing such a church in Freetown. This process, which has been in operation for almost two hundred years, is as old as the very existence of the Krios.

Consequently, for the Krios of Freetown, Christianity is the traditional religion. This partly explains why separatist churches that blend traditional African religions with Christianity have not been a feature of Krio society. This also explains why the majority and the oldest of Christian women's associations in Freetown are comprised of Krio women and serve as an extension of Krio social organization and culture. As a result, this chapter focuses on the influence of these associations on the formation of the ideology of womanhood primarily in Krio society.[5]

Associations developed around churches as an extension of religious fellowship. For women, this was particularly significant as avenues for fellowship with other women. Such associational ties helped to strengthen common beliefs and preserve Christian ideals, especially the ideals of womanhood, monogamous marriage, and motherhood.

These ideals were conservative and not questioned, as marriage and motherhood did not receive the same admonition characteristic of some strands of Western feminism. In fact, many of the women who join organizations like Mothers' Union, and who were interviewed, considered Western feminism as subversive to the family and to Christian values.

Material prosperity was also an ideal supported by the class-conscious nature of Freetown society, by its commercial character, and by the Protestant ethic as understood by Weber.[6] Prosperity was more of an ideal than a reality, for only a few people of Freetown can be described as prosperous. Although there is an effective middle class, the majority is characteristic of most urban dwellers in Africa, and belongs to the lower middle and lower classes.

The standard of living of the majority of the middle class has been eroding as a result of the protracted economic recession and the ten-year rebel war. Historically speaking, prosperity was short-lived. The following perspicacious analysis of the history of the political economy of West Africa by Amin makes an important point.

> The most interesting chapter in the political economy of Sierra Leone cannot, for its part, be separated from the history of Nigeria. It was the trading station at Freetown, where the British navy assembled the freed slaves, which stimulated the formation of the "Krio" bourgeoisie which spread along the whole of the western coast in the nineteenth century and filled the role of a comprador bourgeoisie for British capital. But this class disappeared at the end of the last century, when the British executed their main Krio trading rivals on the pretext that they had taken part in the Temne and Mende revolts.[7]

Today many Krios have a precarious economic existence and few can match the wealth of politicians or Lebanese and European merchants in Freetown. While a number are in the professions, the majority works for wages, or as salaried clerks and artisans.

Krio society is nonetheless marked by class distinctions, which is sometimes reflected in the churches. For example, St. George's Cathedral, the Diocesan seat of the Anglican church, situated in the center of Freetown is the largest and most highly esteemed. It is among the churches that has in its membership some of the most affluent of Freetown's Krio citizens.

Freetown Krio society also exhibits Christian and English characteristics that are apparent in its religious and educational institutions. It must be added, however, that there are also strong retentions of African and Caribbean influences that are even more important.

For example, most rites de passages have strong African elements, as does the complex bilateral kinship structure. Caribbean and African influences are apparent in Krio architecture and music (gumbe), in the traditional dish fufu and female ethnic dress—print en enkintcha, or its forerunner, kabaslot en kotoku. The Krio language, derived largely from English but containing many words from various West African languages and a few words from other European languages, readily represents this cultural blend.[8]

Although the Krios were among the first to practice Christianity in Freetown they only represent one group of Christians. In theory, no church sets out to have a single ethnic composition, but there are churches that carry an ethnic designation and whose services are conducted in the language of the predominant ethnic group.

One reason offered for the establishment of these churches is "the need for the various ethnic groups to worship separately in their own language, so that they may more readily understand the services and offer their worship intelligently."[9]

Mrs. Blanche Benjamin in *print en enkincha* traditional dress. Courtesy: Azania Steady.

Women's associations are increasingly playing an important role in these churches that have increased substantially in Freetown, especially because of internal displacement and refugee flight resulting from the civil wars in both Sierra Leone and Liberia. American churches have been particularly active in promoting services in the local languages. These have included the Assemblies of God, the Evangelical United Brethren, and the Jehovah's Witness. Compared to the well-established Anglican, Methodist, and Catholic churches from Europe, American missions can be regarded as late arrivals.

Over the last two decades, there has been an avalanche of evangelical activities in Freetown, and include several denominations, and tend to attract all ethnic groups. Notable among them are the Church of Jesus Christ of Latter Day Saints (the Mormon Church) primarily from the United States. In addition, the appeal of syncretic churches that are a blend of Christianity and African religions has been increasing and challenging the dominance of earlier syncretic churches such as the Church of the Lord, Aladura (Adejobi), from Nigeria.

Christian Women's Associations

Women's religious associations are widespread among Freetown dwellers, especially Krio women in their middle years. Variously styled as bands, committees, groups, guilds, unions, or societies, these associations are autonomous bodies within each church. A few of them, namely the Mothers' Union and the Women's Volunteers (Anglican), as well as the District Women's Work Committee (Methodist) operate at national and local levels. This is also true of the Women's Society for World Service of the Evangelical United Brethren denomination. Each association has its distinctive uniform (usually a white dress, a sash, and a straw hat with the association ribbon) and membership badges.

Despite their autonomous nature, there is a similarity in their functions. These include mobilizing resources, maintaining the ideology of Christian marriage and motherhood, and philanthropy, promoting healing, facilitating the power of prayer; advancing female charisma and religious leadership and promoting unity.

Mobilizing Resources

Most churches derive their revenue from their members' contributions to the weekly collection that is often insufficient. Additional fund-raising is organized by women's associations. This is done through activities such as luncheon sales, bazaars, fetes, and thanksgiving services. The proceeds go toward the general maintenance and renovation of the church, or toward the purchase of some new particular item such as a silver chalice or a stained glass window.

In some cases, associations such as Women's Guilds and Women's Volunteers have been formed for a special fund-raising project—to purchase a pipe organ or to help the church clear some of its debts. Once the equipment has been purchased or the debt cleared, they may cease to function. Some associations are seasonal—for example, Harvest Communities, which are active only during the period of the harvest festival. As a result, there is much proliferation of short-term women's religious associations, which consequently constitute the largest type of women's associations.

A number of religious associations guarantee the general upkeep of the church by setting aside a maintenance fund for minor repairs and renovations. One association, the Ladies Working Band, is found in almost all Anglican churches and functions in a direct caretaking capacity. The members of each Band regularly clean the church premises, and they decorate it on ceremonial occasions. In the Methodist churches, associations with the names Ladies' Guild, Silent Worker, Ladies Union, Ladies Industrial, and Ladies Auxiliary usually perform similar functions. Members of a number of other associations act as "sideswomen" (ushers) during services.

A portion of the funds raised is donated to charitable organizations, in keeping with Christian teachings about philanthropy. In addition, association members visit hospitals, orphanages, and homes for the handicapped at Christmas to sing carols and to present gifts to the inmates. Members of the Dorcas Association formerly made clothes and purchased food items and religious books for others less fortunate than themselves.

Maintaining the Ideology of Womanhood and Christian Marriage

Religion often fashions rules and regulations that guide gender relations. In the case of women's religious associations, the ideology of womanhood is one that includes monogamous Christian marriage. The responsibility for both is placed in the hands of women.

Another feature is the absence of female secret societies such as Sande that perform initiation rites. For Krio men, freemasonry, an exclusively male secret society has been popular as has hunting societies. In contrast to freemasonry, hunting societies have a few adult female members.

In keeping with Christian doctrine and statutory law, Krio marriage is monogamous and a husband is obligated to support his wife and children. This may explain the relative absence of separatist churches among the Krios, as polygamy is permitted in those churches. Monogamy is frequently regarded not as a Christian institution, but as a specifically European one, lacking spiritual sanction. However it has become a feature of Krio society.

The Mothers' Union, an international association, is the prime example of an association preserving what has come to be regarded as Christian marriage and morality. The Mothers' Union of Sierra Leone has remained virtually the same over the years, as its objectives have remained relatively unchanged. It is organized on Diocesan and local levels throughout the country. Each branch, headed by the "enrolling member," is attached to a church; and at the local level each branch functions separately. All branches meet together once a year to mark the opening of the Mothers' Union year.

The aim of the association is the advancement of the Christian religion in terms of strengthening and preserving monogamous marriage and Christian family life. Although it is part of the Anglican Church, membership is open to other women from Christian denominations. Widowed and divorced women are accepted as members to provide them with spiritual support since, according to their philosophy, their civil status does not reflect their inner spirituality. The following are the main objectives of the association:

- To uphold Christ's teaching on the nature of marriage and to promote its wider understanding.
- To encourage parents to bring up their children in the faith and life of the church.
- To maintain a worldwide fellowship of Christians united in prayer, worship, and service.
- To promote conditions in society favorable to a stable family life and the protection of children.
- To help those whose family life has met with adversity.

It is customary to have sermons and addresses that quote extensively from the Bible, passages that reinforce the home as having a divine origin (Genesis 1:27–28); the importance of creative family relationships (Ephesians 5:21–33); the responsibility of parents to bring up their children in the nurture and admonition of the Lord, (Ephesians 6:4) and nurture their children, and so forth.[10]

For the Mothers' Union, "the home comes before the State, The School and the Church. It is the basic institution of society. The Bible clearly speaks concerning its nature, concerning its purpose and concerning its preservation. A Christian home is a home where Christ is loved and trusted and obeyed sincerely and steadfastly."[11]

At this annual meeting, members reinforce their commitment to uphold the sanctity of Christian marriage and family life, considered prestigious in Freetown society. It is possible that these women guard not morality but their economic security by acting in some respects like a "women's trade union," preserving not only their marriage but also their livelihood.

Maintaining monogamous marriage as an ideal is important as long as family security and continuity depend on the husband being the breadwinner. These associations then promote their dependent status that would make them opposed to feminist ideas that are critical of the family. From my interviews with members of the Mothers' Union, I did not come across anyone that supported the criticism of the family, usually made by some feminists. For many, a stable family is the foundation of a stable society, and they are sure that their theories about the family are the right ones.

During the pioneer days of settlement in Freetown, individual Krio women had a great deal of economic independence and affluent lifestyles, primarily through success in their trading activities. Some of the best properties in early nineteenth-century Freetown were owned by women.[12] Today, many Krio women own property that they either acquired independently or inherited, and have independent incomes from trade and salaried employment.

However, with the exception of a few highly successful women entrepreneurs, career women, and women from well-to-do families, a married woman's income is generally regarded as supplementary to that of her husband. Many household surveys have noted that while female-headed households are common, most of the responses identify men as heads of households.[13]

What is significant is the fact that women's contribution to the household budget is an important supplement since the cost of urban housing, school fees, and transportation

Maroon Church—A historic landmark in Freetown. Courtesy: Christopher Greene.

impose tremendous strains on the husband's income. A working wife, whether self-employed or salaried, is often regarded as an economic asset. Thus, the majority of married women's earnings are seen essentially as supplementary, and women largely seek economic security in marriage because of greater opportunities for male employment within the modern urban structure. Single women heading families usually do so with some difficulty, even when employed, since salaries are generally quite low and inflation high.

The association is very understanding when it comes to women who are divorced. In its view, divorced women should not be abandoned when they need the Church most, since the breakup of their marriage might not be due to their neglect of the family but rather to the recklessness of their husbands or the challenges of life. The Mothers' Union has been filling the need for moral support and advice when marriage and family life seem threatened.

Marriage counseling is also offered by the Mothers' Union Workers and the Enrolling Members. These women are dedicated to the ideals of Christian marriage and family life, and are seen as having invaluable wisdom in helping to resolve marital problems. In addition to holding individual consultations, they frequently address branch meetings on various topics concerning the home and family.

Promoting Healing

Another important function of some types of women's religious associations is comforting the troubled and the sick. The need for this is due in part to underdevelopment and the social malaise of urban life, exacerbated by poor living standards and collective ill health for many. Malaria and other infections and parasitic diseases are prevalent and act synergistically with malnutrition to produce chronic ill health for large sections of the population, particularly children. The trauma of the ten-year war has created many refugees and displaced people who live under difficult conditions in Freetown and in refugee camps and are often in poor health.

Economic hardship and harsh living conditions such as overcrowding, poor housing, poor sanitation, and inadequate municipal services have also contributed to ill health. Stress and chronic anxiety resulting from rapidly changing social and economic conditions aggravate and contribute to hypertension and strokes. Interviews of physicians have revealed high levels of stress-related disease and an increase in the prescription of tension-reducing drugs for women.

Women's religious associations performing the functions of healing have provided the psychological component necessary for mind–body interaction in the healing of the total person. One of the best known was the Martha Davies Benevolent Women's Association, also known as the Jane Bloomer Church. Women with chronic illnesses, and those for whom western and traditional medicine have failed to provide a cure, have turned to divine healing.

Over the years, several charismatic religious leaders, such as Madam Dumbuya who established her own church, have been providing healing services. Part of the appeal of the "Aladura Church" and the "God of Our Light Church" has been due to their healing functions. Newer types of prayer groups similar to the Martha Davies Benevolent Association, and generally called "Struggle" groups, appear from time to time with healing as their main function. Healing covers physical and mental ills as well as social malaise.

Facilitating the Power of Prayer

A number of women who attend these prayer meetings or consult the leaders also seek solutions to social problems, economic hardship, and internal conflicts. The belief in prayer is very strong and is often reinforced during meetings by the leader as follows:

> Does prayer really matter? Is it important to us as individuals? As a family? As a nation? To us as individuals are we too busy that we do not consider it important to set aside specific times to pray? We rush off and leave things undone which need our attention. Is it always easy to say I will do better tomorrow. As Mothers' Union members, we should remember that we are entitled in prayer. To keep Satan out we must be prayer warriors, he trembles when we pray. Our prayers express our love and commitment to Christ. By talking to God we show that we depend on him. Jesus was a great prayer warrior. He prayed without ceasing. We need the supportive power of prayer especially in times of sorrow, sickness, wars, and trauma. We also need to pray however, when we have joy. So let us all share in prayer, upholding one another.[14]

Prayer is often regarded as a powerful force in solving problems and assuaging anxieties. The supplications offered by prayer groups can be seen as an extension of this view. Although prayer is important for all religious groups, it seems to have a particular meaning for members of the Mothers' Union even when the activities are of a more secular nature. Prayer provides women with a weapon in the fight against hardship. They believe that their faith and reliance on God gives them the strength to withstand the "trials and tribulations" of life.

Female Charisma and Religious Leadership

One of the obvious changes in the church is the number of women who are lay preachers. Fewer are ordained ministers, but this is a major improvement for the church. Reflecting the strongly patriarchal family law of the Old Testament, Canon Law institutionalized male dominance. Despite recent changes to diversify the clergy, Church leadership is still a male preserve. This legacy is still reflected in the suspicion of the virtually all-male clergy toward female leadership in the church.

An important function of religious associations therefore is to provide avenues for the development of religious leadership among women, who hold no formal position in the clerical hierarchy in Freetown. Many of the devotional and counseling functions of these associations are conducted by women leaders, and provide an avenue for the development of female religious leadership.

There is evidence from historical sources that in the early days of settlement women often had experience in church leadership. According to Fyfe, they preached and testified in the Nova Scotian churches. One woman had her own congregation in her house at Water Street.[15] However, the development of such female leadership was not encouraged by the Church Missionary Society nor by the Methodist missionaries.

Nor did the African clergy that later assumed the leadership of the missions allow the development of female leaders.

Religious leadership tends to command a large measure of automatic respect even outside the church hierarchy, and this accounts for its importance to men and women alike. Association leadership not only makes a woman an exceptionally good Christian but also adds to her status in the community. Moreover, it secures for her the ultimate glory of a grand funeral.

The greatest honor a Christian can receive after death is to be laid out in church and to have a well-attended funeral spilling over into the churchyard and adjacent streets. A cortege of association members in uniform marching at a woman's funeral is testimony to a life well lived as a Christian. Only very active women, usually leaders of religious associations, receive this great honor.

Women's charismatic qualities are nurtured and find expression in leading others in fellowship and devotion. Meetings usually have periods devoted to the reading of passages from the Scriptures, to Bible study, to prayers, and to hymn singing. Most women enjoy the devotional aspect of meetings because of the pleasure they derive from "feeding the soul" through religious songs, studying the Scriptures, and praying.

Some of the associations, such as the Martha Davies Confidential Benevolent Association and Mrs. Pinkney's Spiritualist association—named after their founders—are essentially prayer groups that have become formalized and have remained active after many years. These two groups offer a revival type of worship, faith healing, and extemporaneous prayer—popular kinds of devotional expression among some Freetown women.

Members of prayer groups in Freetown continue to attend their mainstream churches, unlike separatist groups in other parts of Africa where this type of worship results from a complete break that occurred with the mainstream church. In some cases, separatist churches have their own women priests, but in Freetown members of these prayer groups have not severed links with the mainstream. This may be because mainstream Christianity is already a meaningful religion whose values have become internalized to such an extent that they are an integral part of Krio everyday life. Those desiring a less formalized type of worship have tended to become affiliated with one or more of the American-based evangelical churches in Freetown or the charismatic sects.

Promoting Unity

An attempt made at uniting all Christian women by the United Church Women has been successful. It maintains close cooperation with the Sierra Leone United Christian Council, a non-denominational body. This association aims at encouraging women to come together in a spirit of fellowship:

> To study, speak, and act on issues in the country and in the world that involve moral, ethical and spiritual principles inherent in the Christian gospel.

This exercise has led to a more critical evaluation of women's position in the church, especially with regard to the question of female religious leadership. As noted earlier, many religious women's associations tend to be conservative and maintain existing religious norms and institutional procedures.

The development of a non-denominational multi-ethnic women's association such as the United Church Women on a national level might lead to more profound changes in the future. One of these changes is the increasing number of women who are lay preachers and Ministers. Another is the increasing role of women in development-related activities and in peace building through their church-based activities.

For example, in the Diocese of Bo, the second largest city in Sierra Leone, Mothers' Union members are among those who were admitted to the makeshift Médecins Sans Frontières hospitals to treat wounded war victims. The Freetown branch was represented at the inauguration of the Women's Movement for Peace, whose activities included seminars, workshops, and protest marches for peace.

Christian women regard their associations as essential resources to further augment their faith and devotion. Being a good Christian has high value in their system of merit. They do not make a connection with some feminist interpretations of marriage as an institution that oppresses women. They are more concerned about strengthening their families and raising children who will be good Christians and successful citizens through their devotion to God. For them, everything else is secondary.

Chapter Ten
Comparative Insights at the National Level

The Development–Underdevelopment Nexus

In this chapter, comparisons will be drawn with other countries in Africa, notably Kenya, Nigeria, South Africa, and Algeria, representing the four subregions. In order to examine the themes explored in this book, heavy reliance is placed on studies conducted primarily by women from these countries.[1] In a few instances, this has been supplemented by interviews. In the next chapter, this insight is applied to associations operating at the regional or pan-African level.

Almost everywhere in Africa, women have had a long tradition of organizing for collective action. Despite many challenges, they have achieved a measure of success. It is clear from the case study of Sierra Leone that women's associations play important roles in promoting socioeconomic development and as such, fill a major gap in the inability of the government to meet the country's development needs. At the same time, they try to resist the negative consequences of underdevelopment, most damaging of which is widespread poverty.

Despite the efforts of many governments, gender disparities continue, particularly in the areas of education and employment and in the low representation of women in decision-making bodies or in institutions. Women are also disproportionately affected by poverty and have a high dependency burden. The numbers of women heading households has been increasing steadily and now stands at 40 percent in some countries like Botswana.

The case study of Sierra Leone shows how women's associations have been effectively promoting development, especially in the areas of education, employment, creation, and resource mobilization, as well as political participation for democratization. It also reveals that these associations work to promote gender equality as part of the overall struggle for development and democratization. This is because development assumes priority in countries that face major economic challenges and the destructive forces of corporate globalization.

Examples from other countries in Africa confirm the development–underdevelopment nexus. The Women's Movement in Kenya, sometimes referred to as "the women's group movement" has a tradition of collective work and self-help for development.[2] The largest and best known is Maendeleo Ya Wanawake, which is Swahili and means "progress for women."

Studies have shown that the formation of women's associations in Kenya is improving the situation of land ownership and tenancy in urban areas, where women and children are the majority of slum dwellers. Many of these associations are involved in promoting socioeconomic development through entrepreneurship, community projects, mutual-aid schemes, legal counseling, formal and non-formal education, and self-empowerment programs.

In Nigeria also, women's associations have tried to tackle problems created by economic crises and recession. The National Committee on Women and Development (NCWD) was established in 1982, as a liaison between women NGOs and the government in an effort to promote development and to eliminate gender biases in policies and programs. In 1989, the National Commission for Women drafted a National Policy on women.

Another group, the Women in Nigeria (WIN) association, which has been regarded as being more feminist in orientation and has a membership of both men and women, places greater emphasis on combating discrimination and sexist practices in the family and in the workplace.

In South Africa, women in parliament started the Women's Budget Initiative in March 1996 to promote welfare and the needs of the poorest and most disadvantaged members of society in certain areas, such as education, housing, and social welfare services. Though fairly successful, this initiative has been criticized for functioning well at the national level, but for failing to percolate down to the provincial level.[3]

Examples from North Africa also show that although many associations are led by women of the middle and upper classes, a large number tend to work on development issues such as education and employment, and to challenge underdevelopment and poverty that affect the lower classes. In Egypt, many have sought to raise awareness about the conditions of working and lower-class women and to promote agendas for social and political reform. These reforms appeal to lower-class women and create links with workers and peasant parties.[4]

The role of international NGOs and donors in development has been controversial. Some of the problems of international donors that surfaced in Sierra Leone are evident in other countries. According to Odoul and Kabira, the donor community in Kenya has not done any better than governments in the development arena, since they also undermined women by stressing welfare-type programs and implementing poorly planned and inadequately funded programs. They also brought the patriarchal biases of the international community to their development programs. In addition, many of the international development programs are dominated by men.[5]

Sub-Saharan Africa is the only region of the world where poverty has been steadily increasing during the last two decades. Many studies and reports, including the *Human Development Report* and *The World's Women* have repeatedly confirmed this continuing deterioration in their development indicators. Moreover, Structural Adjustment Programs (SAPs) have not achieved the expected macroeconomic stability, but instead have increased underdevelopment, poverty, and the debt burden. In addition, sub-Saharan Africa accounts for only 3 percent of Foreign Direct Investment (FDI) and 1 percent of world trade.[6]

Despite its improving economic performance, Kenya is facing severe economic problems, including a national debt and the conditionalities of SAPs. In addition, it is being marginalized by the process of globalization, as liberalization and privatization increase the ownership of vital natural resources by non-Kenyans. Per capita incomes continue to decline and the incidence of poverty is almost 50 percent. Most of the households living in poverty are headed by women.[7]

Although women have been entering wage employment in large numbers, and now stand at 21 percent of the formal labor force, they are concentrated in the lower ranks of agricultural industries and forestry, and in the services. In the informal sector, about 78 percent of the enterprises are owned by men and 22 percent by women.[8]

Gender disparities continue to be manifested in education as in Sierra Leone. Despite basic structural changes in the curriculum and in enrollment of girls, there has been a steady decline in the proportion of women entering universities, leveling off at 27 percent in 1993. In institutions of science and technology, female enrollment ranges from 3 to 6 percent. No improvement has been noted for female participation in male-dominated areas. After more than thirty years of independence, girls continue to be overwhelmingly enrolled in secretarial and nursing courses in post-secondary institutions.[9]

Nigeria is one of the largest countries in Africa with a population estimated as approaching one hundred million. Its women have a strong and impressive record of female resistance through collective action against forces of colonialism, underdevelopment, and authoritarianism. Despite these facts, and its enormous wealth in oil, it has some of the most severe problems of poverty.

In Sierra Leone and elsewhere, gender inequality is reinforced by poverty, and women face several constraints in terms of access to resources, education, training, technology, and so forth. SAPs intensify inflation, high interest rates, and debt servicing burdens. Women are increasingly being involved in export production to increase foreign exchange earnings and debt repayments in conformity with the conditionalities of SAPs.[10] To make matters worse, multinational corporations, especially oil conglomorates have been creating havoc for the economy, the environment, and women.

Nonetheless, Nigerian women have been fighting back and have challenged multinational corporations (such as Chevron Texaco and Shell) for promoting underdevelopment in their country. One poignant example of this comes from the Niger Delta. The people of the Niger Delta are among the poorest in Nigeria, despite living on the land that makes the country the sixth largest oil exporter in the world and the fifth biggest supplier of American oil imports.

The lack of development efforts in the Delta has prompted activists to challenge the oil corporations and to focus their demands for roads, water, electricity, and schools from multinational companies that are extracting and appropriating the wealth of the country. The protest of women has received worldwide media attention. In July 2002, Nigerian women staged a siege of an oil terminal to challenge their exploitative practices as well as their destruction of the environment, and to make demands for economic justice. The women wanted the company to provide jobs and promote development for their community.

According to the press, the unarmed women occupied the terminal, stopping exports and trapping about 700 workers inside.

> Other teams of women shut down the docks and the helicopter pads. About 100 police officers and soldiers, armed with assault rifles, were sent into the terminal to protect the facility. They agreed to end their siege after the company offered to hire at least 25 villagers and to build schools and electrical and water systems.[11]

In the situation of South Africa, underdevelopment has significant racial overtones that worsen the problem of poverty. The most profound division in South Africa remains a racial one maintained by de facto Apartheid. It is expressed in patterns of continuing inequality in housing, employment, and education between Blacks and Whites and the persistence of racial exploitation and oppression. Although gender is relevant and has led to the formation of alliances against Apartheid, women's experiences are shaped by the realities of a racial ideology.[12]

In the 1980s, some women's grassroots associations affiliated to the United Democratic Front, tried to organize around consumerist issues that were negatively affecting their livelihood. Foremost among these organizations were the Federation of Transvaal Women, the United Women's Congress in Western Cape and the National Organization of Women.

Black women in South Africa are still struggling to overcome the legacy of colonialism and Apartheid, both of which have severely destroyed their communities. De facto Apartheid is expressed in residential segregation, in depressed township slums, and in infertile homelands. It is a well-known fact that environmental degradation continues through locating waste dumps on land inhabited by Blacks. Toxic chemicals, including waste products like mercury from multinational corporations based in Europe and the United States are dumped on land inhabited by Blacks.[13]

In the case of South Africa, economic and corporate are globalization reinforcing poverty among Blacks, especially among Black women in rural areas. They eke out a living on infertile land of little or no use to the dominant White economy. These *discarded people* are often short of food, clean water, and fuel. Health care, schools, power lines, and other services and facilities are insufficient or absent.

In addition, trade liberalization has resulted in rural women, who constitute the majority, being among the poorest in the country.[14] As Pheko points out, trade liberalization is undermining the efforts of the post-Apartheid government in promoting gender empowerment for women.

> Globalization and liberalization are challenging these rights of women all the time. In South Africa, the entry into the global market is making us more conscious, and apt to question whether these liberal trade regimes have not always used women's socioeconomic status as a bargaining chip in terms of trade relations.[15]

Expansion of the informal sector is a feature of corporate globalization and this trend is evident all over Africa. This is compounded by contraction in the formal public sector and lack of skills for women for jobs in the formal sector. In South Africa, this sector is expanding and shows signs of weakness, instability, and vulnerability to downturns in the economy. Of the 3 million people in this sector, 60 percent are Black women. In the formal labor market, the majority of Black women are domestics, agricultural workers, and low-level industrial workers.

Several associations, characteristic of the mutual-aid associations in Sierra Leone have also been operating among women in South Africa. They help advance their prospects in trade, provide protection from police harassment and assault, and facilitate access to markets and credit. The Self-Employed Women's Union in South Africa, founded in 1994, is an example of this.

In North Africa also, women have fought against the negative effects of underdevelopment. In Algeria, the history of women's mobilization for political and legal participation dates back to their pivotal role in the war of liberation against the French from 1954 to 1962. Since 1984, the passing of the restrictive Family Code has led them to focus their attention on more domestic matters as they lobby for its abolition. Since the later part of the last century, they have repeatedly challenged the violent attacks on women resulting from a fanatical and misogynistic brand of religious fundamentalism.

Embedded in their activism is recognition of the importance of promoting development for women in the areas of education and employment, and of the need to increase women's political participation, legal capacity, and empowerment. Associations such as S.O.S Women help women and children that have been rendered destitute and homeless as a result of divorce or other misfortune.[16]

In Algeria, corporate globalization and the emphasis on liberalization are adding economic pressures through SAPs that result in cutbacks in the social sector, underdevelopment, and dependency. Dependency relations with France have established a pattern of migration, involving men, women, and whole families that has resulted in greater loss of people.[17] Algerian women, including migrant women, are building a network of solidarity and support with international women's groups. Associations such as the Collectif Maghreb Egalité 95 have developed the capacity take on broader international issues.

The Democratization–Authoritarianism Nexus

African women have seized the momentum of the current democratization movement in an attempt to open the engendered political spaces. Yet they face many problems when seeking to promote democratization and challenge authoritarianism. A comparative review reveals similarities as well as differences in their approaches, based on contextual and historical specificities.

The overall goal for democratization tends to resonate with women. These can be discussed under the following headings: authoritarianism, democratization, the role of the state, nationalism and religious fundamentalism.

Authoritarianism is reinforced by corporate globalization, which is undermining the sovereignty of many nation states in Africa and the Global South. It tends to reinforce women's secondary position in society and erode their political participation. There are many examples of women challenging authoritarianism dating from the colonial period. Examples include the famous 1929 women's war in Nigeria, and the participation of women in liberation movements and in the Mau Mau movement of Kenya against the British.[18] Women continued to challenge authoritarian rule in the post-independence regimes of one-party governments and military rule.

With regard to the law, the rights of women have not always been upheld by colonial statutory laws that provided the judicial framework. For example, in Zambia, women have consistently lobbied for the abolition of all gender-based discriminatory laws that date to colonial times.

In Uganda, women's insistence led to the reform of the colonial-inspired constitution and to the establishment of affirmative action, which is still handicapped by the male domination of the political apparatus.[19] In Eritrea also, the quota system has been criticized for not going far enough and does not have women in high-level positions.[20]

In the South African situation, women have been constantly involved in challenging the authoritarianism of colonialism and Apartheid through numerous women's associations and trade unions. The Rural Women's Movement was a grassroots association whose main objectives included resistance to policies of forced removals, and to the Bantustans that deprived Blacks of South African citizenship during the Apartheid era. These women were mainly poor and depended on irregular remittances from male family members working in the mines as migrant laborers. Migrant workers were regularly treated with cruelty and toiled under oppressive conditions and the constant threat of dismissal and loss of income.

African women were among the most victimized under Apartheid and were relegated to the status of minors by law. As a result, they were among the most active in the resistance to Apartheid, both within and outside of Africa. In the early 1920s, the trade unions were the first organizations to provide women in the laundry, clothing, furniture, and baking industries with leadership roles. In 1913, the African National Congress (ANC), the main opponent to the Apartheid regime and now the ruling party, established the ANC Women's League.

The highly centralized and extremely authoritarian police state of the Apartheid regime, represented the most extreme example of White hegemonic dominance on the African continent. It legally sanctioned the institutionalized oppression of Blacks. Ironically, the Apartheid regime was considered democratic and patterned after Western democratic models of constitutional rule. Because of the legacy of racism and authoritarianism, the need for feminist struggles to consolidate a broad-based women's movement is complicated. This is because of the challenges facing women from diverse racial, class, ethnic, and geographical backgrounds. In my interviews with Black South African women, while conducting research in South Africa in 2001, the struggle against racism was regarded as priority, as many institutionalized forms of de facto Apartheid and economic oppression remain intact.

Although Apartheid has been legally abolished, South Africa remains a country of extreme contradictions. The economy is still dominated by Whites, who are in a position to take advantage of corporate globalization and its emphasis on privatization and liberalization. The vast majority of Blacks continue to live in poverty in the townships and rural areas.

Eighty percent of the land is still owned by Whites who constitute 14 percent of the population. Black women are even further marginalized in terms of their limited access to land and other productive resources, in addition to the structural violence evident in the economic oppression of Blacks. This explains why many Black women tend to reject feminism as a White American import that would dilute the Black liberation struggle.[21]

As many scholars have shown, women in Africa continue to campaign for gender equality within the framework of democratization. In Kenya, in addition to campaigning for a repeal of discriminatory laws, such as section 82 of the Kenyan Constitution women have insisted on other changes. These include demands that the government be more committed to affirmative action; that public officials who make derogatory remarks against women be disciplined; and that women be included in key decision-making positions.

Many of these groups, such as the National Council of Women of Kenya and the League of Women Voters organized voter education throughout the country and widely disseminated a booklet on women and democracy and on women's rights. The momentum for democratization also led to the repeal of section 2A of the Constitution in 1991, thus ending authoritarian one-party rule. It also opened the way for multiparty democracy and restored liberal democratic principles as envisaged in the rule of law. These included broad-based political participation, social justice, and the promotion of a free press.

As one of the largest pressure groups, women were able to take their issues to center stage in new policies, plans, and programs for democratization. Their demands included gender equality at all levels, including the government. In 1992, a National Convention was held in Nairobi that influenced almost all political parties to include women's issues in their constitutions and platforms as was the case of Sierra Leone.

Some political parties are regarded by scholars from Kenya as more gender-sensitive than others and might help to improve the situation in future elections. The Democratic Party is regarded as being more inclusive, by having women represented in all bodies of the party and in taking a strong constitutional stand against gender-based discrimination. Commitment to gender issues is one of the criteria for leadership in the party.

A review of women's political participation in Nigeria also shows that the low participation rate of women in politics has been chronic. In the 1998 Constitutional Assembly, there were 14 women elected or appointed as opposed to 565 men. Women have been active in political parties and campaigns and have lobbied for an end to military rule in Nigeria as well as in Sierra Leone.

As demonstrated in the Sierra Leone case, women insisted on linking democratization to conflict prevention and peace building as important aspects of development. However, women's campaigns for democratization and their participation in political parties do not guarantee them positions in government. The example of the 1991 election in Kenya also bears this out. In Algeria also, despite women's participation in political parties, which was viewed as essential for their cause, the 1992 elections yielded only 1.5 percent of female representation in the national legislative government.[22]

Although some progress in democratization is being made, the pace has been slow. In South Africa, implementation of the recommendations of the Commission on Gender Equality has been difficult. This is because the different governmental ministries have different priorities and mandates, and are not often equipped to deal with gender issues. In addition, there is no effective monitoring machinery, and women in parliament who have formed caucuses, often have severe time constraints that prevent them from meeting regularly.

The Role of the State

Since the major agent of development is the state in many countries of the Global South, particularly in Africa, feminist analyses have focused on the role of the state in facilitating or impeding the advancement of women.[23] In most instances, the patriarchal nature of the state has tended to work against women's interests, except to the extent that the state can co-opt women's groups. This has often led to what has been termed bureaucratic feminism, a type of feminism that has official, rather than activist-driven proclivities and that is often promoted by the United Nations process and its "Women's Bureaus."[24]

According to Odoul and Kabira, affiliation of women's associations with the state in Kenya, led to co-optation and domination by the KANU ruling party. The state undermined women's efforts by manipulating those in leadership positions and interfering with their autonomy and agenda-setting priorities.[25] In Kenya and elsewhere, women's bureaus have been criticized for tending to "ghettorize" women's issues.

There has also been a conflict of roles and responsibilities when the Women's Bureau attempts to usurp the position of major women's associations, such as the National Council of Women of Kenya (NCWK) and the Maendeleo Ya Wanawake (MYWO), which were set up to coordinate the work of women's associations.

At the same time, Non-Governmental Organizations (NGOs) are also being criticized for failing to set an agenda for women that would empower them, rather than reinforce gender roles. Tensions continue to exist between the Women's Bureau and some of the major women's associations that center the bureaucratization of the women's movement.

In Nigeria too, the state appears to have been influential in the activities of women's associations, especially through first ladies, a phenomenon that has been referred to by Addullah as "wifeism and activism."[26] The role of the state in agenda setting has been prominent, especially in the National Council of Women's Societies (NCWS), which mandates that its head should be the First Lady of Nigeria.

Other leadership positions tend to be confined to wives of politicians and bureaucratic and other high-ranking women. This has created many conflicts and contradictions within the group, as well as with other women's associations such as the Women in Development group, which was founded in 1983. The NCWS established a policy based on a vision of economic independence, self-reliance, gender equality, and sustainable development. This has been criticized for being less visible on feminist issues, related to reproductive rights, sexual harassment, and so forth.[27]

The case of Sierra Leone represents a departure in terms of co-optation by the government in that the Women's Forum's main objective was to challenge the military government and bring about democratic change through the ballot. Women's wings of political parties on the other hand have tended to be more closely affiliated with the State.

The case of Tunisia offers an interesting deviation. Unlike many countries where the state has not taken the initiative to promote gender equality without strong pressures from women's groups, in Tunisia the state has been a leader and major player in the advancement of women. This is because progress has come about as a result of executive edicts rather than through a social movement, or women's pressure groups.

In 1956, President Habib Bourguiba banned polygamy and insisted on the education of girls. In 1987, President Ben Ali continued this trend by expanding women's rights and empowering them politically and socially. He established the Center for Research Documentation and Information on Women (CREDIF), which collects data and conducts studies on a number of issues affecting women's rights for the purpose of formulating policies and implementing programs. It also had a ministry of women's and family affairs, which was established to develop policies that would positively affect women and their families. It works closely with the National Council for Women and the Family.[28]

The Tunisian case is an interesting study of the advantage of women mobilizing for change versus direct state intervention, which might undermine female activism and reduce the intensity of women's collective action. As a result, Tunisia is often regarded as having a weak women's movement and a weak civil society process. The state tends to discourage women's independent mobilization and to pre-empt their agenda. It has passed several laws that eliminated former restrictions on women.[29]

This raises the question as to whether female emancipation that is given on a silver platter has the same value as one that results from feminist struggles. This point is underscored by the Tunisian film, "The Season of Men," directed by Monfida Tlathi. The film shows that, despite a political and social revolution, women did not achieve liberation.

One scholar has argued that despite efforts to create institutions that are progressive and that attempt to improve the status of women, the government has tended to hinder the actions of women's associations seeking to promote women's rights. In addition, the government does not fully acknowledge or validate many of these associations, thereby forcing women to function on an individual basis or through unofficial associations, and denying them the power to mobilize and to become effective.[30]

Nevertheless, the National Union of Tunisian Women (UNTW) has a membership of over 135,000 women, with subregional and local branches and youth groups. It focuses on the promotion of women's social, economic, scientific, and cultural interests as well as increasing women's participation in decision-making and in the government.

The relationship between female parliamentarians and women's associations has often been complex as several studies show. Tamale's study of Uganda indicates that the relationship between female members of parliament and women outside of parliament was tenuous at best. She insists that "women parliamentarians should especially work with women activists in their efforts to equalize outcomes for women and men in all spheres."[31]

In Nigeria also, Amadiume remarks on the strong contrast between indigenous Igbo women's associations that showed a strong commitment to female solidarity and a unity of purpose, and many of the imported modern associations. She claims that "modern Nigerian history is typified by the exploitation of women's organizational ability by male-dominated political parties and female political careerists."[32]

In South Africa, on the other hand, a number of women in parliament try to hang on to their leadership of women's associations, albeit with great difficulty. They have had recurring problems due to their affiliation with the State, and run the risk of being co-opted and of giving priority to agenda-setting by the State. As Zulu puts it:

> The task facing women in South Africa today is enormous. Nevertheless, a solid foundation has been established. It remains our task to mobilize more women into a united front, encourage debate on the formation of a women's movement, especially now that the Women's National Coalition has fulfilled its original mandate to draft a charter for women's rights. The majority of the women who were active before the formation of the Women's National Coalition are now all in Parliament, and this has left a gap.
>
> We almost have to start from scratch in making sure that we have other women taking the place of those women who are in Parliament today. Many women, now in Parliament, do not want to let go of their positions that they had when they were outside the government. But it becomes difficult for them to split themselves into two and be able to run organizations outside of Parliament.[33]

Controversies involving the State have also revolved around the degree of affiliation of women parliamentarians with political parties. Frene Ginwala, a feminist and speaker of the South African House of Parliament, cautioned against the possible loss of autonomy by being too involved with political parties, and has urged women to build a power base both inside and outside of the ANC, the ruling party.

Women's associations can and do have confrontations with the state. Tarrow defines contentious politics as the type of politics that occurs when ordinary people, often in league with more influential citizens, join forces in confrontation with elites, authorities, and opponents.[34] In Africa, there are many examples of women's collective action that would conform to contentious politics, both during the colonial era, and in more recent times.

However, there are also several examples where women are allied to the state through women's wings of political parties. In addition, affiliation of individual women leaders of associations with the government, as in the case of Constance Cummings-John and the Women's Movement in Sierra Leone, has involved women in helping to shape policies.

Nationalism

As seen in Sierra Leone, the West African Youth League and the Sierra Leone Women's Movement played a seminal role in the decolonization movement in Africa. During the nationalist movement prior to independence in Nigeria, market women were mobilized into powerful organizations and this was a leading factor in a political party's ability to control a certain area. Women's associations in Western Nigeria are among the oldest, largest, and most powerfully organized of all women's associations in the country.[35]

In the postcolonial period, women's associations became involved in nation building and socioeconomic development, but politics remained dominated by men. The marginalization of women in politics was reinforced by a succession of military regimes, which promoted a strong male ethic, punctuated by brief periods of male-dominated civilian regimes that have come to define the State. Ironically, some military regimes in Nigeria have a record of appointing women to high decision-making positions that equals or surpasses the record of civilian governments. On the whole, however, women are marginalized in political parties and in government regardless of whether it is a military or civilian government. Most women who aspire to political office tend to run as independent candidates.

Women's associations in Kenya also protested against British rule and policies that were destructive to the economy and culture. Among their targets were the imposed local government structures and ordinances, forced labor, hut and poll taxes, forced male migration, and the privatization of land. A series of laws and policies were instituted that led to intense resistance of forced labor on plantation, land theft, and violation of human rights by the colonial state.

Organized protests were held by African men and women, culminating in the Mau Mau guerrilla war of independence in the 1950s. Up to 5 percent of the forest guerrilla fighters of the Mau Mau were women and a larger number helped the war efforts by providing administrative and logistical support.[36]

The protests against colonial rule strengthened women's movements in other ways. They contributed to a high degree of female militancy. Many women, as well as men, lost their rights to land. Many became heads of households with fewer resources and a heavier workload.

In order to reduce the tensions brought about by the anti-British liberation war of the Mau Mau, wives of colonial officers formed the MYWO in 1952. It later became a major women's association in mobilizing women for nation-building and socioeconomic development in the post-independence period. Women's groups became a vital component of the mobilization of the population's potential and strengths for rapid development. About 80 percent of Kenya's cash economy has been attributed to cooperatives and women's groups.

In South Africa, nationalism took the form of the struggle against Apartheid. Women became active members of anti-Apartheid groups, such as the African National Congress, the Pan African Congress of Azania, the Campaign for Defiance Against Unjust Laws, Trade Unions, the Black Consciousness Movement, and so forth.

In addition, they set up their own associations, such as The Black Women's Federation. Their political efforts included not only campaigns against Apartheid, but also against Bantu education which prepared Africans for low level jobs. They also lobbied against the pass laws and forced removals. African women were actively involved in the Cato Manor uprisings, the Crossroads protests, the Soweto uprisings, and numerous other protests and campaigns. They were among the victims of the Sharpeville Massacre of 1960.

That same year, liberation movements such as the ANC and the PAC were banned, a move that severely restricted the political activities of African women.[37] In 1990, the Apartheid regime lifted the ban on liberation movements and entered an era of negotiations aimed at dismantling the Apartheid system.

This presented an opportunity for the opening up of new political spaces and building awareness, not only about national liberation, but also about gender issues. This momentum inspired what has been described as "a burgeoning women's movement" through which women insisted on putting gender issues on the national agenda for the new South Africa.

The Apartheid system had reinforced patriarchal values held by all groups in many respects. In his address to parliament in 1994, the year that marks the end of de jure Apartheid, President Nelson Mandela recognized this and made the following statement:

> The objectives of the Reconstruction and Development Program will not have been realized unless we see in visible and practical terms, that the condition of women has radically changed for the better and that they have been empowered to intervene in all aspects of life as equal members of our society.

In 1994, the Women's National Coalition, which is made up of ninety different women's associations, adopted the Women's Charter for Effective Equality. This attempt at unity was unreal, since Apartheid had divided women according to racial groups for over two generations. This called into question the very notion of "sisterhood." The experience of most African women with White women has been as exploited domestic servants.

The attempt at alliance building should not be interpreted as an example of female solidarity, or as an attempt at universalizing the feminist struggle. Rather, it was part of a process of mobilization against Apartheid, within an international climate that promoted a feminist agenda. It fuelled consciousness about feminist issues, which the emerging South Africa could not afford to ignore. The adoption of feminist issues at this opportune time was politically strategic, since the aim of the new South Africa was to eliminate racial discrimination. The institutional racism imposed by Apartheid was anchored on a patriarchal ideology supporting White supremacy.

The achievement of the Women's National Coalition was nonetheless spectacular in terms of lobbying and campaigning for female representation in government. The Women's League of the ANC was also influential in mobilizing women from other organizations to ensure a new constitutional order that would meet women's demands. Although much more needs to be done to achieve full equality, there have been some achievements. The first Speaker of the House of Representatives was a woman, Dr. Frena Ginwala.

The progress has continued and South Africa today is one of the most advanced countries in terms of female representation in high levels of government and in parliament. It was included among the 16 countries that have more than 25 percent of women in government in 1999. The percentage of women in parliament was 30 percent and the trend has continued.[38] South Africa became the fourteenth country in Africa to have women serve in the position of head of state, acting head of state, or deputy heads of state, through the appointment of a woman, Phumzile Mlambo-Ngouka, as Deputy President.

In the new South Africa, women are guaranteed equality under the law and have seized the opportunity in the democratization process to ensure full citizenship and a political process in which women are represented at all levels of political decision-making. In addition, institutions have been established to implement and monitor gender equality.

These include the Commission on Gender Equality, a focal point for women that is located in the office of the President of the country. The country has also enacted legal reforms to prevent discrimination in employment, provide maternity and other benefits, and protect women from domestic violence.

Despite these developments, implementation has been difficult, due to resource constraints and the lack of mechanisms to effectively monitor the gains made by women. At the local level, there is a quota of 50 percent representation by women designated by the ANC. However, this does not guarantee effective policies and programs for the advancement and empowerment of women. In addition, the establishment of the Office of the Status of Women is likely to be faced with problems of implementation or lead to the de facto ghettorization of women's issues in only one ministry.

In North Africa also, women advanced the nationalist cause and challenged colonial rule, the best known of which was the involvement of women in the Algerian war of liberation against France from 1954 to 1962. Almost 11,000 women took part in the war as combatants. The war was an opportunity for women to participate in a movement that has been described by Lazreg as a rational response to an otherwise irrational historical situation.[39]

The struggle for national liberation, however, does not always guarantee women equality in the aftermath of the struggle. After the war, women were relegated to the background as traditional patriarchal attitudes re-emerged after the war. Nationalism did not always protect women. Despite their war efforts and the establishment of the Union of Algerian Women in 1963, the restrictive Family Code was passed in 1984.

Many scholars have observed that the Family Code contradicts both the national laws, which guaranteed equality to women, and the traditional laws, which ensured their protection. Instead, the Family Code established patriarchal dominance of men over women that included legalizing polygamy and establishing gender-based discrimination in matters relating to divorce, inheritance, matrimonial guardianship, and fundamental freedom. Since its enactment, many women's associations, political parties, human rights lawyers, and activists have been campaigning for its abolition.[40]

The National Women's Coordination serving as an umbrella association has as its priority, the abolition of the Family Code. Other objectives include promoting the equal rights of women in education, employment, and political participation.

Religious Fundamentalism and Democratization

In northern Nigeria, where Muslims are predominant, women received the franchise only in 1978. Yet, in many ways, Islam provided opportunities for female political participation at an unprecedented scale. One party, the People's Redemption Party (PRP) appealed to the masses and women and stressed social justice, liberation, and human dignity, based on the principles of the Koran.

The party attracted a category of women comprising prostitutes and concubines known as "Karuwai" and was able to win the 1979 elections for Kano State, thereby setting in motion, women's political participation in the North. In the 1983 elections, it was the only party to name a woman as a presidential running mate.

Its radical platform raised awareness about the need for gender equality and the advancement of women. Its aims included programs for introducing legislation to abolish child marriage; to give protection to all women married under all laws; and to ensure that all female children have equal rights and opportunity in educational institutions.[41]

The Federation of Muslim Women's Associations in Nigeria (FOMWAN), founded in 1985, is an independent forum, from which Muslim women could examine gender issues relating to religion and women's roles in society in general. It emphasizes education and skills acquisition and conducts classes in adult literacy and primary education with an Islamic bias. It also runs nursing schools and centers for vocational education. It has pioneered a pedagogical approach and a methodology called "each one, teach one," which has been effective in building women's skills.[42]

Religion has also played an important role in mobilizing women in North Africa. In general terms, these associations tend to have a strong feminist agenda aimed primarily at laws deriving from the interpretations of the Sharia and the Koran.

These laws have strong patriarchal proclivities, particularly in relation to family law and inheritance.

Although women's associations have tended to be class based, certain issues of common interest to women, such as campaigns against discriminatory family laws and the "family code" have mobilized all classes of women. Women have also taken collective action against religious fundamentalism.

Algerian women are among the best known for challenging religious fundamentalism that threatens to deprive them of their rights. They continue to fight a moral war against Islamic fundamentalism that targets women and advances a strict patriarchal order that is restrictive to women. A number of women have been subject to violent attacks and murders at the hands of religious fundamentalists.

As a result, "women have been forced into the forefront of fighting violence because many have been victims, either as specific targets or as a result of the killing of members of their family."[43] Some associations have been formed specifically to resist acts of violence and to help the victims. These include the Association of Solidarity and Support to Families and Victims of Terrorism.

Nationalist struggles and resistance to fundamentalism have usually been regarded as bearing some similarities, in that both are based on ideological beliefs and notions of superiority. A study by Amrane-Minne argues for a clear continuity between the struggle for independence and the mobilization of women against Islamic fundamentalism. She maintains that through their courage, Algerian women have become a symbol of resistance to religious fundamentalism. In this way, they use a variety of methods, including demonstrations, feminist writings, and the use of the media to get their message across.[44]

According to Fades, women's activism appears to be transcending Algerian society in ways that defy the status quo and challenge the religious imperatives of the legislative process. She maintains that by calling for the repeal of the Family Code and discriminatory laws, and by demanding recognition of their status and rights as citizens, women are implicitly advancing their desirability for a secular state. According to her:

> By aspiring to equality, women fundamentally refuse to accept oppression and necessarily pose the question of fundamental liberties; for the redistribution of power by institutional means, in a nutshell, the rule of law.[45]

Summary

This chapter has shown how the main themes of this book have implications for other countries in Africa. The countries selected for special focus, have all experienced colonial rule and have been impacted negatively by the global political economy. Women's collective action continues to inspire efforts toward development and democratization. Underdevelopment and authoritarianism still pose many challenges that impede development and the quest for democratization. The situation of Sierra Leone is not unique in this regard. However, it has the added dimension of a ten-year civil war that occurred fairly recently and that has tended to intensify the challenge for women. The comparative review shows that the other countries have all faced armed conflict at some point of their history. However, even when a country is at peace, it can be confronted with ominous economic and political problems.

The war in Sierra Leone was not an isolated event. It was a phenomenon that had international dimensions, fuelled by conflicts over diamonds. Like many of the countries in Africa, Sierra Leone is also facing the oppressive conditions imposed by the global economy through the neoliberal agenda of corporate globalization and the conditionalities of SAPs.

The struggle to promote development and democratization has required collective action by women in many countries of Africa that go beyond gender concerns. As long as countries remain in protracted economic crisis fuelled by the global economy, the women's movement in Africa will be conditioned by exigencies that undermine society as a whole.

Chapter Eleven
Comparative Insights at the Regional/Pan-African and International Levels

Introduction

Significant developments in regional and international cooperation have inspired women's collective action at the pan-African level, and this study would not be complete without a discussion of them. This is because of their impact on agenda setting and alliance building across the continent, and their link to the international women's movement. This chapter will explore the main themes of the book at the regional pan-African level and conclude with a reflection on the contribution of African women to the international women's movement.

The African Union (AU) places emphasis on the important role of women in promoting many of the objectives of the New Partnership for African Development (NEPAD). Although NEPAD has been criticized for not involving grassroots organizations, especially women's associations, in its formulation, women have been lobbying for decision-making roles in the implementation of NEPAD and in the AU.

By operating at the regional level, pan-African women's associations hope to maximize their reach, amplify their objectives, and leverage larger resources and networks. They have developed strategies to overcome barriers and constraints imposed by both the authoritarianism of some governments and of corporate globalization that impact on development and the prospects for democratization.

The approach of intergovernmental bodies, such as the AU and the United Nations, to gender equality and the advancement of women, has been criticized for being too bureaucratic and government-controlled. Nonetheless, many of these pan-African associations work with United Nations organizations as a result of the women's world conferences and because of the funding opportunities provided by the United Nations. In addition, the United Nations, in general, especially the United Nations Development Program (UNDP) is increasingly stressing the goal of building partnerships with NGOs for international cooperation.

According to Viola Morgan, head of the Gender Program of the Regional Bureau for Africa of the UNDP:

> The strength of these associations lies in their dynamism, commitment and resourcefulness. Many of them have evolved in response to some pressing issues that complement

what the governments may or may not be doing. Their ability to network at the national, regional, and international level and to mobilize resources are additional assets. Governments are often faced with competing priorities, some of which relate to urgent security issues. NGOs, especially women, are closer to the people and to the problems of development and are more inclined to act.[1]

The Development–Underdevelopment Nexus

It is clear that pan-African women's associations have sprung up to face the challenges of development and underdevelopment. Without doubt, women continue to be the mainstay of most rural economies and the predominant players in the informal economy of the urban areas. The vulnerability of African economies to powerful global economic forces and faulty development policies become magnified at the regional level. These associations have played an important role in promoting women's economic participation and in challenging underdevelopment and environmental degradation. They also strive to promote development, especially in areas of entrepreneurship development, education, training, research on global and local political economies, access to technology, and information dissemination.

For example, the African Federation of Women Entrepreneurs (AFWE) facilitates cooperation among women entrepreneurs in Africa and internationally. It has affiliates in 45 countries and its membership consists of manufacturers and exporters of goods and services. Another association, ABANTU for development works in 19 African countries and focuses on development problems by building partnerships between women and men to overcome these problems. It aims to eliminate gender inequality by eradicating the cultural, legal, and political obstacles to women's attainment of equality and economic independence before the law and in society. Among its major activities are training in the area of policy analysis, economics, health care, leadership, the media, and environmental conservation.

The Greenbelt Movement founded in 1977 by Professor Wangari Maathai, an internationally renowned Kenyan scientist and Nobel Laureate for Peace, is one of the most well-known environmental women's associations in Africa. Its impact has been felt all over the world and it has spurned similar women-led tree-planting environmental associations and programs in other parts of Africa. It was founded in response to the plight of rural women in Kenya, where environmental degradation was rapidly leading to deforestation, forcing women to travel long distances for firewood. Soil erosion was also affecting the yield from their crops, leading to malnutrition. To make matters worse, pollution was contaminating their drinking water.

Twenty-six years after its founding, women of the Greenbelt Movement in Kenya had planted 15 million trees; run tree nurseries that generated income; and provided effective forestry management for their communities. According to Dr. Maathai and other leaders, the Greenbelt Movement has served as a mechanism for the empowerment of women.[2] It is credited with providing a forum for women to become creative and effective leaders and to change their environments for the better. They have also promoted public awareness about the linkages among food, energy, and health. Children learn about the Greenbelt Movement at school and small farmers have become more aware of the importance of agro-forestry.

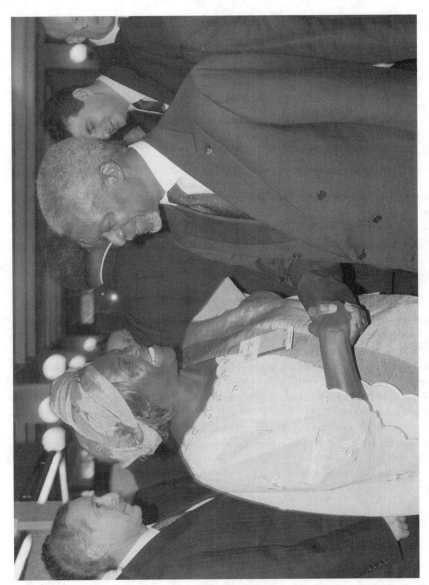

Dr. Wangari Maathai of Kenya, winner of the 2005 Nobel Peace Prize, being congratulated by Kofi Annan, Secretary-General of the United Nations, a fellow Nobel Peace Prize Laureate. Courtesy: United Nations.

The Movement has also been politically active and has often come into conflict with the government's schemes for urban expansion that involve building of concrete structures in parks and commons. The Movement is well known for its passive resistance to such developments and has staged marches and protests that have sometimes led to arrests and police brutality. One such event involved the planting of trees at the Karura Forest, near Nairobi.

The Greenbelt Movement promotes the notion of planting trees as an act of promoting democracy. In 1998, Dr. Maathai stood as a candidate for the presidency of Kenya. Although she did not win, she established a place for women in Kenyan politics that linked nation building with ecologically sound and sustainable development. She is currently the deputy minister for environment.

A number of pan-African associations promote development through the media and communication. Some have increased women's access to information through various channels, including the Internet. The African Women's Development and Communication Network (FEMNET), which has a high international profile, is one such association. FEMNET was founded to strengthen the role of women NGOs in promoting gender equality and women's empowerment, through gender-sensitive policies and planning, equitable development, and capacity building. Comprised of a large number of women professionals, particularly in media, it publicizes its activities through the mass media, award ceremonies, exhibitions, and publications.[3]

A large number of associations promote development through education, skills development, and research. This is intended to overcome problems of underdevelopment related to illiteracy, the slow pace of educational equity, and lack of gender-relevant data for development. Although the gender gap is closing at the primary and secondary school levels in all regions of the world, women still lag behind in some countries of Africa and South Asia. It has been projected that the majority of the 866 million illiterates will consist of women living in Africa and South Asia in 2005. At the tertiary level and in decision-making positions in education, men outnumber women in most countries of the world.[4]

One of the associations working toward gender equality in education is the Forum for African Women Educationalists (FAWE), which operates in 31 countries. Its goal is to promote grassroots activism and mobilization to increase the demand for substantial improvements in female education. Its membership is comprised of professional women involved in education as teachers and administrators, and policy makers. Its advocacy involves challenging gender-based discrimination in education, which includes stereotypical beliefs that investing in women is a waste of money.

> Educating girls and women is the single most important investment that yields maximum returns for development. The infant mortality rates decrease, children have a higher probability of getting a good education and most importantly, women become generators, which increased the economic base of the family. The most important issue in any country is the number of girls that have access to education and the quality of education they receive as measured by levels of retention and performance.[5]

FAWE conducts research on the impact of official policies on female education and tries to evaluate ways of using resources more efficiently, thus enabling more girls to receive adequate education. It is engaged in advocacy for educational reform and for an end to gender-based discrimination through its publications, the media, and scholarships. It also gives awards for innovative approaches to the promotion of female education. FAWE has established links to other women's associations engaged in similar activities as well as with international organizations such as the United Nations Educational, Social and Cultural Organization (UNESCO).

The association also works with many partners, including the Female Education in Mathematics and Science in Africa Association (FEMSA) in increasing the levels of education of women in Africa. The aim of FEMSA is to improve the participation and performance of girls in science, mathematics, and technology at the primary and secondary levels. Its activities include reforming the curriculum, teacher education, training, and improvement in testing to ensure greater participation and better test scores in these subjects for girls as well as boys.

The performance record increasingly shows that although boys, in general, perform better than girls, the failure rate was high for both genders. Many of the schools did not have the basic necessities for a good education and often lacked facilities and equipment, as well as textbooks in science and mathematics. According to FEMSA, poverty is a major factor in the failure of students in these subjects.[6] This underscores the fact that the underdevelopment and impoverishment of many African countries is a major factor in poor educational performance of pupils, especially girls.

Some associations are engaged in conducting research and in analyzing the problems of development and its gender implications. The Association of African Women for Research and Development (AAWORD) analyzes the problems of underdevelopment from a historical and global perspective. AAWORD's mission includes decolonizing research, which for the most part has served and responded to needs outside of Africa and centering African priorities. It also advocates for the transformation of gender relations and the social conditions of underdevelopment.

It has organized continent-wide research projects on democratization, globalization, the impact of the HIV/AIDS epidemic on women, and the role of young women in research and development. The ultimate aim is to build a powerful women's movement that links human rights to the theory and practice of development and making visible and effective, the contribution of African women to development. AAWORD regards feminism as an ideological, social, and political movement, whose ultimate goal is the fundamental transformation of the existing social structures based on patriarchal values.

AAWORD also seeks women's empowerment and full participation in all stages and at all levels of the political power structure. It regards politics as an all encompassing and democratic process. To this end, it emphasizes the fact that feminism can be an agent of economic and social change and can advance our understanding of the relationship between gender subordination and human development.[7]

Akina Mama wa Africa which is Swahili for "solidarity among African women," is a pan-African association that emphasizes training for development and policy reform. It aims "to create space for African women to organize autonomously, identify issues of concern to them and speak for themselves. It also seeks to provide solidarity, support, awareness and to build links with African women active in the areas of their own development."[8]

It is based in London with a regional secretariat in Uganda, and has branches in other parts of Africa. The association established the African Women's Leadership Institute (AWLI) to encourage and train sufficient numbers of women for informed and enlightened leadership that will ultimately promote a progressive African women's development agenda. Training is provided in organizational skills, feminist leadership, and policy-related advocacy. It is convinced that the development of a feminist constituency among the next generation of African women leaders is essential to the future success of the African women's movement.

The Democratization–Authoritarianism Nexus

Associations operating at the pan-African level give women a chance to maximize their influence on political processes in the African region as a whole. Many of them promote the empowerment of women by seeking to advance the process of democratization and to influence policies toward greater social and gender equality in political decision making. A number of them have established alliances with international women's associations and regularly attend international meetings.

One of the most prominent pan-African women's associations working to advance democratization and to challenge authoritarianism is the Women, Law and Development in Africa (WILDAF), based in Harare. It is a women's network dedicated to promoting democratization and human rights for women. It also seeks to strengthen strategies which link law and development with an increase in women's political participation.

The network brings together over 150 associations in 22 countries. It also advocates and supports programs on violence against women. WILDAF played an important role in mainstreaming women's rights into the African Charter on Human and People's Rights.

The Pan-African Women' Liberation Organization (PAWLO) aims to create awareness about the role of pan-Africanism and African women in liberation struggles. It is an outcome of the Congress of Pan African Organizations held in Kampala, Uganda in 1994. It develops networks with similar associations in Africa and in the African Diaspora and conducts workshops in training and skills building for political participation.[9]

The Study and Research Group on Democracy and Economic and Social Development in Africa (GERDDES) is based in Benin, with branches in 21 other countries. It is charged with promoting and consolidating democracy in Africa. It conducts training seminars aimed at developing the capabilities of women politicians, especially at the local level. It has been credited with the large turnout of women rural voters in the Senegalese presidential election of 2000. GERDDES has also had some success in Mali, a country that has been subject to military rule for more than two decades.

Although all of these associations have development objectives some have taken on the challenge of the two themes in a decisive manner. Included in their activities is political participation to promote democratization, conflict prevention, and peace building. They also advocate and have activities for economic empowerment, gender budgeting, media advocacy, communication, and research for development and education. On the international level, they are building alliances with other women's associations to combat problems stemming from international economic policies that promote underdevelopment.

Although the Forum for Women in Democracy (FAWODE), based in Uganda, has a national origin, its impact has been significant outside of Uganda. It emerged out of a two-decade armed struggle, based on the belief that women's political participation was essential to the ending of conflicts in Uganda and the subregion.

After two decades of war, 51 women who were part of the 284 members of the new Constituent Assembly working on a new Constitution, formed a Women's Caucus. They joined forces with youth, workers, and disabled persons in pushing for equality between men and women under the law, in the new Constitution and its successive bills adopted by the Ugandan Parliament. In time, The Women for Action for Development Organization joined them to form a new association called the FAWODE.

The association has held numerous "Gender Dialogues" ranging from agricultural subjects to health and education. It gives voice to grassroots women through the media and in bringing them in touch with policy makers. It aims to strengthen the skills and capabilities of women for effective political participation, electioneering, and decision-making. Its activities include guaranteeing women one-third of the local council seats; offering courses for women councilors on the legislative process; and training of parliamentarians.

It also organizes workshops for trainers at the national and subregional levels. Emphasis is placed on policy reform relating to land tenure and family law. These are areas of special concern to women, who have been restricted by gender-biased policies and laws from contributing fully in to shaping new national democracy.[10]

FAWODE's impact on the subregion has been significant. For example, its candidates' training package, combined with its Parliamentarian's Training Project have been successfully applied to winning seats for women from all socioeconomic groups in Uganda and in other countries as well. This training package is now being used by women political candidates in countries of the subregion, such as Malawi, Namibia, Rwanda, South Africa, Tanzania, Zimbabwe, and Botswana.

Among the problems cited by FAWODE are resource constraints that have hampered its activities at networking in the subregion. Despite these problems, it is forging ahead, strengthening old activities, and initiating new ones. Although it is difficult to assess its full impact, especially of new activities, it shows a generally positive trend in building women's capacity for political participation toward democratization.

The Women's Development Foundation (WDF) is a South African-based NGO that also aims to increase women's political participation leadership in decision-making bodies. While the main focus of its activities are in South Africa, it has also worked with women public officials throughout East and Southern Africa, including Uganda, Swaziland, Tanzania, Zimbabwe, Botswana, and Kenya. It also works with men and holds training sessions for both male and female local councilors and parliamentarians in constituency building, assertiveness, and gender sensitization.[11]

The association has been innovative in the concept, plan, and implementation of electoral rules, procedures, and management. For example, it seeks to alleviate cultural obstacles and stereotypes about women's roles in society that impede their participation in politics by using indigenous methods of communication as well as the media. It also tries to surmount practical obstacles to women's participation, such as transportation to polling stations and organizes child-care services on the day of elections. The Foundation initiated the 50/50 quota system for political parties to promote gender equality in the selection of political candidates. It also lobbies the South African Independent Electoral Commission to make political parties accountable in meeting this target.

Despite the difficulties of finding an appropriate framework for regional corporation, it has had some successful collaborative activities with other countries. This resulted in some of the gains by women in elections in Botswana, Tanzania, and Uganda.

Emang Basidi is another important political association with impact that goes beyond its national boundaries. Although based in Botswana, it has successfully collaborated with women's associations in Namibia, South Africa, and Tanzania. Bolstered by its motto, "Stand Up for Women," it works to heighten awareness of gender-specific issues and to promote the political empowerment of women. Since 1986, it has been a strong advocate for women's legal rights and has aggressively lobbied for legal reform.[12]

Prior to the 1994 parliamentary elections, Emang Basidi elevated and expanded its activities to include a broad-based and multifaceted "Political Education Project." The project aimed to increase the level of women's representation in elected office and include women's rights and issues on political platform of parties. It also supported activities to educate women on the link between voting and improving their living conditions and to promote the awareness of women's under-representation in politics.

Its "Vote for Women" campaign has successfully mobilized public opinion in favor of gender equality in government and sparked the interest of political parties as well. Its voter education workshops have been held frequently, involving sessions in which voters are able to question candidates and learn the true meaning of democracy and accountability which go beyond mere voting. It produced a "Democracy/Voter Education Manual" in conjunction with the Women's Development Foundation of South Africa.

Training programs are organized for women local councilors and members of parliament; women's wings of political parties and women decision-makers in government, parastatals, and in the private sector. Its curriculum includes the following: learning about the structures and procedures of local government, including the relationship between local and national governance; lobbying, coalition building, gender analysis, and gender budgeting; delivering constituency services and organizing programs with the media.

As a result of its activities, the number of women Members of Parliament in Botswana has doubled from 9 to 18. As such it has contributed to enlarging the pool of role models for young women, not only in Botswana, but also in the subregion.

Conflict Prevention and Peace Building

Women form a substantial percentage of the people involved in peace movements in Africa. As a result of the proliferation of armed conflicts in many countries in Africa,

Sarah Daraba Kaba of the Mano River Women's Peace Network of West Africa (second from left) with other members of the Mano River Women's Peace Network. Courtesy: UN Photo/Stephenie Hollyman, New York.

a number of women's associations have emerged to promote conflict prevention and peace building at the pan-African level. They often work behind the scenes and away from the limelight of the media, which tends to focus on the activities of men in situations of conflict and in peace negotiations.

As one of the most affected victims of war, women have a vested interest in peace. In addition, the task of caring for the sick and wounded imposes additional responsibilities on women that continue long after the wars have ended. Women's associations and peace networks have been active in almost all the areas of conflict, especially in Rwanda, Liberia, Sierra Leone, Mali, Ethiopia, Mozambique, Angola, South Africa, Uganda, and the Democratic Republic of the Congo.

According to some of the leaders of associations interviewed, peace building ideally should encompass a preventive and sustained strategy applicable before, throughout, and after the conflict. Despite the tendency to exclude women from peace-building efforts, there are several encouraging examples of women's associations working for peace at a pan-African level.

Among these is the Federation de Reseau des Femmes Africaines pour la Paix (FERFAP) founded in response to the genocide in Rwanda that left 60 percent of the adult female population widowed. It promotes education for peace through workshops, adult literacy, and non-formal education for children, whose education had been affected by armed conflict. The association also works through traditional community and council procedures to see that justice is done and that the 125,000 war prisoners awaiting trial in Arusha pay for their crimes against humanity.[13]

Femmes Africa Solidarité (FAS) was founded by African women leaders after the 1995 World conference on Women in Beijing. Based in Geneva, Switzerland, it works on peace-building programs throughout Africa and in close collaboration with the AU, the Economic Commission for Africa, and women's associations. It was created in response to "the explosion of violent conflicts tearing apart the fabric of society in Africa" and to advance the implementation of the Beijing Platform for Action. Their objectives include giving value to women's initiatives and enhancing their capabilities and rights as fully fledged participants in peace making and peace building. It has supported women's peace efforts in Burundi, Sierra Leone, and Liberia.[14] The Mano River Women's Peace Network of West Africa (MARWOPNET) which operates in Sierra Leone and Liberia is affiliated to FAS.

An international conference was organized in South Africa in 1997 on "Leadership in the Burundi Peace Process." Women from Burundi had the opportunity to discuss and analyze the historical causes and consequences of the conflict. From their perspective, the conflict was a result of colonial rule that instituted notions of ethnic superiority and discrimination.

While large-scale ethnic killings did not seem to have occurred during the colonial period, the advent of German and Belgian rule cultivated the notion of superiority of the Tutsi over the Hutu. Because all those in management positions and in the universities during the colonial administration were Tutsi, the groundwork was laid for the mechanism that perpetuated this discrimination following independence.[15]

FAS identifies poverty as a major factor in preventing the development of pluralism and democracy from becoming a reality. Poverty is viewed as being responsible for the social injustices that provoke war. FAS sees economic vitalization as an essential complement to the peace process. It considers the war in Burundi as "war against poverty."[16]

This appears to be in keeping with a study by UNIFEM that argues that although women and men may bring their interests and ideological proclivities into the peace process, their perspectives and impacts are different. The study argues that women's understanding of social justice and of the deleterious impact of wars on development can influence peace negotiations so that they can produce more constructive, inclusive, and sustainable outcomes.

> The absence of women from the process results in setbacks to the development of society at large and undermines democracy.[17]

There are many factors contributing to armed conflict in Africa of which underdevelopment and poverty play a major part. Also of importance is the unjust economic world economic system being propelled by corporate globalization that is destroying many of the economies of Africa. This has become reinforced by the sale of arms and illegal drugs used by rebels and insurgents.

The Contribution of African Women to the International Women's Movement

One of the most important ways in which African women have contributed to the international women's movement is by emphasizing and expanding the agenda of development. This included giving emphasis to the negative impact of the global political economy on African countries and bringing it to the attention of the international community.

In addition they have helped to promote a rights-based approach to development that is now considered a sine qua non to the advancement of women. The right to development is now accepted as a basic human right. This has reinforced the need to address the negative impact of the international political economy on African countries. This is particularly marked by the continuing de facto debt burden, SAPs, and corporate globalization. Many Plans of Action and Declarations emanating from International meetings have sought to address the negative impact of the global economy on many countries, especially in Africa.

Concerns about the impact of the global economy have also led to African women forming alliances with other associations from the Global South, since they have not been regarded as priority by the majority of feminists from the Global North. As a result, development problems emanating from the global economy have tended to assume a low profile in the international women's agenda.

Some associations of women from the Global South, such as the AAWORD have made the challenge of the global economy their primary focus. They have contributed significantly to revising the paradigms used to study gender relations and women's location in the international economy.

AAWORD emphasizes the neocolonial nature of research on African women and analyzes the negative impact of the global political economy associated with the

Nkosazana Dlamini Zuma, Minister for Foreign Affairs of the Republic of South Africa, addressing participants at the Forty-Ninth Session of the Commission on the Status of Women, today at UN Headquarters. Courtesy: UN Photo/Mark Garten, New York.

burdens of debt, SAPs, and neoliberal policies, such as privatization.[18] It has organized conferences against the negative effects of globalization. It urges African women to become actively involved in the assessment of the Highly Indebted Poor Countries Initiative (HIPC) of international financial institutions:

> It is thus time for us to launch an African women's movement that goes beyond monitoring the World Bank's commitment to women, but engages itself with monitoring the international financial institutions that are actively contributing to mortgaging women's well-being. In so doing, we may need to recall that the Bretton Woods Institutions—the World Bank, the International Monetary Fund (IMF) and the General Agreements on Tariffs and Trade (GATT) founded in 1944, are indeed agencies of the United Nations. For that reason, we would make them accountable to women's needs, despite the fact that unlike the United Nations General Assembly, they do not function on the principle of "one country, one vote" but on the basis of "one dollar, one vote."[19]

Other pan-African women's associations, such as ABANTU for Development, have also criticized the inimical consequences of corporate globalization. In a statement issued at a regional meeting held in Cape Town, South Africa in 1996, the association called for joint campaigns against the World Bank and the IMF and other financial institutions:

> to restructure their economically unsound, undemocratic and dehumanizing policies and programs.[20]

Development Alternatives for Women in a New Era (DAWN) was founded in 1984 as an alliance of women activists, researchers, and policy makers of women in the "Third World" or Global South. It was inspired by AAWORD.[21] It is an international feminist network that promotes critical perspectives on development and challenges the unequal and androcentric model of development that is propelled by market forces that impoverish many nations, social groups, and women.[22]

DAWN is committed to developing an alternative framework and method to attain the goals of economic and social development, based on the principles of justice and freedom from all forms of oppression by gender, class, and nation. It maintains that existing structures of domination are embedded in a colonial heritage that perpetuate the objectives of powerful nations and social classes.

As founding members of DAWN, African women have helped to develop a position that views existing global economic and political structures as producing inequalities between classes, genders, and ethnic groups, at international, regional levels, and national levels. As researchers and activists, they seek to build alliances with women at the grassroots level. DAWN has branches in Africa, holds regular research meetings on topics dealing with subjects like globalization, the feminist movement, the state, and governance.

Much of DAWN's global advocacy involves building partnerships with other associations and networks to achieve reform of international financial institutions such as the World Bank and the IMF. It also advocates for the accountability of governments to the commitments made at international meetings and Plans of Action on gender equality, as well as on poverty alleviation.

African women have also contributed to the decolonization project in general, as delegates to international conferences and as members of movements for nationalism and independence. The struggle against Apartheid was another important contribution. International solidarity among feminists, inspired by African women, contributed in part to bringing down the Apartheid regime in South Africa. The United Nations Commission on the Status of Women established an agenda item on "Women and Apartheid," which has led to several agreements opposing de jure and de facto Apartheid and proposing guidelines for racial equality. The United Nations Conference Against Racism held in South Africa in 2001 also bolstered the struggle of women against de facto Apartheid.

Individual African women have also made important contributions to the international agenda and the international women's movement. [Dr. Professor Wangari Maathai, winner of the 2004 Nobel Prize for Peace, is not only the first African woman to win the Nobel Prize, but is also an icon for both intellectual and activist leadership in the international women's movement.] She is committed to women's empowerment at the grassroots level and works to protect the environment against degradation, which can be a central point of conflict. In the area of legal equality, one of the first female Supreme Court justices in the world, Justice Annie Jiagge from Ghana was among the pioneers who contributed to the drafting of the Convention on the Elimination of All Forms of Discrimination Against Women (CEDAW).

In the 1980s and 1990s, a number of African women involved in the work of the United Nations for gender equality and the advancement of women made noteworthy contributions. For the fourth World Conference on Women held in Beijing in 1995, Mrs. Gertrude Mongella of Tanzania served as the Secretary-General of the conference. She is now the president/speaker of the parliament of the AU. Ambassador Marvat Tallawy of Egypt was a member of the Commission on the Status of Women that drafted the Declaration on Violence Against Women. Mrs. Shafika Selami-Meslem from Algeria served as Director of the Division for the Advancement of Women and deputy secretary-general of the Third World Conference on Women in Nairobi in 1985.

Several other African women have contributed in a number of ways to the work on the advancement of women within the United Nations System and have also contributed to improving the situation for women within these organizations. These include Dr. Achola Pala Okeyo of Kenya, Ms. Thelma Awori of Uganda, Dr. Felicia Ekejiuba of Nigeria, Ms. Philomena Kintu of Tanzania, Ms. Yasine Fall of Senegal, Ms. Rachel Mayanja of Uganda, Dr. Laketch Dirasse of Ethiopia, Ms. Viola Morgan and Dr. Filomina Steady, both of Sierra Leone.

Despite these international and collective efforts, government-inspired policies for gender equality and for the advancement of women have their limitations. African female scholars are among those that have criticized state-oriented and intergovernmental approaches that can limit the effectiveness of International Plans of Action and Conventions. These international approaches are increasingly being viewed as examples of bureaucratic and state sponsored feminism that tends to accommodate rather than challenge patriarchy.[23]

For this reason, the emphasis on development alternatives and the challenge of the global economic system by African women and women from the Global South have been important for the international women's agenda. Without doubt, the international women's movement would not be the same without the contributions of women from Africa.[24] Conceptual and theoretical advancement have been influenced both by the contributions of AAWORD, WILDAF, FEMNET, FAS, and DAWN, among others.

The contributions to the agenda on development, human rights, and the elimination of racism by African women's associations and individual women from Africa have been significant. Through collective action and individual effort, women in Africa are contributing to the transformation and empowerment of women's lives, not only in Africa but also internationally.

Summary

Women's associations at the pan-African level expand the opportunities for collective action by women. They reflect the importance of the themes of this study at the regional level, since some of the problems of underdevelopment and authoritarianism transcend national boundaries. They also facilitate the development of formal and informal networks and strategies to overcome barriers and other constraints that might impede action at the national level. Many of the pan-African associations promoting peace, such as FAS are able to avoid polarizations and tensions at the national level and can offer neutral and safe spaces for resolving conflict.

A number of problems transcend national boundaries. These include cross-border conflicts, environmental degradation, cross-border trade, the HIV/AIDS pandemic, and corporate globalization. As has been demonstrated, conflict prevention and peace building have gender implications, some of which can best be addressed at the pan-African level. Regional associations can also expand the potential for development and democratization by maximizing their resources and exchanging ideas and information that can promote a more active role for women in economic and political decision-making.

Chapter Twelve
Conclusion

Engendering and Enhancing Development and Democratization

This book has sought to answer the following questions: What is the significance of women's collective action in Africa? What does this tell us about the relationship between development and underdevelopment? What is the role of women in promoting democratization, given the legacy of authoritarianism at national and international levels? How does the social fabrication of women's positions and roles determine collective identities, the nature and function of women's associations, and their institutional manifestations? How do women seek to empower themselves collectively and for what reason? What is the contribution of African women to the international women's movement?

To answer these questions, I have used a conceptual model based on two explanatory themes, namely the development–underdevelopment nexus and the democratization–authoritarianism nexus. The link between these processes represents a *nexus of action and reaction*. I argue that it is this nexus that becomes the main theater for their collective action. It is in the dynamic interface of the two themes that the symbolic and material significance of women's collective action are expressed.

A review of relevant theoretical frameworks showed that no single theoretical approach would suffice. What is required is an eclecticism that draws insights from development theories, especially dependency and theories of the political economy, feminist theory, and social movement theory.

Dichotomous models were found limiting and required more flexible interpretations. This was particularly true of the public–private debate and rigid social distinctions implied by the term "gender." For example, rather than being a representation of the devaluation of women's roles in social reproduction, "motherhood," an integral function of the private sphere, can be a mobilizing force for development and democratization. Studies from Sierra Leone, Nigeria, and Kenya bear this out.

Similarly, the private sphere does not constitute an exclusive female domain. In fact, it is constantly being invaded by the "public" hand of the state in determining laws that regulate marriage and family matters. Colonial-based statutory laws are notoriously discriminatory against women and their repeal has been a significant factor in female mobilization in many countries.

A "collective empowerment index" appears to be the missing link for bridging the gap between what Longwe and Clarke describe as a lack of correlation between a "women's self-reliance index" linked to individual capacity and "women's empowerment index" that measures levels of female representation in political and managerial positions.[1]

Empowerment through collective action tends to facilitate the movement of women from self-reliance to empowerment in terms of political and managerial positions, not only in Sierra Leone, but also in other countries and at the pan-African level. However, the gender bias in the body politic and women's generally low numbers in government can prevent them from being effective after they get into political office.

In exploring both the essentialist and the post-modernist paradigms, it could be argued that, given the hegemonic and universalizing nature of corporate globalization, essentialist and modernist realities continue to be relevant. Despite their diverse origins and social locations, women of the Global South have tended to act collectively and in an essentialist manner when resisting the negative, universal, and totalizing impact of the global political economy.

They have been among the activists in the anti-globalization movement and in challenging the authoritarian and undemocratic proclivities of multinational corporations and the international financial institutions, such as the World Bank, the International Monetary Fund (IMF), and the World Trade Organization (WTO). Through collective action, associations like Association of African Women for Research and Development (AAWORD) and Development Alternatives for Women in a New Era (DAWN) have sought to promote alternative approaches to development that are more humane, egalitarian and that are based on advancing the welfare of people, rather than accumulating profits.

The domination of economic structures, and the exploitative nature of the neo-colonial post-independence period are clearly being reinforced by corporate globalization. These events construct gender relations in a manner that has reinforced women's disadvantages in the economic, political, and social spheres. This leads to the further exclusion of women from the structures of power and to the devaluation of their contributions.

Much of women's activism through collective action is in response to the phenomenon of global economic domination. As this study has shown, women have been challenging structures of power and demanding access to decision making for many years. In addition, they have been constructing alternative avenues for power through leadership in associations; developing alternative mechanisms for capital accumulation; and promoting democratization and a culture of peace. In some cases, they have had success but in most instances they are struggling against formidable odds.

As has been demonstrated, these associations have the capacity to transform themselves into politically active entities, individually or collectively to challenge exploitative economic conditions and gender-based hierarchies. Almost always, however, their agenda is not limited to women or gender issues, but have a larger socio-centric objective to transform society through development and democratization.

With regard to Sierra Leone, the in-depth study of over eighty women's associations in Freetown supports the thesis that women's collective action seeks to accomplish two main overlapping objectives. One is to promote development while resisting underdevelopment, and the other is to promote democratization while challenging authoritarianism. In this process, women seek to empower themselves collectively as well as to transform their societies.

Mobilization and agenda setting in Sierra Leone are influenced to a large extent by primary group ties of kinship, ethnicity, and community. However, associations operating at the national level as umbrella associations, federations, and spontaneous movements for development and democratization, tend to cut across primary group ties and to be representative of a more diverse group.

The concern of women over larger development-oriented goals and issues of democratization and citizenship has taken priority over narrow feminist goals. Feminist quests for gender equality in terms of jobs and political positions can sometimes be viewed as being too individualistic and self-centered and as a luxury, given the major challenges of underdevelopment and poverty. Instead, objectives are broadly defined and generally in the main goals of promoting development and democratization. The country has ratified the *Convention on the Elimination of All Forms of Discrimination Against Women*, making women's legal rights guaranteed by law in a technical sense. However, compliance remains a challenge, due largely to problems of underdevelopment. Translating international laws into domestic legislation and enforcing them can be hampered by the paucity of resources.

Sierra Leone's recent history of political upheavals and a rebel war has helped to shape the priorities of these associations and has intensified the quest for development and democratization, both of which are viewed as interrelated. Women have been in the forefront of the movements for both development and democratization that incorporate, rather than centralize, gender equality and the advancement of women.

The real and potential contributions of women's associations have been significant in many areas. Among these is their function as shadow development agencies contributing to many development projects, including large-scale ventures, such as the building of schools, orphanages, collectives, and clinics.

They have also helped to promote an informed and active female citizenry and have kept women abreast of civic matters and of events on the political landscape. Through these efforts, they have increased awareness about the important role of women in development and have influenced many policy changes toward the establishment of a National Policy for Women. Women's needs, concerns, and aspirations are now articulated at several ministerial levels and in National Development Policies and Plans.

The development of female human resources and the building of women's skills and legal capacity to a large extent have been due to the work of women's associations. They have also enriched the social, cultural, and recreational life of Freetown through their fund-raising activities and public events. These associations serve as the conscience of society and have often advocated against a number of social problems including the high cost of living, unemployment, crime, and militarization. They have established programs to combat the increasing school dropout rate for girls and the rehabilitation of women and girls affected by the war.

Regardless of size or membership composition, almost all associations pursue interests that pertain to society as a whole. These *socio-centric interests* can and often include "practical women's interests" and "strategic gender interests." They often take the form of political, economic, and educational empowerment; alleviating poverty; providing mutual aid; promoting philanthropic objectives for the less fortunate; and improving their social, cultural, spiritual, and personal well-being.

Their weakness lies primarily in their generally low level of funding, which imposes resource constraints and limits their effectiveness. Cultural attitudes toward women and gender biases also tend to impede their work, especially when dealing with entrenched bureaucracies and the military. Some associations are plagued with structural weaknesses that stem from role conflicts between the leadership and the rank and file, and personality conflicts among the leadership. Internal problems of management have also surfaced from time to time, especially in the larger associations, leading in a few cases to factions being formed or splinter groups separating from the main body.

By far the greatest challenge is that which operates within the constraints of a colonial legacy and the hostile international political economic system led by corporate globalization. These forces work against development and democratization and undermine the welfare and well-being of African peoples of both genders.

The explanatory themes were then examined in other countries, namely Nigeria, Kenya, South Africa, and Algeria, representing the four subregions of Africa. In particular, similarities were found in terms of the impact of corporate globalization and protracted recession in creating and reinforcing underdevelopment and poverty. To a large extent, women employed comparable strategies of resistance to underdevelopment. All the countries face similar challenges to democratization. These were apparent in the relationship with the state, the role in nationalism, and the impact of religion and fundamentalism on democratization.

This comparative insight was further extended to women's associations operating at the regional/pan-African level. These continent-wide associations maximize the objectives, activities, and resources of those that operate at the national level. They also help to coordinate and harmonize agendas that are of priority to the African region. An analysis has also been made of the contribution of African women to the international women's movement.

One striking feature of the pan-African women's associations is their tendency to have an agenda that gives a higher profile to feminist concerns in the narrow sense of promoting gender equality. Although they are also promoting development and democratization, they tend to work more closely with international feminist movements and international organizations that are more inclined to promote gender equality above other issues. Since much of their funding comes from international sources that sometimes impose their own priorities, these associations have been inclined to align their agenda-setting priorities with those of their international donors.

The driving force behind the themes of this book is a critique of the dominant model of development, based on economic and political liberalization. Instead of producing development and democratization, it has fostered underdevelopment and authoritarianism. It is the tension between the ideals and realities of the international political economy and its national and local manifestations that has defined, promoted, and circumscribed women's collective action.

Through this process women seek to empower themselves and to develop their own brand of African feminism, with its altruistic and humanistic overtones. In this feminism factors such as colonialism, Apartheid, structural racism, and the international political economy are central.

Women engage in collective action to maximize the social capital generated, not only from new opportunities, such as the international women's movement and the new momentum for democratization, but also from their traditional modes of mobilization. In addition, because the nature of power introduced through colonialism excluded them, they mobilize to capitalize on the new political spaces opened by the global momentum for democratization.

The study illustrates one of the main arguments of the book, that women's associations are primarily shadow development agencies, providing goods and services that are either minimally provided by governments or not provided at all. This is a characteristic symptom of underdevelopment, inspired by a colonial legacy that exploited Africa's resources, but failed to build structures for development. The postcolonial period has continued this legacy into the new millennium. It has been marked by dependent economic relations and the marginalization of Africa in the process of corporate globalization.

The objectives and actions of women's associations are not limited to gender issues. Given the reality of oppressive global economic forces and the mounting poverty faced by the majority of women, narrow feminist concerns pale in the face of the material and existential conditions of women and their families that continue to deteriorate.

My study has shown that women's interests are socio-centric and that women's rights include the rights to development. The challenge for development is underdevelopment that profoundly undermines not only women's rights, but also the rights of all members of society. Democratization is essential in restoring these rights and is a critical aspect of development, since it involves challenging authoritarianism, including the authoritarianism of the global economy.

As this study has shown, African women have been in the forefront of the international women's movement. Furthermore, they have made substantial contributions as governmental delegates, NGOs, international civil servants and as individuals in the development of international Plans of Action, Conventions and Declarations. Their contribution has been particularly marked in the areas of decolonization and in the struggles against racism, especially Apartheid.

African women brought their own theories, critiques, wisdom, and experiences to the international agenda-setting process and have contributed to a truly global and humanistic revisionist project. They have also helped to refine and advance the postmodernist vision, by insisting on multiple truths, histories, and realities, while also recognizing the continuing relevance of essentialist paradigms. Above all, it is worth reiterating that African women have helped to advance the agenda on development and democratization and have joined in the global resistance to forces of underdevelopment and authoritarianism, emanating primarily from the global economic system.

Women's associations in Africa are an exploding phenomenon changing the African landscape and representing a dual interactive process. On the one hand, they act collectively to promote objectives of development and democratization, which include an empowered female citizenry. One the other hand, they aim to resist the forces of underdevelopment, globalization, and authoritarianism.

Pursuing development is an integral part of a brand of African feminism and an African feminist agenda. While encompassing struggles for gender equality, it often goes beyond controversies about male dominance, sexuality, reproductive rights, and the quest for person fulfillment. This is in keeping with the holistic nature of African feminism that of necessity has to take into account the colonial legacy, the trans-Atlantic Slave Trade, structural racism in Africa and in the African Diaspora, underdevelopment, and the impact of the global political economy on both men and women. The collective consciousness reflects a feminism that is socio-centric and humanistic, with concerns that apply to society as a whole.

According to the positions explored in this book, African feminism transcends individualism and tends to involve women as a group in struggles against oppression of all kinds, based on economic, political, and cultural domination. It is more holistic and humanistic and has a greater potential for social transformation. In this regard, it conforms to the definitions of African Feminism first articulated in 1981 in *The Black Woman Cross-Culturally* and in later publications.

African feminism was defined there as a feminism in which sexism cannot be isolated from the larger political and economic forces responsible for the exploitation and oppression of both men and women of Africa and of the African Diaspora. It also seeks to promote these objectives within in a human-centered context.[2]

Judging from the proliferation of women's associations over the last quarter of a century in Africa, one can say with some certainty that a type of gender revolution is taking place in Africa. This revolution has more to do with the economic challenges of underdevelopment and poverty and with political instability than with narrow feminist concerns. It is a revolution against the economic injustice of the global economy that is wreaking havoc on African countries and on the social fabric and well-being of African peoples. It is also a revolution against authoritarianism, at both the national level and at the level of the global political economy.

Collective action by women will have to involve building alliances with other movements within and outside Africa, which are seeking to humanize and democratize the unjust international economic system. This calls for an international feminist agenda that gives priority to promoting economic justice at the global level. Based on this study of women's collective action at national, regional, and international levels, one can conclude with some certainty that it is only through effective transformation of the global political economy that feminist concerns can assume center stage in Africa.

Notes

Chapter One The Challenge and Conceptual Framework

1. Steady, 1981, 2002; AAWORD, 1985; Sen and Grown, 1986; Dembele, 1999, 2002; Fall, 1999, among others.
2. Beauvoir, 1989 edition; Rosaldo and Lamphere, 1974; Kaplan, 1992; Katzenstein and Mueller, 1987 among others for various interpretations of this approach.
3. Lebeuf, 1963; MacCormack (Hoffer), 1974; Aidoo, 1981; Okonjo, 1981; Sudarkasa, 1981.
4. See for example, Steady, 1981; Amadiume, 1987, 1997; Oyewumi, 1997; Nnaemeka, 2003.
5. Collins, 1990 (2000 edition); Mohanty, Russo, and Terres, 1991; Basu, 1995; Nnaemeka, 1998. See also Steady, 1981.
6. Sivard, 1977; See AAWORD, 1985; Sen and Grown, 1986; Reno, 1998; Starr, 2000; Rowbotham and Linkogle, 2001.
7. Jayawardena, 1986; Moghadam, 1994; Wieringa, 1992, 1995; Alexander and Mohanty, 1997; Rosander, 1997; Nnaemeka, 1998; Tripp, 2000, among others.
8. Rowbotham, 1975; Beneria, 1978, 2003; Young, Wolkowitz, and McCullagh, 1981. For development see Boserup, 1989; Sen and Grown, 1986, among others. See also Rosaldo and Lamphere, 1974 for one of the earliest cultural interpretations.
9. Caplan and Budjra, 1982.
10. See Urdang, 1979, 1989; Lewis, 1976; Wipper, 1984; Kabira *et al.*, 1993; Khasiani and Njiro, 1993; Rosander, 1997; Woodford-Berger, 1997; Tripp, 2000, 2001.
11. Wipper, 1984.
12. See Abdullah, 1995.
13. For a variety of views, see Stroebel, 1976; Caplan and Budjra, 1982; Kabira *et al.*, 1993; Khasiani and Njiro, 1993; Abdullah, 1995; Aubrey, 1997.
14. Aubrey, 1997, p. 38.
15. Steady, 1981, 1987.
16. Copestake and Welled, 1993; Aubrey, 1997.
17. Sen and Grown, 1986, pp. 90–93 and Wipper, 1995 for classifications.
18. Rosander, 1997, p. 9.
19. Dworkin, 1976, 1989; Daly, 1978, 1985. Ecofeminists also see a universal essence that links women to nature and its preservation and condemns patriarchal values that lead to the domination and destruction of nature. See Reuther, 1975; Shiva, 1989; Diamond and Orenstein, 1990 for expressions of this view.
20. See Foucault, 1984; Robinow, 1984.
21. Ray and Korteweg, 1999.
22. Molyneaux, 1985. Some scholars have not supported Molyneaux's model, based on their research in countries of the Global South. These include Agarwal, 1992.
23. See Steady, 1981, 1987, 2000 for African Feminism and Basu, 1995 for local feminisms.

24. Steady, 2002a.
25. Amadiume, 1987; Oyewumi, 1997.
26. Amadiume, 1987, 1997; see also Oyewumi, 1997.
27. Okonjo, 1981. See also Lebeuf, 1963; Paulme, 1963; Sudarkasa, 1987 among others, for women's participation in politics and in the public sphere.
28. See Amadiume, 1987 for the Igbo of Nigeria; Stamp, 1986 for Kenya and Steady, 1975 for Sierra Leone.
29. Awe, 1997.
30. Rai, 2000.
31. Zulu, 1998.
32. Cohen and Rai, 2000.
33. Zirakzadah, 1997 for the first position; Stienstra, 2000 and Cockburn, 2000 for the second position.
34. Rai, 1994.
35. Alvarez, 1990, p. 29.
36. Lloyd, 1999.
37. Tarrow, 1998. Parpart and Staudt for some perspectives on women and the state in Africa.
38. *Wisconsin State Journal*, July 16, 2002 among others.
39. Amadiume, 1987, p. 196.
40. Wolfensohn, 2000.
41. UNAIDS, 2003.
42. The life expectancy rate is 54 years; adult female literacy 34%; maternal mortality 500 per 100,000 births; infant mortality 92 per 1000 live births; and the majority of refugees and people living with HIV/AIDS are women.
43. Sarris and Shams, 1991; Dembele, 1999, 2002; McBride and Wiseman, 2000.
44. Ake, 1990.
45. See Rostow, 1967 for modernization theory in general. Boserup, 1989, for one of its earliest critiques in terms of its application to women of the Global South; Snyder and Tadesse, 1995.
46. Pettman, 1996; Fall, 1999; Khor, 2000; Starr, 2000; Dembele, 1999, 2002; Steady, 2002.
47. Bundlender and Hewitt, 2002; UNDP, 2002; Morgan and Tropps, 2000; Council of Europe, 1998.
48. Mama, 1997.
49. United Nations, 2000, p. 42.
50. United Nations, 1995b.
51. Also referred to as "The Third World" and "Developing Countries." See Etienne and Leacock, 1980 for a study of women and colonialism.
52. See Bangura, 1992, 1994; Falk, 1999; Sethi, 1999; Fall, 1999; Rupert, 2000; Temple, 2000; Steady, 2002.
53. Khor, 2000.
54. Frank, 1969; Amin, 1973; Frank, 1981; were leading examples of this theoretical viewpoint. The feminist version was well articulated by Sen and Grown, 1986; Fall, 1999.
55. See Steady, 2002 for a study of the impact of globalization on Black Women in Africa and the African Diaspora.
56. Tarrow, 1998.
57. Ake, 1990, 1996.
58. Tarrow, 1998, p. 3.
59. Staudt, 1986.
60. AAWORD, 1985; Sen and Grown, 1986 for the articulation of these positions.
61. Sen and Grown, p. 19, following similar views expressed earlier in *AAWORD Newsletter—Feminism in Africa*, vol. II/III, 1985.
62. Longwe and Clarke 1999. See also Longwe, 1990.
63. Longwe and Clarke, 1999. See also United Nations, 1995a.

Chapter Two   The Context and Background

1. Rodney, 1981.
2. See Steady, 2001 for a study of the culture of this group with a focus on women.
3. Reno, 1995; Hirsch, 2001.
4. Sierra Leone Government, 1965, 1968, 1992. The last census done in 1985 (published in 1992) listed the population as 3.5 million. The latest census was conducted in December 2004 and the Provisional Results give the total population as approximately five million.
5. McCulloch, 1950 and updated data from censuses, household surveys, and interviews.
6. See Davies, Davies, Gyorgy and Kayser, 1992 for interviews with women in various occupations.
7. Provinces Land Act, Chapter 122 of the Laws of Sierra Leone was enacted to regulate the use, ownership, and occupation of Provincial land.
8. Koroma, 2003.
9. Most of the references for the historical background are from the following sources: Mota, Teixeira, 1625, translated by Paul Hair, 1977; Fyfe, 1962; Fyle, 1981. See also Foray, 1977.
10. Steady, 1985 for a study of the role of polygamy, among other things in women's work in two rural areas in Sierra Leone.
11. Combahee River Collective Statement, 1986.
12. Banton, 1957; Little, 1965.
13. Dworzak, 1994.
14. Dworzak, 1994. See also, Dworzak, 1990; Barlay, 1990; Sierra Leone Government, 1994; Department of Economic and Development Planning, 1993. Non-Governmental Organizations (NGOs) is a term used by the United Nations that has become popularized to include all women's association. NGOs usually have some international contacts and collaboration.
15. United Nations, 1985.
16. Abdullah, 1995 for first ladies' associations in Nigeria.
17. *Ibid.*
18. Preamble of the constitution of the National Federation of Sierra Leone Women.
19. Williams, 2003.
20. Personal communication, 1994.
21. Field research was conducted in 2002 in conjunction with the Human Resource Development Organization of Freetown (HURDO).

Chapter Three   Collective Action for Political Empowerment

1. MacCormack, 1974.
2. Lucan, 2004.
3. The title of Mayor in Freetown was replaced by "Chairman of the Committee of Management." However, this has reverted to Mayor following a move to decentralization and greater autonomy in local government in 2004.
4. Steady, 1975.
5. *Ibid.*
6. Steady, 1975, p. 11.
7. See Day, 1997 for a profile of Kadi Sesay.
8. Weeks, 1992.
9. Akhigbe, 1999, p. 5
10. Women's Forum, 1997, p. 34.
11. Jusu-Sheriff, 2000, p. 2.
12. Women's Forum, 1997, p. 86.
13. Women's Forum, 1997.

14. Personal communication, June 8, 1998.
15. Women's Forum, 1997, pp. 92 and 93.
16. Women's Forum, 1997, p. 98.
17. National Commission for Democracy in Sierra Leone, 1996.
18. See French, 1997a for further discussion of this issue and the events surrounding it.
19. Women's Forum, 1997, pp. 89–90.
20. Jusu-Sheriff, 2000.
21. Jusu-Sheriff, 2000, p. 4.
22. Field interviews and Femmes Africa Solidarité website, 2004.
23. Darbor, 2003. See also Campaign for Good Governance, 2004.
24. UNFPA unpublished documents and those of the Ministry of Health confirm this.
25. Movement for Progress Party, 2002.
26. Jusu-Sheriff, 2000, p. 5.

Chapter Four   Collective Action for Economic Empowerment

1. Sierra Leone Government, 1992.
2. White, 1987, among others.
3. United Nations, 2000, p. 144.
4. United Nations, 2000, p 130.
5. IPAM, 1991.
6. *Ibid.*
7. *Ibid.*
8. Concord Times, 2004.
9. Personal communication with members of the association during several field research encounters.
10. Cummings-John, 1995, pp. 92–93.
11. Cummings-John, 1995; Denzer, 1981.
12. Cummings-John, 1995, p. 89.
13. Cummings-John, 1995.
14. Cummings-John, 1995; see also obituary by Fyfe, 2000.
15. See Ardener, 1964; Geertz, 1962.
16. Geertz, 1962.

Chapter Five   Collective Action for Educational and Occupational Empowerment

1. Numerous studies exist on this subject. See Bloch *et al.*, 1998; Chanana, 2001 and Kwesiga, 2002.
2. Fyfe, 1962, p. 132.
3. Fyfe, 1962, p. 102.
4. *Ibid.*
5. Fyfe, 1962, p. 514.
6. Mrs. Lati Hyde-Forster, personal communication.
7. Hamilton, 1971; see also Cromwell, 1989.
8. UNDP, 1996, 1997.
9. See also United Nations, 2000, p. 85.
10. *Ibid.*, p. 94.
11. Many studies have documented these trends. Among these are May-Parker, 1986 and Beoku-Betts, 1998.
12. Cummings-John, 1995.
13. Cummings-John, 1995, p. 78.

14. See Steady, 1985.
15. J. Dworzak, personal communication, 1996.
16. Thorpe, 2000.
17. *Ibid.*
18. *Ibid.*
19. *Ibid.*
20. United Nations, 1985.
21. African Development Bank, 1990. Reinforced in subsequent publications of the Africa Development Report, especially in 1996 and 1997.
22. Assie-Lumumba, 1995; Beoku-Betts, 1998.
23. Beoku-Betts, 1998, p. 173.

Chapter Six "Traditional" Associations

1. Data and information on traditional secret societies and on traditional religious societies, in general, are derived from several published and unpublished sources, interviews, and discussion group sessions. Published sources include; Margai, 1948; Porter, 1953, 1963; Banton, 1957; Van Gennep, 1960; Kenyatta, 1938; Little, 1965; D'Azevedo, 1962, 1980; Peterson, 1968, 1969; MacCormack, 1974, 1977; McCulloch, 1950; Richards, 1975; Steady, 1975; Williams, 1979; Bledsoe, 1980; Wilson, 1981; Boone, 1986; Harrell-Bond, 1976; Cohen, 1971; Day, 1998.
2. Cohen, 1971.
3. D'Azevedo, 1962.
4. Richards, 1975.
5. Williams, 1979.
6. Information on the organization of Sande are from fieldwork interviews, general knowledge living in Freetown, and from studies, in particular, Richards, 1975.
7. MacCormack, 1974, 1977.
8. Van Gennep, 1960.
9. Margai, 1948.
10. Data based on information from informants and supplemented by Richards, 1975.
11. See Phillips, 1979 for an analysis of the iconography of the Mende Sowei mask. See also Boone, 1986 for a study by an art historian of the Sande/Bondo mask and other symbols of the Sande/Bondo society and ritual.
12. Bledsoe, 1980.
13. Published information on the Poro Society is difficult to obtain. It is almost impossible for a woman to conduct research on this society. Most of the information is derived from Little, 1951.
14. Little, 1951.
15. Margai, 1948.
16. Peterson, 1969, p. 298.
17. Banton, 1957.
18. Williams, 1979.
19. Williams, 1979, p. 14 and Koso-Thomas, 1987.
20. Koso-Thomas, 1987.
21. Dreifus, 1977.
22. Baker-Bensfield, 1977.
23. See JENDA, 2000 for a critique of the quest for "designer vaginas" through plastic surgery in relation to criticisms of genital mutilation. See Weitz, 1998, for extensive and rare discussions on the subject from a feminist perspective. See also Robertson, 1996; James and Robertson, 2002.
24. Williams, 1979.

25. Oladipupo, 2003.
26. French, 1997b.
27. Haja Sally Sasso, personal communication.
28. *Ibid.*
29. Foday, 1996.
30. Dugger, 1997.
31. French, 1997a.
32. French, 1997b.
33. Babatunde, 1997.
34. *Ibid.*
35. Steady, 1975.

Chapter Seven   Mutual-Aid Associations

1. Social Welfare Department Unpublished Reports, several years.
2. See Porter, 1953, 1963; Banton, 1957.
3. Banton, 1957; Little, 1965.
4. Harrell–Bond, 1976.
5. A Yoruba custom popularized in Freetown by the Krios.

Chapter Eight   Islam and Women's Associations

1. Fyle, 1981.
2. Peterson, 1969; Fyle, 1981.
3. See Blyden, 1967.
4. Bassir, 1954, p. 252.
5. Trimingham, 1961, p. 46.
6. Mernissi, 1975, p. 81.
7. Stroebel, 1976.
8. Trimingham, 1961.
9. Stroebel, 1976.

Chapter Nine   Christianity and Women's Associations

1. See Steady, 1976, 1978.
2. Selena Hastings, Countess of Huntingdon (1707–1991) was the central figure in the evangelical revival of the eighteenth century in England and founder of the Countess of Huntingdon's Connexion. Countess of Huntingdon Churches were founded in Sierra Leone by Blacks from Nova Scotia who landed in Freetown in 1792 after being granted freedom following the American Revolutionary War.
3. See Fashole-Luke, 1968.
4. Fashole-Luke, 1968, p. 132.
5. See Steady, 2001.
6. Weber, 1930.
7. Amin, 1973.
8. See Fyfe, 1962; Porter, 1963; Peterson, 1969; Wyse, 1980, 1991; Steady, 2001 among others.
9. Fashole-Luke, 1968.
10. Mothers' Union, 1996.
11. Mothers' Union, 1996, pp. 14–15.
12. Fyfe, 1962.

13. Sierra Leone Household Surveys and Censuses were conducted at relatively frequent intervals before the war. The latest census was conducted in December 2004.
14. Mothers' Union, 1996.
15. Fyfe, 1962.

Chapter Ten   Comparative Insights at the National Level

1. The case study of Kenya relies on Kameri-Mbote and Kibwana, 1993; Khasiani and Njiro, 1993; Nzomo, 1993a, b; Odoul and Kabira, 1995; Kabira, Odoul, and Nzomo, 1993; Stamp, 1986; Wipper, 1984 and United Nations documents. Studies of Nigeria relied on Okonjo, 1981; Mba, 1982; Ekejiuba, 1991; Abdullah, 1991, 1995; Okeke, 2000, 2002. Studies of South Africa relied on Magubane, 1979; Rivkin, 1981; Walker, 1991; Zulu, 1998, 2000; Pheko, 2002. Studies on Algeria and the North Africa relied on Bouatta, 1997; Fades, 1994; Fernea, 2000; Karem, 1997; Lazreg, 1990; Lloyd, 1999; Ziai, 1997.
2. Abdullah, 1991.
3. Several studies are aware of this phenomenon, including Zulu, 1998, 2000.
4. Fernea, 2000.
5. Odoul and Kabira, 1995.
6. See UNDP, *Human Development Report*, several years and UN, *The World's Women*, several years. UNDP, 2002. Enterprise Africa, Regional Bureau for Africa. See Lancaster, 1991 for a study of SAPs.
7. Government of Kenya, 1990; See also Nzomo, 1993a, b and Khasiana and Njiro, 1993.
8. Government of Kenya, 1990.
9. *Ibid.*
10. Abdullah, 1995.
11. *Wisconsin State Journal*, July 16, 2002.
12. Most of the references for post-Apartheid South Africa come from Zulu, 1998 and 2000; presentations at the World Conference Against Racism in Durban in 2001 and my interviews with South African women in 2001.
13. Several grassroots movements staged demonstrations against the dumping of mercury and other industrial waste products from corporations in the United States on lands inhabited by Blacks, particularly Zululand. Clor Chemicals, which dumps mercury on Zululand, is one of the culprits.
14. Zulu, 1998, p. 147. Garba and Garba, 1999, p. 16.
15. Pheko, 2002, p. 102.
16. Lloyd, 1999.
17. See Raissiguier, 2002 for a study of African women migrants in France.
18. Urdang, 1979; Van Allen, 1972, 1976; Nzomo, 1993; Odoul and Kabira, 1995.
19. Tamale, 1999.
20. Hale, 2000.
21. Zulu, 1998 expresses this, as do many African women's associations interviewed in South Africa during the World Conference against Racism in Durban, 2001.
22. See Kameri-Mobote, 1993.
23. Parpart and Staudt, 1989.
24. See discussions of bureaucratic feminism in Alvarez, 1990; Oyewumi, 1997.
25. Odoul and Kabira, 1995.
26. Abdullah, 1995.
27. *Ibid.*
28. Ziai, 1997.
29. *Ibid.*
30. *Ibid.*
31. Tamale, 1999, p. 200.

32. Amadiume, 1987, p. 183; also Mba, 1982.
33. Zulu, 1998, p. 156.
34. Tarrow, 1998.
35. Studies of women's associations in Nigeria cited here include, Mba, 1982; Ekejiuba, 1991; Abdullah, 1995 and United Nations documents.
36. Odoul and Kabira, 1995.
37. References for South Africa are taken from several sources that include Rivkin, 1981, Lapchik, 1981; Zulu, 1998, 2000.
38. United Nations, 2000.
39. Lazreg, 1990.
40. See Saadi, 1991; Bouatta, 1997.
41. A number of studies have discussed these developments, see Ekejiuba, 1991; Okonjo, 1981.
42. Ekejiuba, 1991 among others.
43. Lloyd, 1999, p. 6.
44. Amrane-Minne, 1999.
45. Fades, 1994, p. 61.

Chapter Eleven Comparative Insights at the Regional/Pan-African International Levels

1. Viola Morgan, personal communication, 2003.
2. Data on the Greenbelt Movement is derived from several sources including the association's own documents, personal communication with the leaders, especially Professor Maathai, site visits to projects in Kenya, and the Internet.
3. Data and information on FEMNET are derived from its documents and interviews with its members, especially the executive. Of importance also is AMWA/FEMNET/WILDAF, unpublished Report, 2000.
4. United Nations, 2000.
5. FAWE, 2001.
6. FEMSA Project, 2001.
7. Expressed in a number of AAWORD documents, reports, and its journal *ECHO*.
8. Data on Akina Mama Wa Africa is derived from unpublished annual reports, proposals, speeches, UNDP reports, Morgan and Tropps, 2000, Akina Mama website, and interviews with members, especially the executive members.
9. Morgan and Tropps, 2000.
10. Much of the data on regional associations is derived from Morgan and Tropps, 2000, UNDP documents, NGO documents and reports and interviews with leaders of some of these associations.
11. Morgan and Tropps, 2000. Most of the data on WDF is derived from UNDP libraries, documents from the association and interviews.
12. UNDP, 2001; Morgan and Tropps, 2000; Documents of the association and interviews.
13. Morgan and Tropps, 2000 and UNDP documents from the Gender Program.
14. Femmes Africa Solidarité (FAS), 2001; Data used for FAS include interviews, reports from the organization, reports from countries, and associations working with FAS, UNDP, and the United Nations library. Also Morgan and Tropps, 2000; UNIFEM, 2000.
15. Femmes Africa Solidarité (FAS), 2001, p. 15.
16. *Ibid.*
17. UNIFEM, 2000.
18. AAWORD, 1998.
19. AAWORD, 1998, p. 8.
20. ABANTU, 1996, p. 5.

21. Much of the revisionist scholarship, which inspired Third World Feminist writings can be traced to AAWORD; its publications; meetings; journal *ECHO* and its contributions to UN meetings and publications, especially during the United Nations World Conference on Women held in Nairobi, Kenya in 1985.
22. Sen and Grown, 1986.
23. Oyewumi, 1997; Alvarez, 1990.
24. Snyder and Tadesse, 1995.

Chapter Twelve   Conclusion

1. Longwe and Clarke, 1999.
2. Steady, 1981, 1987, 2000, 2002.

Bibliography

AAWORD (1985) *Newsletter, Feminism in Africa*, Vol. II/III.
AAWORD (1998) "Editorial." *ECHO: Journal of the Association of African Women for Research and Development.* Dakar: AAWORD.
ABANTU (1996) *Abantu for Development Newsletter*, Vol. 2.
Abdullah, H (1991) Women in Development: A Study of Female Wage Labour in Kano Manufacturing Sector, 1945–1990. Unpublished Ph.D. thesis, University of Hull.
Abdullah, H (1995) "Wifeism and Activism: The Nigerian Women's Movement" in A Basu (ed.), *The Challenge of Local Feminisms: Women's Movements in Global Perspective.* Boulder: Westview Press.
Afshar, H (ed.) (1993) *Women in the Middle East: Perceptions, Realities and Struggles for Liberation.* New York: St. Martin's Press.
Africa Development Bank (1990) *Africa Development Report.* Abidjan: ADB.
Agarwal, B (1992) The Gender and Environment Debate: Lessons from India. *Feminist Studies*, 18(1): 119–158.
Aidoo, A (1981) "Asante Queen Mothers in Government and Politics in the Nineteenth Century" in F C Steady (ed.), *The Black Woman Cross-Culturally.* Cambridge: Schenkman Publishing Company.
Ake, C (1990) Democracy and Development. *West Africa*, March 1990.
Ake, C (1996) *Democracy and Development in Africa.* Washington, DC: The Brookings Institution.
Akhigbe, L (1999) Sierra Leone: The Inside Story—Why Peace has been so Elusive. *New African*, July 25, 1999.
Alexander, MJ and CT Mohanty (1997) "Introduction: Genealogies, Legacies, Movements" in MJ Alexander and CT Mohanty (eds) *Feminist Genealogies, Colonial Legacies, Democratic Futures.* New York: Routledge.
Allen, A (1997) Who Needs Civil Society? *Review of African Political Economy*, 73: 329–337.
Allen, K, D Budlender, and R Sharp (2000) *How to do a Gender-Sensitive Budget Analysis: Contemporary Research and Practice.* Sydney and London: Australian Agency for International Development and the Commonwealth Secretariat.
Alvarez, S (1990) *Engendering Democracy in Brazil: Women's Movements in Transition Politics.* New Jersey: Princeton University Press.
Amadiume, I (1987) *Male Daughters, Female Husbands: Gender and Sex in an African Society.* London: Zed Books.
Amadiume, I (1997) *Reinventing Africa: Matriarchy, Religion and Culture.* London: Zed Books.
Ames, V and P Parman (1984) Challenging Imperial Feminisms. *Feminist Review*, July 1984.
Amin, S (1973) *Neocolonialism in West Africa.* New York: Monthly Review Press.
Amin, S (1974) *Accumulation on a World Scale: A Critique of the Theory of Underdevelopment.* New York: Monthly Review Press.
Amrane-Minne, D (1999) Women and Politics in Algeria from the War of Independence to our Day. *Research in African Literatures*, Indiana University Press Journals, Austin.
Anonymous (1999) No Glass Ceiling for Women. *New African*, March 1999.

Ardener, EW (1972) "Belief and the Problem of Women" in J La Fontaine (ed.), *The Interpretation of Ritual*. London: Tavistock Publications.
Ardener, S (1964) The Comparative Study of Rotating Credit Associations. *Journal of the Royal Anthropological Institute*, 94: 201–229.
Assie-Lumumba, N (1995) Gender and Education in Africa: A New Agenda for Development. *Africa Notes* (April) 1–4.
Aubrey, L (1997) *The Politics of Development, Cooperation, NGOs, Gender and Partnership in Kenya*. London: Routledge.
Awe, B (1977) "The Iyalode in the Traditional Yoruba Political System" in A Schlegel (ed.), *Sexual Stratification: A Cross Cultural View*. New York: Columbia University Press.
AWMC (1999) African Women's Media Center Booklet. Dakar: AWMC.
Babatunde, E (1997) The need to take a broader view of African cultural practices. Commentary. *The Philadelphia Inquirer*.
Bangura, Y (1992) "Authoritarian Rule and Democracy in Africa: A Theoretical Discourse" in F Gibbon *et al*. (eds), *Authoritarianism, Democracy and Adjustment: The Politics Economic Reform in Africa*. Uppsala: The Scandinavian Institute of African Studies.
Bangura, Y (1994) Economic Restructuring, Coping Strategies and Social Change: Implications for Institutional Development in Africa. *Discussion Paper DP 52*. Geneva: UNRISD.
Banton, M (1957) *West African City: A Study of Tribal Life in Freetown*. London: Oxford University Press.
Barker-Benfield, GJ (1977) "Sexual Surgery in Late Nineteenth Century America" in D Dreifus (ed.), *Seizing Our Bodies: The Politics of Women's Health*. New York: Vintage Books.
Barlay, K (1990) *An Analysis of Programmes to Integrate Sierra Leonean Women in Development*. Freetown: National Population Commission.
Bassir, O (1954) Marriage rites among the Aku (Yoruba) of Freetown, *Africa*, Vol. xxiv, 251–155.
Basu, A (1992) *Two Faces of Protest: Contrasting Modes of Women's Activism in India*. Berkeley: University of California Press.
Basu, A (ed.) (1995) *The Challenge of Local Feminisms: Women's Movements in Global Perspective*. Boulder: Westview Press.
Beauvoir, S de (1989 edn) *The Second Sex*. New York: Vintage.
Beneria, L (1978) *Reproduction, Production and the Sexual Division of Labor*. Geneva: International Labor Organization.
Beneria, L (2003) *Gender, Development and Globalization: Economics as if People Mattered*. New York: Routledge.
Beoku-Betts, J (1998) "Gender and Formal Education in Africa: An Exploration of the Opportunity Structure at the Secondary and Tertiary Levels" in M Bloch *et al*. (eds), *Women and Education in sub-Saharan Africa: Power, Opportunities and Constraints*. Boulder: Lynne Rienner.
Bledsoe, C (1980) *Women and Marriage in Kpelle Society*. Stanford: Stanford University Press.
Bloch, M, J Beoku-Betts, and B Tabachnick (eds) (1998) *Women and Education in sub-Saharan Africa: Power, Opportunities and Constraints*. Boulder: Lynne Rienner.
Blyden, EW (1967) *Islam, Christianity and the Negro Race* (1887, reprinted). Edinburgh: The University Press.
Boone, S (1986) *Radiance from the Waters: Ideals of Feminine Beauty*. New Haven: Yale University Press.
Boserup, E (1989) *Women's Role in Economic Development*. London: Earthscan Publications Ltd.
Bouatta, C (1997) *Evolution of the Women's Movement in Contemporary Algeria: Organization, Objectives and Perspectives*. Helsinki: UNU World Institute for Development Economics Research (UNU/WIDER).
Budlender, D and G Hewitt (2002) *Gender Budgets Make Cents: Country Studies and Good Practice*. London: Commonwealth Secretariat, Gender Section.

Campaign for Good Governance (2004) *Silent Victims, Young Girls at Risk: An Evaluation of Post-War Rape and the Response to Rape in the Provinces of Sierra Leone*. Freetown, Campaign for Good Governance.
Caplan, P and J Budjra (1982) *Women United, Women Divided: Comparative Studies of Ten Contemporary Cultures*. Bloomington: Indiana University Press.
Chafetz, J and G Dworkin (1986) *Female Revolt: Women's Movement in World and Historical Perspective*. Toronto: Rowan and Allenheld.
Chanana, K (2001) *Integrating Women's Education: Bounded Visions, Expanding Horizons*. New Delhi: Rawat Publications.
Cockburn, C (2000) "The Women's Movement: Boundary-Crossing on Terrains of Conflict" in R Cohen and S Rai (eds), *Global Social Movements*. London: Athlone Press.
Cohen, A (1971) The Politics of Ritual Secrecy, Man, n.s. Vol. 6, 427–448.
Cohen, R and S Rai (eds) (2000) *Global Social Movements*. London: Athlone Press.
Collins, P (2000 edn) *Black Feminist Thought: Knowledge, Consciousness and the Politics of Empowerment*. New York: Routledge.
Combahee Women's River Collective Statement (1986) *Black Feminists Organizing in the Seventies and Eighties*. New York: CWRC.
Concord Times (2004) "In Sympathy with Our Poor Sierra Leone Mothers." Column. *Concord Times*, February 23, 2004.
Copestake, J and K Welled (eds) (1993) *Non-Governmental Organizations and the State in Africa: Rethinking Roles in Sustainable Agricultural Development*. New York: Routledge.
Council of Europe (1998) *Gender Mainstreaming: Conceptual Framework, Methodology and Presentation of Good Practice*. Strasburg: Council of Europe.
Cromwell, A (1989) *An African Woman Feminist: The Life and Times of Adelaide Smith Casely-Hayford*. Washington, DC: Howard University Press.
Cummings-John, C (1995) *Memoirs of a Krio Leader*. Ibadan, Nigeria: Sam Bookman for Humanities Research Center.
Cutrufelli, M (1985) *Women of Africa: Roots of Oppression*. London: Zed Books.
Daly, M (1978) *Gyn/Ecology: The Metaphysics of Radical Feminism*. Boston: Beacon Press.
Daly, M (1985) *Beyond God the Father: Toward a Philosophy of Women's Liberation*. Boston: Beacon Press.
Darbor, M (2003) Educating for a secure future. *Human Security and Dignity: Fulfilling the Promise of the United Nations*, DPI/NGO Conference.
Davies, C, A Davies, A Gyorgy, C Kayser (eds) (1992) *Women in Sierra Leone: Traditional Voices*. Freetown: Partners in Adult Education. Women's Commission.
Day, L (1997) Dr. Kadi Sesay: Sierra Leone Feminist and Advocate for Democracy. *Issue: A Journal of Opinion*, African Studies Association, Vol. 27, 19–22.
Day, L (1998) "Rites and Reason: Precolonial Education and its Relevance to the Current Production and Transmission of Knowledge" in M Bloch, J Beoku-Betts and R Tabachnick (eds), *Women and Education in sub-Saharan Africa: Power, Opportunities and Constraints*. Boulder: Lynne Rienner Publishers.
D'Azevedo, W (1962) Some Historical Problems in the Delineation of a Central West Atlantic Region. *New York Academy of Sciences*, 96: 512–538.
Dembele, DM (1999) "The Political Economy of Debt Adjustment and Globalization in Africa" in Y Fall (ed.), *Africa: Globalization, Gender and Resistance*. Dakar: AAWORD Book Series, no. 1.
Dembele, DM (2002) "Trade Liberalization and Poverty in sub-Saharan Africa" in FC Steady (ed.), *Black Women, Globalization, and Economic Justice*. Rochester, Vermont: Schenkman Books.
Denzer, L (1981) Constance Cummings-John of Sierra Leone: Her Early Political Career. *Tarikh* 25: 20–32.
Department of Development and Economic Planning (1993) *Experiences of Local Communities/NGOs: Institutional Issues and Constraints*. Freetown: Sierra Leone Government.

Diamond, I and G Orenstein, (eds) (1990) *Reweaving the World*. San Francisco: Sierra Club Books.
Dreifus, C (ed.) (1977) *Seizing Our Bodies: The Politics of Women's Health*. New York: Vintage Books.
Dugger, CW (1997) "Tug of Taboos: African Genital Rite vs. American Law." A T & T WorldNet series.
Dworkin, A (1976) *Our Blood: Prophecies and Politics in Sexual Politics*. New York: Harper Row.
Dworkin, A (1989) *Letters from a War Zone: Writings 1976–1989*, New York: E.P. Dutton.
Dworzak, J (1990) "Women in Sierra Leone" in G Ashworth (ed.), *Change*. London.
Dworzak, J (1994) *Directory and Profile of NGOs Preparing for Dakar Preparatory Conference*. Freetown: The Women's Unit, IPAM.
Economic Commission for Africa (ECA) (1995) *Social Development*. Addis Ababa: ECA.
Economic Commission for Africa (ECA) (1998) *Economic Report on Africa*. Addis Ababa: ECA.
Economic Commission for Africa (ECA) (2000) *Economic Report on Africa*. Addis Ababa: ECA.
Ekejiuba, F (1991) Participation in the Democratization Process in Nigeria. *African Studies Working Papers*. Boston University, Boston.
Etienne, M and E Leacock (eds) (1980) *Women and Colonialism: Anthropological Perspectives*. New York: Praeger.
Fades, F (1994) Les associations femmes Algeriennes face à la menace Islamiste. *Nouvelles Questions Feministes*, 15(2): 51–65.
Falk, R (1999) *Predatory Globalization: A Critique*. Cambridge: Polity Press.
Fall, Y (ed.) (1999) *Africa: Gender, Globalization and Resistance. Africa: Globalization, Gender and Resistance*. Dakar: AAWORD Book Series, no. 1.
Fashole-Luke, EW (1968) "Religion in Freetown" in C Fyfe and ED Jones (eds), *Freetown: A Symposium*. Freetown: Sierra Leone University Press.
FAWE (2001) *Federation of African Women Educationalists Booklet*. New York: UNDP.
Femmes Africa Solidarité (2001) *Engendering the Peace Process in Burundi*. UNDP Series: Women's Best Practices. Geneva: FAS.
Femmes Africa Solidarité (2003) "MARWOPNET awarded 2003 United Nations Prize in the field of Human Rights." Press Release, Geneva, December 3, 2003.
FEMSA (2001) FEMSA Project. New York: UNDP.
Fernea, E (2000) The Challenges of Middle-Eastern Women in the Twenty-First Century. *Middle East Journal*, Washington, Spring 2000.
Foday, HK (1996) "The Test of Womanhood" in K Sesay (ed.), *Burning Words, Flaming Images*, Vol. 1. London: SAKS Publications.
Foray, CP (1977) *Historical Dictionary of Sierra Leone*. London. (Cited in A Wyse 1991.)
Foucault, M (1984) *Foucault Reader* edited by Paul Rabinow. New York: Pantheon Books.
Frank, G (1969) *Capitalism and Underdevelopment in Latin America: Historical Studies of Chile and Brazil*. New York: Monthly Review Press.
Frank, G (1981) *Reflections on the World Economic Crisis*. New York: Monthly Review Press.
French, HW (1997a) "Africa's Culture War: Old Customs, New Values," A T & T WorldNet Service. February 2, 1997.
French, HW (1997b) "Sierra Leone Journal: Disfiguring Ritual Bonds Women to the Past," Grafton, Sierra Leone. A T & T WorldNet Service, February 2, 1997.
Fyfe, C (1962) *A History of Sierra Leone*. London: Oxford University Press.
Fyfe, C (2000) Obituary—Constance Cummings-John, *Expo Times*, March 17, 2000, reprinted in FC Steady, *Women and the Amistad Connection*, pp. 267–269.
Fyle, CM (1981) *The History of Sierra Leone: A Concise Introduction*. London: Evans Brothers, Ltd.
Gallie, W (1962) "Essentially Contested Concepts" in M Black (ed.), *The Importance of Language*. London: Prentice Hall.

Garba, A and K Garba (1999) "Trade Liberalization, Gender Equality and Adjustment Policies in sub-Saharan Africa" Y Fall (ed.), *Africa: Gender, Globalization and Resistance*. Dakar: AAWORD.
Geertz, C (1962) The Rotating Credit Association: A Middle Rung in Development. *Economic and Cultural Change*, 10(3).
Gordon, A (1996) *Transforming Capitalism and Patriarchy: Gender and Development in Africa*. Boulder: Lynne Rienner.
Government of Kenya (1990) Central Bureau of Statistics. Nairobi: Government of Kenya.
Haavio-Mannila, E et al. (eds) (1985) *Unfinished Democracy: Women in Nordic Politics*. Oxford: Oxford University Press.
Hale, S (2000) "The Soldier and the State: Post-Liberation Women, The Case of Eritrea" in M Waller and J Rycenga (eds), *Frontline Feminisms: Women, War and Resistance*. New York: Garland Publications
Hamilton, P (1971) "Women of Africa," paper presented at the Extra-Mural Studies Seminar on Community Development, Fourah Bay College, Freetown.
Harrell-Bond, BE (1976) *Modern Marriage in Sierra Leone: A Study of the Professional Group*. The Hague: Mouton.
Hay, J and S Stichter (1995) *African Women South of the Sahara*. New York: John Wiley and Sons.
Hearn, J (2001) The "Uses and Abuses" of Civil Society in Africa. *Review of African Political Economy* (London) March 2001.
Hirsch, J (2001) *Sierra Leone: Diamonds and the Struggle for Democracy*. International Peace Academy, Occasional Paper Series. Boulder: Lynne Rienner.
Imam, A, A Mama, and F Sow (eds) (1998) *Engendering African Social Sciences*. Dakar: CODESRIA Book Series.
IPAM (1991) *Women in Management and Decision-Making Positions in Freetown*. Freetown: The Women's Unit, IPAM.
James, S and C Robertson (2002) *Genital Cutting and Transnational Sisterhood: Disputing the U. S. Polemics*. Urbana: University of Illinois Press.
Jayawardena, K (1986) *Feminism and Nationalism in the Third World*. London: Zed Books.
Jellicoe, MR (1955) Women's Groups in Sierra Leone. *African Women*, 1: 35–43.
JENDA: Journal of Culture and African Women's Studies (2000). Genital Landscaping, Labia Remodeling and Vestal Vaginas: Female Genital Mutilation or Genital Cosmetic Surgery: Readings on Designer Vaginas. Electronic Journal, *Jenda*, 1(1).
Jusu-Sheriff, Y (2000) Sierra Leonean Women and the Peace Process. *Accord: An International Review of Peace Initiatives*, September 2000.
Kabira, W, J Odoul and M Nzomo (eds) (1993) *Democratic Change in Africa: Women's Perspectives*. Nairobi: AAWORD.
Kameri-Mobote, A and K Kibwana (1993) "Women, Law and the Democratization Process in Kenya" in W Kabira et al. (eds), *Democratic Change in Africa: Women's Perspectives*. Nairobi: AAWORD-Kenya and ACTS Gender Institute.
Kaplan, G (1992) *Contemporary Western European Feminism*. New York: New York University Press.
Karem, A (ed.) (1997) "Women, Islamisms and State: Dynamics of Power and Contemporary Feminisms in Egypt" in *Muslim Women and Politics of Participation: Implementing the Beijing Platform*. Syracuse: Syracuse University Press.
Katzenstein, M and C Mueller (eds) (1987) *The Women's Movement in the United States and Western Europe*. Philadelphia: Temple University Press.
Kenyatta, J (1938) *Facing Mount Kenya*. London: Secker and Warburg.
Khasiani, SA and EI Njiro (eds) (1993) *The Women's Movement in Kenya*. Nairobi: AAWORD.
Khor, M (2000) *Globalization and the South: Some Critical Issues*. Geneva: UNCTAD.
Koonz, C (1987) *Mothers in the Fatherland: The Family and Nazi Politics*. London: Jonathan Cape.

Koroma, S (2003) *The Discrimination Facing Non-Natives under CAP 122 (Sections 3 & 4).* Freetown: University of Sierra Leone.

Koso-Thomas, O (1987) *The Circumcision of Women: A Strategy for Eradication.* London: Zed Books.

Kuper, H (1968) *The Swazi: An African Aristocracy.* Pasadena: Holt, Rinehart and Winston.

Kwesiga, J (2002) *Women's Access to Higher Education in Africa: Uganda's Experience.* Kampala: Fountain Publishers Ltd.

Lancaster, C (1991–1992) Democracy in Africa's Foreign Policy. *Foreign Policy,* 85, Winter.

Lapchik, R (1981) "The Role of Women in the Struggle Against Apartheid in South Africa" in FC Steady (ed.), *The Black Woman Cross-Culturally.* Cambridge: Schenkman Publishing Company.

Lazreg, M (1990) Gender and Politics in Algeria: Unraveling the Religious Paradigm. *Signs* 15(4): 755–758.

Lebeuf, A (1963) "The Role of Women in the Political Organization of African Societies" in D Paulme (ed.), *Women of Tropical Africa.* London: Routledge and Kegan Paul.

Lewis, B (1976) "The Limitation of Group Action among Entrepreneurs. The Market Women of Abidjan, Ivory Coast" in N Hafkin and E Bay (eds), *Women in Africa: Studies in Social and Economic Change.* Stanford: California University Press.

Little, J (1994) *Gender, Planning and the Policy Process.* London: Pergamon Press.

Little, K (1951) *The Mende of Sierra Leone.* London: Routledge and Kegan Paul.

Little, K (1965) *West African Urbanization: A Study of Voluntary Associations in Social Change.* Cambridge: Cambridge University Press.

Lloyd, C (1999) Organizing Across Borders: Algerian Women's Associations in a Period of Conflict. *Review of African Political Economy,* no. 82.

Longwe, S (1990) From Welfare to Empowerment: The Situation of Women in Development in Africa. Working Paper, no. 204. East Lansing, Michigan State University, Office of Women in International Development.

Longwe, S and R Clarke (1999) Towards Improved Leadership for Women's Empowerment in Africa: Measuring Progress and Improving Strategy. Lusaka: African Leadership Forum.

Lucan, T (2004) *The Life and Times of Paramount Chief, Madam Ella Koblo Gulaman.* Freetown: Penpoint Publishers.

MacCormack, CM (1974) "Madam Yoko: Ruler of the Kpa Mende Confederacy" in M Rosaldo and L Lamphere (eds), *Women, Culture and Society.* Palo Alto: Stanford University Press.

MacCormack, CM (1977) Biological Events and Social Control. *Signs: Journal of Women, Culture and Society,* Autumn. Chicago: Chicago University Press.

Magubane, B (1979) *The Political Economy of Race and Class in South Africa.* New York: Monthly Review Press.

Mama, A (1997) "Sheroes and Villains: Conceptualizing Colonial and Contemporary Violence Against Women in Africa" in M Alexander and C Mohanty (eds), *Feminist Genealogies, Colonial Legacies, Democratic Futures.* New York: Routledge.

Margai, MAS (1948) Welfare Work in a Secret Society. *African Affairs,* 48(3): 223–233.

Margolis, D (1993) Women's Movements Around the World: Cross-Cultural Comparisons. *Gender and Society,* 7(3), September 1993.

May-Parker, J (1986) *Women's Employment Patterns, Discrimination and Promotion of Equality in Africa: The Case of Sierra Leone.* Geneva: International Labour Office.

Mba, N (1982) *Nigerian Women Mobilized: Women's Political Activity in Southern Nigeria, 1900–1965.* Berkeley: Institute of International Studies, University of California.

McBride, S and J Wiseman (eds) (2000) *Globalization and Its Discontents.* New York: St. Martin's Press.

McCulloch, M (1950) *Peoples of the Sierra Leone Protectorate.* London: International African Institute.

Mernissi, F (1975) *Beyond the Veil: Male-Female Dynamics in a Modern Muslim Society.* Cambridge: Schenkman Books.

Milbrath, L (1965) *Political Participation*. Chicago: Chicago University Press.
Moghadam, VM (1994) *Identity Politics and Women*. Boulder: Westview Press.
Mohanty, CT, A Russo, and L Terres (eds) (1991) *Third World Women and the Politics of Feminism*. Bloomington: Indiana University Press.
Molyneux, M (1985) Mobilization Without Emancipation? Women's Interests, State and Revolution in Nicaragua. *Feminist Studies*, 11(2): 227–253.
Molyneux, M (2001) *Women's Movements in International Perspective: Latin America and Beyond*. New York: Belgrave.
Morgan, R (1984) *Sisterhood is Global: The International Women's Movement Anthology*. Garden City: Anchor Books.
Morgan, V and S Tropps (2000) *Looking Around and Looking Ahead: Dynamics of Gender Partnerships*. New York: UNDP.
Mota, AT da (1625) *An Account of Sierra Leone and the Rivers of Guinea of Cape Verde*. Lisboa: Junta de Investigacoes Cientificas Do Ultramar.
Mothers' Union, The (1996) *The Mothers' Union Magazine* (70th Anniversary Special Edition). Freetown, August 1996.
National Commission for Democracy in Sierra Leone (1996) Background Document. Freetown: NCDSL.
NEPAD (2000) *New Partnership for African Development*. Pretoria: African Union.
Nnaemeka, O (ed.) (1998) *Sisterhood, Feminisms and Power: From Africa to the Diaspora*. Trenton: Africa World Press.
Nnaemeka, O (2003) Nego-Feminism: Theorizing, Practicing and Pruning Africa's Way. *Signs*, 29(2).
Nzomo, M (1993a) "Engendering Democratization in Kenya" in W Kabira *et al.* (eds), *Democratic Change in Africa: Women's Perspectives*. Nairobi: AAWORD-Kenya and ACTS Gender Institute.
Nzomo, M (1993b) "The Kenyan Women's Movement in a Changing Political Context" in SA Khasiani and EI Njomo (eds), *The Women's Movement in Kenya*. Nairobi: AAWORD-Kenya.
Odoul, J and W Kabira (1995) "The Mother of Warriors and Her Daughters: The Women's Movement in Kenya" in A Basu (ed.), *The Challenge of Local Feminisms: Women's Movements in Global Perspective*. Boulder: Westview Press.
Okeke, P (2000) Recognizing Tradition: Women's Rights and Social Status in Contemporary Nigeria. *Africa Today*, 47(1), Winter 2000.
Okeke, P (2002) "Women's Movements" in T Zeleza and D Eyoh (eds), *Encyclopedia of Twentieth Century African History*. London: Routledge.
Okonjo, K (1981) "Women's Political Participation in Nigeria" in FC Steady (ed.), *The Black Woman Cross-Culturally*. Cambridge: Schenkman Publishing Company.
Oladipupo, S (2003) "Outlaw Female Circumcision Now," *P.M. News*. Lagos, June 19, 2003.
Ovis, S (2001) Civil Society in Africa or African Civil Society? *Journal of Asian and African Studies* (Leiden).
Oyewumi, O (1997) *The Invention of Women: Making an African Sense of Western Gender Discourses*. Minneapolis: University of Minnesota Press.
Papart, J and K Staudt (eds) (1989) *Women and the State in Africa*. Boulder: Lynne Rienner.
Paulme, D (ed.) (1963) *Women of Tropical Africa*. London: Routledge and Kegan Paul.
Peterson, J (1968) "The Sierra Leone Creole: A Reappraisal" in C Fyfe and ED Jones (eds), *Freetown: A Symposium*. Freetown: Sierra Leone University Press.
Peterson, J (1969) *Province of Freedom*. London: Faber and Faber.
Pettman, J (1996) *Worlding Women: A Feminist International Politics*. London: Routledge.
Phillips, R (1979) *The Sande Society Masks of the Mende of Sierra Leone*. Ph.D. dissertation, University of London, School of Oriental and African Studies.
Pheko, M (2002) "Privatization, Trade Liberalization and Women's Socioeconomic Rights: Exploring Policy Alternatives" in FC Steady (ed.), *Black Women, Globalization and*

Economic Justice: Studies from Africa and the African Diaspora. Rochester: Schenkman Books.

Porter, AT (1953) Religious Affiliation in Freetown, Sierra Leone. *Africa*, 23: 3–14.

Porter, AT (1963) *Creoledom: A Study of the Development of Freetown Society.* London: Oxford University Press.

Rai, S (1994) "Gender and Democratisation: Or What Does Democracy Mean for Women in the Third World," *Democratisation*, Vol. 1, No. 2.

Rai, S (2000) "International Perspectives on Gender and Democratization" in S Rai (ed.), *International Perspectives on Gender and Democratisation.* New York: St. Martin's Press; London: Macmillan Press.

Raissiguier, C. (2002) "Scattered Markets, Localized Workers: Gender and Immigration in France" in FC Steady (ed.), *Black Women Globalization and Economic Justice: Studies from Africa and the African Diaspora.* Rochester: Schenkman Books.

Ray, R and AC Kortweg (1999) Women's Movements in the Third World: Identity, Mobilization and Autonomy. *Annual Review of Sociology*, Palo Alto, 1999.

Reno, W (1995) *Corruption and State Politics in Sierra Leone.* New York: Cambridge University Press.

Reno, W (1998) *Warlord Politics and African States.* Boulder: Lynne Rienner Publishers.

Reuther, R (1975) *New Woman, New Earth.* New York: The Seabury Press.

Richards, JVO (1975) Some Aspects of the Multivariant: Socio-Cultural Roles of the Sande of the Mende. *Canadian Journal of African Studies* 9(1): 103–113.

Rivkin, E (1981) "The Black Woman in South Africa: An Azanian Profile" in FC Steady (ed.), *The Black Woman Cross-Culturally.* Cambridge: Schenkman Publishing Company.

Robertson, C (1996) Grassroots in Kenya: Women's Genital Mutilation and Collective Action, 1920–1990. *Signs* 21(3): 615–642.

Robertson, C and I Berger (eds) (1986) *Women and Class in Africa.* New York: Africana Publishing Company.

Robinow, P (1984) *The Foucault Reader:* New York: Pantheon.

Rodney, W (1981) *How Europe Underdeveloped Africa.* Washington, DC: Howard University Press.

Rosaldo, M and L Lamphere (eds) (1974) *Women, Culture and Society.* Palo Alto: Stanford University Press.

Rosander, E (ed.) (1997) Transforming Female Identities: Women's Organizational Forms in West Africa. *Seminar Proceedings* no. 31. Uppsala: Nordiska Afrika Institutet.

Rostow, WW (1967) *The Stages of Economic Growth: A Non-Communist Manifesto.* Cambridge: Cambridge University Press.

Rowbotham, S (1975) *Hidden from History: Rediscovering Women in History from the 17th Century to the Present.* New York: Pantheon Books.

Rowbotham, S and S Linkogle (eds) (2001) *Women Resist Globalization: Mobilizing for Livelihood and Rights.* London: Zed Books.

Rupert, M (2000) *Ideologies of Globalization: Contending Visions of a New World Order.* New York: Routledge.

Saadi, N (1991) *La Femme et la Loi en Algerie.* London: UNU/WIDER, Editions Bouchene.

Sarris, A and H Shams (1991) *Ghana Under Structural Adjustment: The Impact of Agriculture and the Rural Poor.* New York: New York University Press.

Sen, G and C Grown (1986) *Development, Crises and Alternative Visions: Third World Women's Perspectives* (DAWN) New York: Monthly Review Press.

Sethi, R (ed.) (1999) *Globalization, Culture and Women's Development.* New Delhi: Rawat Publications.

Shiva, V (1989) *Staying Alive: Women, Ecology and Development.* London: Zed Press.

Sierra Leone Government (1965) *1963 Population Census of Sierra Leone. Vols. I and II.* Freetown: Central Statistics Office.

Sierra Leone Government (1968) *Household Survey of the Western Area, Advance Report.* Freetown: Central Statistics Office.

Sierra Leone Government (1987) *Household Survey of the Western Area, Final Report*. Freetown: Central Statistics Office.
Sierra Leone Government (1992) *1985 Census of Population and Housing. Vol. 1*, Summary Results. Freetown: Central Statistics Office.
Sierra Leone Government (1994) *Report of a Workshop on Government and Non-Governmental Organizations: Partners in Sustainable Development*. Freetown.
Sivard, R (1977) *World Military and Social Expenditures*. Leesburg: WMSE Publications.
Snyder, M and M Tadesse (1995) *African Women and Development: A History*. London: Zed Books.
Stamp, P (1986) "Kikuyu Women's Self-Help Groups" in C Robertson and I Berger (eds), *Women and Class in Africa*. New York: Africana Publishing Company.
Starr, A (2000) *Naming the Enemy: Anti-Corporate Movements Confront Globalization*. New York: Zed Books.
Staudt, K (1986) Women, Development and the State: On the Theoretical Impasse. *Development and Change*, 17(2): 325–333.
Steady, FC (1975) *Female Power in African Politics: The National Congress of Sierra Leone*. Munger Africana Library Monograph, Pasadena: California Institute of Technology.
Steady, FC (1976) "Protestant Women's Associations in an African City" in N Hafkin and E Bay (eds), *Women in Africa*. Palo Alto: Stanford University Press.
Steady, FC (1978) "The Role of Women in the Churches of Freetown, Sierra Leone" in E Fashole-Luke *et al.* (eds), *Christianity in Independent Africa*. London: Rex Collings.
Steady, FC (ed.) (1981) *The Black Woman Cross-Culturally*. Cambridge: Schenkman Publishing Company.
Steady, FC (1982) African Women, Industrialization and Another Development: A Global Perspective. *Development Dialogue*, nos. 1–2.
Steady, FC (1985) "Women's Work in Rural Cash Food Systems: The Tombo and Gloucester Development Projects in Sierra Leone" in S Muntemba (ed.), *Rural Development and Women: Lessons from the Field*. Geneva: International Labour Office.
Steady, FC (1987) "African Feminism: A Global Perspective" in R Terborg-Penn *et al.* (eds), *Women of the African Diaspora: An Interdisciplinary Perspective*. Washington: Howard University Press.
Steady, FC (2000) "African Feminism" in L Code (ed.), *Encyclopedia of Feminist Theories*. London: Routledge.
Steady, FC (2001) *Women and the Amistad Connection: Sierra Leone Krio Society*. Rochester: Schenkman Books.
Steady, FC (ed.) (2002) *Black Women, Globalization and Economic Justice*. Rochester: Schenkman Books.
Steady, FC (2002a) "Engendering Change Through Egalitarian Movements: The African Experience" in C Murphy (ed.), *Egalitarian Politics in the Age of Globalization*. New York: Palgrave.
Stewart, S (1997) Happy Ever After in the Marketplace: Non-Governmental Organizations and Uncivil Society. *Review of African Political Economy*, 71: 11–34.
Stienstra, D (2000) "Making Global Connections among Women: 1970–99" in R Cohen and S Rai (eds), *Global Social Movements*. London: Athlone Press.
Strobel, M (1976) "From Lelemama to Lobbying: Women's Associations in Mombasa, Kenya" in N Hafkin and E Bay (eds), *Women in Africa*. Stanford: Stanford University Press.
Sudarkasa, N (1981) "Female Employment and Family Organization" in FC Steady (ed.), *The Black Woman Cross-Culturally*, Cambridge: Schenkman Publishing Company.
Sudarkasa, N (1987) "The Status of Women in Indigenous African Societies" in R Terborg-Penn and A Rushing (eds), *Women in Africa and the African Diaspora*. Washington, DC: Howard University Press.
Tamele, S (1999) *When Hens Begin to Crow*. Kampala: Fountain Publishers.
Tarrow, S (1998) *Power in Movement: Social Movements and Contentious Politics*. Cambridge: Cambridge University Press.

Temple, C (2000) "What is Globalization?" in S McBride and J Wiseman (eds), *Globalization and its Discontents*. New York: St. Martin's Press.
Terborg-Penn, R and A Rushing (eds) (1987) *Women in Africa and the African Diaspora: A Reader*. Washington, DC: Howard University Press.
The Caucus (1993) *The Neglected Arms Race: Weapons Proliferation in the 1990s: A Research Report on Global Weapons Proliferation*, Washington, DC: Arms Control and Foreign Policy Caucus.
Thomas, AC (1985) *The Population of Sierra Lone: An Analysis of Population Census Data*. Freetown: Fourah Bay College, Demographic Research and Training Unit.
Thorpe, C (2000) A Gender Perspective on the Culture of Peace in Post-Conflict Situation, Presentation at a Conference on Higher Education for Peace, Tromso, Norway, May 4–6, 2000.
Trimingham, JS (1961) *Islam in West Africa*. Oxford: Clarendon Press.
Tripp, A (2000) Rethinking Differences: Comparative Perspective from Africa. *Signs*, Spring, 2000.
Tripp, AM (2001) Women's Movements and Challenges to Neopatrimonial Role: Preliminary Observations from Africa. *Development and Change* 32: 33–54.
UNAIDS (2003) Report, UNAIDS. Geneva.
UNCTAD (1999) *Trade and Development Report*. Geneva and New York: United Nations.
UNCTAD (2000a) *The Least Developed Countries Report*. Geneva and New York: United Nations.
UNCTAD (2000b) *World Investment Report*. Geneva and New York: United Nations.
UNDP (1996) *Human Development Report*. New York: UNDP.
UNDP (1997) *Human Development Report*. New York: UNDP.
UNDP (2001) *The Gender Program Review*. New York: UNDP/RBA.
UNDP (2002) *Enterprise Africa*. New York: UNDP/Regional Bureau for Africa.
UNESCO (1992, 1993, 1994) *Statistical Yearbook*. Paris: UNESCO.
UNESCO (1994a) *Women and the Democratization Process in Africa*. Dakar: UNESCO.
UNESCO (1994b) *Donors to African Education (DAE)*. Paris: UNESCO.
UNIDO (1995) *Women in Manufacturing Patterns, Determinants and Trends—Case Study*.
UNIFEM (2000) *Women at the Peace Table: Making a Difference*. New York: UNIFEM.
United Nations (1985) The Nairobi Forward-Looking Strategies for the Advancement of Women.
United Nations (1989) *World Survey on the Role of Women in Development*. New York: United Nations.
United Nations (1995a) *The World's Women: Trends and Statistics*. New York: United Nations.
United Nations (1995b) *The Beijing Platform for Action*. New York: United Nations.
United Nations (1997) Statement of the Ghanaian Representative at the Commission on the Status of Women, New York.
United Nations (2000) *The World's Women: Trends and Statistics*. New York: United Nations.
UNRISD (1995) *Adjustment, Globalization and Social Development*. Report of the UNRISD/UNDP International Seminar on Economic Restructuring and Social Policy.
Urdang, S (1979) *Fighting Two Colonialisms: Women in Guinea-Bissau*. New York: Monthly Review Press.
Urdang, S (1989) *And Still They Dance: Women, War and the Struggle for Change in Mozambique*. New York: Monthly Review Press.
Van Allen, J (1972) Sitting on a Man: Colonialism and the Lost Political Institutions of Igbo Women. *Canadian Journal of African Studies*, 6(2): 165–181.
Van Allen, J (1976) " 'Aba Riots' or 'Igbo Women's War'? Ideology, Stratification and the Invisibility of Women" in N Hafkin and E Bay (eds), *Women in Africa: Studies in Social and Economic Change*. Stanford: California University Press.
Van Gennep, A (1960) *The Rites of Passage*, translated by M Vizedom and G Caffee. Chicago: University of Chicago Press.
Walker, C (1991) *Women and Resistance in South Africa*. New York: Monthly Review Press.

Wanjiku, J and M Nzomo (eds) (1993) *Democratic Change in Africa: Women's Perspective*. Nairobi, Kenya: AAWORD.
Ward, K (ed.) (1990) *Women Workers and Global Restructuring*. Ithaca: ILR Press, Cornell University.
Weber, M (1930) *The Protestant Ethic and the Spirit of Capitalism*. New York: Charles Scribner's Sons.
Weeks, J (1992) *Development Strategy and the Economy of Sierra Leone*. London: St. Martin's Press.
Weitz, R (1998) "A History of Women's Bodies" in R Weitz (ed.), *The Politics of Women's Bodies: Sexuality, Appearance and Behavior*. New York: Oxford University Press.
White, F (1987) *Sierra Leone Women Traders: Women on the Afro-European Frontier*. Ann Arbor: University of Michigan Press.
Wieringa, S (1985) *The Perfumed Nightmare: Some Notes on the Indonesian Women's Movement*. The Hague: Institute of Social Studies.
Wieringa, S (1992) Ibu or the Beast: Gender Interests in Two Indonesian Women's Organizations. *Feminist Review*, 41: 989–113.
Wieringa, S (1995) *Subversive Women: Women's Movements in Africa. Asia, Latin America and The Caribbean*. London: Zed Books.
Williams, B (1979) *The Traditional Birth Attendant: Training and Utilization*. Freetown: Government of Sierra Leone.
Williams, G (ed.) (2003) *A Call to Love for Development*. Freetown: Ro-Morong International, Ltd.
Wilson, J (1981) *Women Paramount Chiefs in National Politics: Madam Ella Koblo-Gulama of Kaiyamba Chiefdom, a Case Study*. Bachelor's thesis, University of Sierra Leone-Fourah Bay College.
Wipper, A (1984) The *Maendeleo ya Wanawake* Movement in the Colonial Period. *Rural Africana* 29: 195–214.
Wipper, A (1995) "Women's Voluntary Associations" in MJ Hay and S Stichter (eds), *African Women: South of the Sahara*, 2nd edn. New York: Longmans Scientific and Technical.
Wolfensohn, J (2000) Rethinking Development: Challenges and Opportunities. Bangkok, Thailand: UNCTAD X, February 12–19.
Women in Nigeria Editorial Committee (1985) *Women in Nigeria Today*. London: Zed Books.
Women's Forum (1997) *Women's Participation in the Peace Process in Sierra Leone*. Freetown: Women's Forum.
Woodford-Berger, P (1997) "Associating Women: Female Linkage, Collective Identities and Political Ideology in Ghana" in E Rosander (ed.), *Transforming Female Identities: Organizational Forms in West Africa*. Uppsala: Nordiska Afrikainstitutet.
World Bank (2002) *Genderstats*. Washington, DC: World Bank.
Wyse, A (1980) Searchlight on the Krios of Sierra Leone: An Ethnographical Study of a West African People. *African Studies*, Occasional Paper No. 3, Freetown: Fourah Bay College.
Wyse, A (1991) *The Krio of Sierra Leone: An Interpretive History*. Washington, DC: Howard University Press.
Young, K, C Wolkowitz and R McCullagh (eds) (1981) *Of Marriage and the Market: Women's Subordination in International Perspective*. London: CSE Books.
Ziai, P (1997) "Personal Status Codes and Women's Rights in the Maghreb" in A Karem (ed.), *Muslim Women and the Politics of Participation: Implementing the Beijing Platform*. Syracuse: Syracuse University Press.
Zirakzadah, C (1997) *Social Movements in Politics: A Comparative Study*. London: Longman.
Zulu, L (1998) The Role of Women in the Reconstruction and Development of the New Democratic South Africa. *Feminist Studies* 24 (1): 46–75.
Zulu, L (2000) "Institutionalizing Changes: South African Women's Participation in the Transition to Democracy" in S Rai (ed.), *International Perspectives on Gender and Democratisation*. London: Macmillan Press, Ltd.

Index

Page numbers in bold indicate figures.

AAWORD, 19, 161, 167, 169, 171, 174
ABANTU, for Development, 169
Accord
   Abidjan, 55
   Conakry, 55
Adejobi, *see* Church of the Lord
adult literacy, *see* literacy, adult
African Federation of Women Entrepreneurs (AFWE), 158
African feminism, 177, 178
African Methodist Episcopal (AME), 84
African scholars, 15
African Union (AU), 12, 157, 170
African women, history of, *see* history of African Women
African women scholars, 2, 15
African Women's Development and Communication Network, *see* FEMNET
African Women's Leadership Institute (AWLI), 162
Akina Mama wa Africa, 161
Aladura, *see* Church of the Lord
Algeria, Algerian, 8, 18, 141–153
All People's Congress (APC), 39–40, 71
alma mater, 84
alumnae associations, 91
ancestors, 98
Annie Walsh Memorial School, *see under* School
anti-Apartheid groups
   African National Congress (ANC), 18, 146, 151
   Pan African Congress of Azania (PAC), 18, 151
anti-globalization movement, 174

anti-Structural Adjustment Riots, *see* riots, anti-Structural Adjustment
Apartheid, 3, 18, 146, 151–152, 170, 177
Arab Fund for Development, The, 121
armed conflict, 3, 42
"ashuobi," 115
Asians, 25
Association of African Women for Research and Development, *see* AAWORD
Australia, 86
authoritarianism, 3, 39, 173
Awe, Bolanle, 7

Bangura, Zainab, 39, 57
Bantustans, 146
Benin, 162
biological systems, 6
birth attendant, 97, 103
Bondo, 29, 31, 95–107; *see also* Sande
Botswana, 141, 163, 164
British grammar school model, 84
Bullom, 27
businesses
   Asian, 24
   European, 24
   Lebanese, 24
   Syrian, 24

Campaign for Good Governance, 53
Canada, 22
Caribbean, 131
Cease-Fire Agreement, 55
CEDAW, 35
chiefdoms, 25
Chief Medical Officer, 79
chiefs, female, 37

Christianity, 25, 129, 130
Christian marriage, *see* marriage: Christian
Christian missionaries, 75, 104, 129
Christian women's associations, 129–140
Church Missionary Society, *see* CMS
Church of the Lord, Aladura, 133
circumcision, female, 103–109
civil society, 8, 50
Civil Society Organizations (CSOs), 5
class, 131
collective empowerment index, 174
CMS, 75, 76, 129–130
collective identities, 4, 5–6
colonial heritage, 5, 6, 7, 12
colonialism, 6, 7, 18, 177
colonial rule, 2
Combahee River Collective, The, 30
Commission on the Status of Women (CSW), 170
Conakry Accord, 55
conflict, armed, 42
conflict resolution, 89
Conteh, Memuna, 39
Convention on the Elimination of all forms of Discrimination Against Women, *see* CEDAW
Cooperatives, 69–73
corporate globalization, 1, 2, 5, 6, 12, 18, 144, 171, 174
cottage industries, 60
coup d'état, 38, 54
credit associations, rotating, 66–69
Crowther, Adjai, 76
culture of peace, 84, 89–91
Cummings-John, Constance, 34, 37, 38, 63, 64, 79, 84, 150
customary law, 113
customary marriage, 113

Dancing Kompins, 101
Danquah, Mabel Dove, 62
Davies, Clarice, 34
DAWN, 19, 169, 171, 174
de Bouvoir, Simone, 2
debt burden, 92
democracy/democratization, 5, 6, 15, 173
democratization–authoritarianism nexus, *see under* nexus

development, 5, 173
development, socioeconomic, 7
Development Alternatives for a New Era, *see* DAWN
development issues, 4
development–underdevelopment nexus, *see under* nexus
Dillsworth, Florence, 38
displaced people, 24
division of labor, *see* labor, division of
domestic workers, 59
Dumbuya, Madam, 137

Economic Community of West African States, *see* ECOWAS
economic roles, 7
ECOMOG, 54, 56
ECOWAS, 54
ECOWAS Monitoring Group, *see* ECOMOG
education
 female, 75–93
 formal, 1, 75–93
 Islamic, 76, 119
 non-formal, 1, 91
 technical, 83
 universal primary, 92
 university, 83
 vocational, 84–88
 Western, 75, 119
Egypt, 170
Eid-ul-Fitri, 121, 125
Emang Basidi, 164
empowerment, 2, 18
England, 76
environment/environmental conservation, 3, 158
environmental degradation, 144
European, 24, 131
Ethiopia, 170

family, nuclear, 91
Family Code, 153, 154
FAS, 166, 167, 171
FAWE, 88–91, 160–161
FAWODE, 163
Federation of Muslim Women Associations in Nigeria, 154

female activism, 50, 53
female chiefs, 37
female circumcision, 103–109
female education, 83
female essence, 6
Female Institution, The, 76
female leadership in church, 138
female migration, 111
female nature, 6
female solidarity, 4, 62, 109
feminism/feminisms, 2, 6, 7
   African, 177, 178
   Western, 130
feminist economists, 4
feminist issues, 5, 6
feminist scholars, 4, 6
feminist scholarship, 2
feminists, Western, 108
Femmes Africa Solidarité (FAS), 166
FEMNET, 160, 171
FEMSA, 161
fertility, 97, 98, 109
Fewry, Oredola, 80
First Ladies, 4
formal education, *see under* education
formal market, *see under* market
formal sector, *see under* sector
Foulah, 119
Fourah Bay College, 76
free trade, 13
Fourth World Conference on Women, Beijing, 44
Freetown Secondary School for Girls, *see under* School for Girls
Fulani, 25

Gambia, The, 17, 42, 63
gara, 65–66
Gbujama, Shirley, 34, 39
gender-based discrimination, 85
gender bias, 53–54
gender budgeting, 11
gender disparities, 80
gender division of labor, 4
gendered education, 75–78
gender equality/equity, 2, 58
gender gap, 83
gender in development, 11

gender inequality, 12
gender mainstreaming, 12, 47
gender roles, 90
gender-specific tasks, 29
gender vulnerabilities, 55–56
GERDDES, 162
Ghana, 53
Ginwala, Frena, 152
Girls' Industrial School, *see under* School
Girls' Vocational School, *see under* School
global capital, 3, 4
global economy, 9, 12, 19
Global North, 167
global political economy, 3, 5
Global South, 3, 4, 6, 11, 12, 18, 19, 167, 170, 174
God Ngewo, 98
grassroots movements, *see* Movements: grassroots
grassroots women's associations, 19
Great Britain, 22
Greenbelt Movement, the, 158
Gross Domestic Product (GDP), 111
50/50 Group, the, 56, 57
Guinea, 65, 66
Gulama, Madam Ella Koblo, 37, 45

Haja, 106, 123, 127
Hajj, 123
healing, 137–138
HIPC (Highly Indebted Poor Countries), 169
history of African women, 6
HIV/AIDS, 11, 14, 55–56, 161, 171
Holst-Roness, June, 38
"Holy Johnson," 130
Human Development Report, 22
humanistic feminism, 9
human resource development, 92
Hyde-Foster, Lati, 79

IMF, 13, 169, 174
identity politics, 30
Igbo society, 7
Index
   Collective Action (CAI), 19
   Collective Empowerment, 174
   Women's Empowerment (WEI), 19
   Women's Self-Reliance (WSI), 19

industrialization, 60
influences
  Caribbean, 131
  African, 131
informal sector, 59
initiation, 98–99, 105
Institute
  Sierra Leone Technical, 85
  Vocational, 86
interests, socio-centric, 6
interests, strategic, 6
international donors, 176
International Monetary Fund, *see* IMF
International Women's Movement, *see under* movement
Islam, 25, 119, 124
Islamic fundamentalism, 154

Jihad, 119
Jonah, James, 49

Kabbah, Ahmad Tejan, President, 54, 121
Kankalay, 79, 88, 125, 126, 127, 128
Karefa-Smart, John, Dr., 39
Kenya, Kenyan, 7, 18, 141–155, 158, 163, 173
kinship, 4, 127, 131
kinship structure, 131
Kono, 99
Koran, 122, 123
Koso-Thomas, Olayinka, Dr., 103
Krio, Krios, 24, 129, 130, 131, 134
Krio language, 25
Krio society, 130, 134
Krio women, 135

labor, division of, 20
labor market formal, 60
Lagos Plan of Action, 92
land, communally owned, 27
Land Act, 27
Least Developed Countries (LDCs), 21
Lebanese, 24, 65, 66, 131
Liberated Africans, 24
Liberia, 17, 96, 166
Limba, 25
literacy, adult, 84, 86, 87
Lokko, 25, 99

Longwe, Sarah, 19

Maathai, Wangari, 158, 159, 170
Maendaleo Ya Wanawake, 4, 141
mainstreaming gender, 47
Malawi, 163
Mammy Yoko, Madam Yoko, 37, 98
management positions, 60
Mandingo, 25, 119
Mano River Women's Peace Network of Africa (MARWOPNET), 55, 166
manufacturing, 60, 61
Margai, Milton, Dr., 101
marriage, 96–97, 98
  Christian, 133, 134, 135
  customary, 113–114
  statutory, 113–114
Mau Mau movement, 18, 145, 151
mediation, 89
medical officer, chief, 79
Mende, 24, 95, 96, 99
merchants
  European, 131
  Lebanese, 66, 131
  see also traders
Methodist Girls' High School, *see under* School
micro-credit, 66–69
midwife, 97
militarism, 12, 55–56
Millennium Declaration, 11, 12
mining companies, 24
model, women and collective action, 9, **10**
modernization, 11
Molyneaux, 6
Momoh, Joseph Saidu, President, 38, 40
Morgan, Viola, 157
motherhood, 173
mothering, multiple, 114–115
Mothers' Union, 133, 134, 138, 140
movement, international women's, 2, 104, 167–171
Movement for Progress Party (MOP), 57
Movements
  grassroots, 3
  women's, 3, 6
multinational corporations, 9, 12–13
multiple mothering, *see* mothering

Muslim, Muslims, 75, 119, 121, 122, 124
Muslim women and girls, 122, 125, 128
Muslim women's associations, 119–128
mutual aid, 111, 126
mutual-aid associations, 3, 111–117
mutual-aid schemes, 142
mutual-aid societies, 115

Nairobi Forward-Looking Strategies for the Advancement of Women (NFLS), 35
Nairobi World Conference on Women *see under* World Conference on Women
Namibia, 163, 164
National Commission for Democracy, 49
National Conferences on Democratization, 49–50
National Policy for Women, 34, 175
National Provisional Ruling Council, *see* NPRC
NEPAD, 12, 18, 92, 157
New Agenda for the Development of Africa, *see* UN-NADAF
New Partnership for African Development, *see* NEPAD
New Zealand, 86
nexus
  democratization–authoritarianism, 12–18, 162–164, 173
  development–underdevelopment, 10–15, 158–162, 173
  of action and reaction, 1, 9, 173
NGO/NGOs, 5, 8, 31, 32, 33, 34, 48, 61, 158, 63, 177
Nigeria, Nigerian, 18, 107, 131, 133, 141–155, 173
Nobel Prize for Peace, 159, 170
non-governmental organizations, *see* NGOs
NPRC, 38, 40, 48
nuclear family, 91

Oku, 119, 121–123
Old Girls' Associations, 91
Osusu, 66–69, 127
Oyewumi, 7

Pan-African women's associations, 158
paramount chieftaincies, 25
patriarchal control, 5, 124

patriarchal family, 8
patriarchal lineage structure, 7
patriarchal societies, 2
patriarchal values, 93
PAWLO, 162
peace, culture of, 89
peace, peacekeepers, 4, 5, 42, 54–56, 57
philanthropy, 134
Pilgrimage to Mecca, 121, 127
political parties, women's wings of, 34, 117
political party/parties, 127
political spheres, 7
politics of representation, 2
polygamous, polygamy, 124
Poro, 96, 99–100
poverty, 59
poverty eradication, 12
private sphere, 2, 8
privatization, 13
Provinces Land Act, 27
public-sphere/private-sphere debate, 2, 273
"pul bondo" ceremonies, 103, 112

racism, 3, 170, 171
racism, structural, 3, 177
Racism, United Nations Conference Against, 170
rebel war, 21, 24, 36, 42, 175
refugees and displaced people, 24
regional/pan-African level, 157, 176
religion, 154
religions, African traditional, 122
religious fundamentalism, 153–154
religious leadership, 139
Reuben-Johnson, Lydia, 79
Revolutionary United Front, *see* RUF
rights-based approach, 167
riots, anti-Structural Adjustment, 8
rites of passage, 98
Rodney, Walter, 22
Roosevelt, Eleanor, 84
Roosevelt Preparatory School for Girls, *see under* School for Girls
Rotating Credit Associations, 66–69
RUF, 40, 54, 55
Rwanda, 17, 163

Saint George's Cathedral, 131
Saint Joseph's Secondary School, *see under* School for Girls
Sande, 29, 31, 96, 111; *see also* Bondo
Sankoh, Foday, 21
Sapes, 27
SAPs, 1, 24, 59, 80, 92, 143, 155, 169
Sasso, Sally, 106
School
   Annie Walsh Memorial, 78, 79
   Girls' Industrial, 78
   Girls' Vocational, 79
   Kankalay, 79
   Methodist Girls' High, 78
   Reuben Johnson Memorial, 79
   Saint Edwards Boys' Secondary, 78
School for Girls
   Freetown Secondary, 91
   Roosevelt Preparatory, 79, 84
   Saint Joseph's Secondary, 38, 78, 91
SCSL, 55
secret societies, traditional, 3, 95–104, 111, 117
sector
   formal, 59
   informal, 59, 144
   service, 59
Security Council, 56
Sembene, Ousman, 18
Senegal, Senegalese, 8, 17, 18, 170
Sesay, Kadi, 39
sexual asymmetry, 2
sexuality, 97
shadow development agencies, 175, 177
Sharia, 122
Sharpville Massacre, 151
Shell Oil Company, 9
Sherbro, 95
Sierra Leone, 7, 18, 21–36, 147, 148, 150, 155, 166, 173
Sierra Leone People's Party (SLPP), 40, 56, 64
Sierra Leone Technical Institute, 85
Sierra Leone Women's Movement, 29, 31, 38, 61–65
"Sisterhood is Global," 2
SLANGO, 34
Slave Trade, Trans-Atlantic, *see* Trans-Atlantic Slave Trade
Smythe, Amy, 39

social change, 105
social divisions, 2
social movement theory, 8–9
social transformation, 2
sociocentric interests, 6, 175, 177
sociocentric objective, 174
solidarity, female, 4
South Africa, South African, 8, 12, 17, 18, 53, 141, 142, 144, 149–153, 163, 164, 170
South African Budget Initiative, 12
Soviet Union, 15
Soweto uprisings, 151
Special Court for Sierra Leone, *see* SCSL
State, the role of, 148–150
Steady, Henry Metcalfe, The Reverend, 78, 84
Steele, Nancy, 38
Stevens, Siaka Probyn, President, 38, 40
Strasser, Valentine E.M., Captain, 38, 40
strategic interests, *see* interests, strategic
Structural Adjustment Programs, *see* SAPs
structural inequalities, 2
structural racism, 12, 177
Sunni, 119
Swahili, 141
Swaziland, 163
Switzerland, 86
Syrian, 24, 65

Tanzania, 8, 18, 163, 164
tasks, gender-specific, 29
Taylor, Charles, 24, 40
technical education, *see under* education
Technical Institute, Sierra Leone, 85
Temne, 24, 96, 99, 115
ten-year war, *see* rebel war
Thomas, Abator, 39
Thorpe, Christiana, 80, 88, 89
Togo, 54
trade liberalization, 144
traders/merchants
   Lebanese, 65, 66
   Marakas, 65
   Syrian, 24, 65
   Yoruba, 65
trade union, 64
trading, 59
traditional associations, 3, 95–109
traditional birth attendants/midwife, 97, 103

traditional sanctions, 113
Trans-Atlantic Slave Trade, 24, 119, 129
Tunisia, Tunisian, 148, 149

UNAMSIL, 25
Uganda, 17, 146, 163, 170
underdevelopment, 6, 15, 18, 177
UNDP, 157
UNIFEM, 167
United Nations (UN), 157
United Nations Mission in Sierra Leone, see UNAMSIL
United Nations Development Program, see UNDP
United Nations World Conferences on Women, 3
United States of America, 22, 53, 86
university education, 83
UN-NADAF, 13, 14, 16
UN Security Council, 56

Vai, 99
Van Gennep, 98
Vocational education, see under Education
Vocational Institute, 86

Wallace-Johnson, Isaac T.A., 37
war in Sierra Leone, 155
West African Youth League, 37, 64, 150
Western feminism, see feminism/feminisms: Western
Western feminists, see feminists, Western
WILDAF, 162, 171
Williams, Belmont, Dr., 103
Williams, Jeredine, 39
womanhood, 98, 99
Women, Law and Development in Africa, see WILDAF
women in the United States, 6
Women's Bureau/Bureaus, 32, 35
Women's Congress, 38
Women's Development Foundation (WDF), 163
Women's Forum, 43–49, 57
women's liberation, 6
Women's Movement, 3; see also Sierra Leone Women's Movement
women's movements, 3
Women's National Coalition, 152

Women's National Commission, 8
women's rights, 1
Women's War, 145
Women's wings of political parties, see under political parties
World Conference on Women, 3, 48
  Beijing, 44
  Nairobi, 32
World Bank, 13, 169
World Trade Organization (WTO), 13, 174

Yoruba, 7, 65, 119
Young Women's Christian Association, see YWCA
Youth League, 37, 64, 150
YWCA, 85–88

Zambia, 8, 146
Zimbabwe, 17, 163

Index of People's Names

Awe, Bolanle, 7

Bangura, Zainab, 39, 57, 107
Benkah-Coker, Hannah, 62, 79
Bloomer, Jane, 137

Caseley-Hayford, Adelaide, 79
Caulker, Honoria Bailor, 37
Conteh, Memuna, 39
Crowther, Adjai, 76
Cummings-John, Constance, 34, 37, 38, 63, 64, 79, 84, 150

Danquah, Mabel Dove, 62
Davies, Clarice, 34
de Bouvoir, Simone, 2
Dillsworth, Florence, 38
Dumbuya, Madam, 137

Fewry, Oredola, 80

Gbujama, Shirley, 34, 39
Ginwala, Frena, 152
Gulama, Madam Ella Koblo, 37, 45

Holst-Roness, June, 38

Johnson, "Holy," 130

Kabbah, Ahmad Tejan, 54, 121
Karefa-Smart, John, 39
Koso-Thomas, Olayinka, 103

Longwe, Sarah, 19

Maathai, Wangari, 158, 159, 170
Mammy Yoko, Madam Yoko, 37, 98
Margai, Milton, 101
Momoh, Joseph Saidu, 38, 40
Morgan, Viola, 157

Oyewumi, Oyeronke, 7

Reuben-Johnson, Lydia, 79
Rodney, Walter, 22
Roosevelt, Eleanor, 84

Sankoh, Foday, 21
Sasso, Sally, 106
Sembene, Ousman, 18
Sesay, Kadi, 39
Smythe, Amy, 39
Steady, Henry Metcalfe, 78, 84
Steele, Nancy, 38
Stevens, Siaka Probyn, 38, 40
Strasser, Valentine E.M., 38, 40

Taylor, Charles, 24, 40
Thomas, Abator, 39
Thorpe, Christiana, 80, 88, 89

Wallace-Johnson, Isaac T.A., 37
Williams, Belmont, 103
Williams, Jeredine, 39

Index of Place Names

Algeria/Algerian, 8, 141, 145, 147, 153
Angola, 18
Australia, 86

Benin, 162
Botswana, 141, 163, 164
Burundi, 17

Canada, 22
Central African Republic, 17

Egypt, 170
England, 76
Ethiopia, 170

Gambia, The, 17, 42, 63
Ghana, 53
Great Britain, 22
Guinea, 65, 66
Guinea-Bissau, 17, 18

Kenya/Kenyan, 7, 18, 141, 142, 143, 145, 147, 151, 158, 163, 173

Lesotho, 17
Liberia, 17, 96, 166

Malawi, 163
Mozambique, 17, 18

Namibia, 163, 164
New Zealand, 86
Nigeria, 18, 107, 131, 133, 141, 142, 145, 147, 148, 149, 173

Rwanda, 17, 163

São Tomé and Principé, 17
Senegal/Senegalese, 8, 17, 18, 170
Sierra Leone, 7, 9, 18, 21–36, 147, 148, 150, 155, 166, 173
South Africa/South African, 8, 12, 17, 18, 53, 141, 142, 144, 149–153, 163, 164, 170
Soviet Union, 15
Swaziland, 163
Switzerland, 86

Tanzania, 8, 18, 163, 164
Togo, 54
Tunis/Tunisian, 148, 149

Uganda, 17, 146, 163, 170
United States of America, 22, 53, 86

Zambia, 8, 146
Zimbabwe, 163

CAUSATION IN NEGLIGENCE

The principal objective of this book is simple: to provide a timely and effective means of navigating the current maze of case law on causation, in order that the solutions to causal problems might more easily be reached, and the law relating to them more easily understood. The need for this has been increasingly evident in recent judgments dealing with causal issues: in particular, it seems to be ever harder to distinguish between the different 'categories' of causation and, consequently, to identify the legal test to be applied on any given set of facts. *Causation in Negligence* will make such identification easier, both by clarifying the parameters of each category and mapping the current key cases accordingly, and by providing one basic means of analysis which will make the resolution of even the thorniest of causal issues a straightforward process. The causal inquiry in negligence seems to have become a highly complicated and confused area of the law. As this book demonstrates, this is unnecessary and easily remedied.

Causation in Negligence

Sarah Green

·HART·
PUBLISHING
OXFORD AND PORTLAND, OREGON
2015

Published in the United Kingdom by Hart Publishing Ltd
16C Worcester Place, Oxford, OX1 2JW
Telephone: +44 (0)1865 517530
Fax: +44 (0)1865 510710
E-mail: mail@hartpub.co.uk
Website: http://www.hartpub.co.uk

Published in North America (US and Canada) by
Hart Publishing
c/o International Specialized Book Services
920 NE 58th Avenue, Suite 300
Portland, OR 97213-3786
USA
Tel: +1 503 287 3093 or toll-free: (1) 800 944 6190
Fax: +1 503 280 8832
E-mail: orders@isbs.com
Website: http://www.isbs.com

© Sarah Green, 2015

Sarah Green has asserted her right under the Copyright, Designs and Patents Act 1988,
to be identified as the author of this work.

Hart Publishing is an imprint of Bloomsbury Publishing plc.

All rights reserved. No part of this publication may be reproduced, stored in a retrieval system, or
transmitted, in any form or by any means, without the prior permission of Hart Publishing, or as
expressly permitted by law or under the terms agreed with the appropriate reprographic rights
organisation. Enquiries concerning reproduction which may not be covered by the above should be
addressed to Hart Publishing Ltd at the address above.

British Library Cataloguing in Publication Data
Data Available

ISBN: 978-1-84946-331-7

Typeset by Hope Services, Abingdon
Printed and bound in Great Britain by
CPI Group (UK) Ltd, Croydon CR0 4YY

BLACKBURN COLLEGE LIBRARY	
BB 61201	
Askews & Holts	25-Mar-2015
UCL346.4103 GRE	

For my boys

PREFACE

Half my lifetime ago, I was asked in an interview a question about duplicative causation (as I would now recognise it). The answer seemed obvious to me at the time and, a few weeks later, I surmised from the outcome of that interview that my answer had been correct. The intervening years have taught me how naïve that conclusion was, but have not moved me from my instinctive response. This book is an extended account of that response.

But For the following, however, it is a book which would not have been written: St Hilda's College, for being a congenial and supportive place to work, and for providing some of the most glorious contemplative scenery in Oxford. Particular thanks are due to my colleagues Konstanze von Papp, Dev Gangjee, Andrea Dolcetti, Selina Todd, Rachel Condry, Susan Jones, Katherine Clarke, Lucia Nixon, Suzie Hancock, Jonathan Williams and Julia Bradley; to Sheila Forbes for being as empathetic a Principal as one could wish for, and to Peter Tullett, John Hamilton, Mike Newell, Andy Oakley, Maggie Bunting, Linda Inness, Elaine Sumner, Richard Kirkland, Graham Smith, Selina Collingwood and Jamie Franklin for making the place what it is. I also owe a debt of gratitude to several of my students, who not only put up with hearing about this book over the course of a couple of years, but were also willing to discuss various causal problems above and beyond those usually posed by undergraduate study: Ben Foster, Michael Poolton, Akshay Chauhan and Christian Goulart Mc Nerney deserve particular mention for the insights they were willing to share.

Working in the University of Oxford Law Faculty is a privilege and a pleasure, and thanks are due to many fellow members, past and present: Louise Gullifer, Ben McFarlane, Roderick Bagshaw, Donal Nolan, Simon Douglas, Imogen Goold, Becca Williams, Mindy Chen-Wishart, James Goudkamp, Paul Cowie, Joshua Getzler, Sandy Meredith, Timothy Endicott, Caroline Norris, Michelle Robb, Margaret Watson, Paul Burns and Bento de Sousa for making my life either easier, or more interesting, or both. Lord Hoffmann, Paul S Davies, Andy Burrows and Rob Stevens were all good enough to field specific questions, and to refuse to agree with me on several issues. It is no exaggeration to say that this manuscript would have been submitted many months later, and would have made any copy-editor run for the hills, had it not been for the genius and attention to detail of the indefatigable Jodi Gardner.

Beyond Oxford, I have gained much from discussions with Richard Wright, John Murphy, Alan Beever, Ken Oliphant, Jenny Steele, Leigh-Ann Mulcahy, Adam Heppinstall, Sir Jeremy Stuart-Smith, Sandy Steel, Jonathan Morgan, Jamie Lee, Claire McIvor, Nick McBride, Graham Virgo, Matt Dyson and Bob Sullivan.

I am especially grateful to Alex Broadbent for his intellectual generosity, to Mark Ingham and Daniel Horsley for checking my maths, to Robert Miles for his ongoing friendship and correspondence, and to Daniel Magnowski for (always) providing the soundtrack. Thanks also to Rachel Turner and to the team (new and old) at Hart/Bloomsbury.

Richard Hart was a huge source of encouragement at the inception of this project, and provided invaluable advice and reassurance throughout: he remained apparently convinced that this was a good idea, even when I wasn't.

I owe much to Jane Stapleton and to Paul Davies for first asking me the question that I am still thinking about nearly two decades later, and for the inspiration and guidance that they have provided, both during my years at Balliol, and ever since.

Nowhere in this book am I as fearful of omission as I am here, and this list is inevitably incomplete.

Finally, thank you Alfie and Benjamin, for reminding me that this book is no more than that. And thank you Al, for everything.

CONTENTS

Preface	vii
Table of Cases	xi
Table of Legislation and Official Publications	xv

1 Introduction 1
    Structure of the Book 5

2 The Necessary Breach Analysis and But For Causation 8
    Why But For Causation? 8
    The Balance of Probabilities 10
    Aggregation 13
    Specific Concept of Cause 15
    Counterfactual to Factual 17
    Current Perspectives on But For Causation 19

3 Basic Principles 32
    Basic Causal Principles 32
    'Operative': the Second Stage of the NBA 43
    The Significance of a Risk Which Has Actually Eventuated ... 47

4 Duplicative Causation (Real and Potential): Overdetermination and Pre-emption 58
    Factual Basis 58
    Overdetermination (Real Duplicative Causation) 59
    What Constitutes an Overdetermined Event? 61
    Double Omissions 65
    Combination of Tortious and Non-Tortious Factors 66
    Pre-emption 77
    Moral Luck 87

5 Material Contribution to Injury 94
    Factual Basis 94
    Medical Negligence 109
    The 'Doubling of the Risk' Test 113

6 Material Increase in Risk 123
    Factual Basis 123
    The Necessary Breach Analysis and Evidentiary Gaps 131
    Single Agent 148

7	Lost Chances	152
	Factual Basis	152
	Type 1 Cases Explained	154
	Type 2 Cases Explained	158
	How Far Does Hypothetical Third Party Action Take Us?	170
8	Concluding Thoughts	173
Index		177

TABLE OF CASES

AB v Ministry of Defence [2012] UKSC 9, [2013] 1 AC 78................................35–36, 114–15
Accident Compensation Corp v Ambros [2007] NZCA 304, [2008] 1 NZLR 340............170
Allied Maples v Simmons & Simmons [1995] 4 All ER 907 (CA) 155–57, 159–60, 164–65, 168–70
Amaca Pty Ltd v Booth [2011] HCA 53, (2011) 246 CLR 36..125
Amaca Pty Ltd v Ellis [2010] HCA 5, (2010) 240 CLR 111 ....2, 119–20
Amaca Pty Ltd (formerly James Hardie and Co Pty Ltd) v Hannell [2007] WASCA 158, (2007) 34 WAR 109...123, 128
Anderson v Hedstrom Corp 76 F Supp 2d 422 (NY 1999)...66
Anderson v Minneapolis, St Paul & Sault-Ste Marie Railway 179 NW 45 (Minn 1920) 67
Athey v Leonati [1996] 3 SCR 458 (SCC).. 106–7
Bailey v Ministry of Defence [2007] EWHC 2913 (QB), [2009] 1 WLR 1052;
 [2008] EWCA Civ 883, [2009] 1 WLR 1052 .... 4, 67, 94, 98, 107–9
Baker v Willoughby [1970] AC 467 (HL)...16–17, 39–43, 46, 51–52
Baldwin & Sons Pty Ltd v Plane (1998) 17 NSWCCR 434 (NSWCA).................................97
Barker v Corus [2006] UKHL 20, [2006] 2 AC 572 .... 35, 97, 128–29, 131–34, 139–40, 144–50
Bendix Mintex Pty Ltd v Barnes (1997) 42 NSWLR 307 (NSWCA)...........................97, 123
Benton v Miller [2005] 1 NZLR 66 (NZCA) 170
Bolton Metropolitan Borough Council v Municipal Mutual Insurance Ltd [2006]
 EWCA Civ 50, [2006] 1 WLR 1492 132
Bonnington Castings Ltd v Wardlaw [1956] AC 613 (HL)........................ 6, 34–35, 63, 94–98
Borealis AB v Geogas Trading SA [2010] EWHC 2789 (Comm), [2011]
 1 Lloyd's Rep 482 42
Browning v War Office [1963] 1 QB 750 (CA)...86
Chaplin v Hicks [1911] 2 KB 786 (CA) 164, 168–69
Chappel v Hart [1998] HCA 55, (1998) 195 CLR 232 156
Chester v Afshar [2004] UKHL 41, [2005] 1 AC 134 .... 4, 7, 16, 45, 154–56, 159, 166–70
Clements v Clements [2012] SCC 32, [2012] 2 SCR 181................................ 12, 120, 127, 145
Collett v Smith [2009] EWHC Civ 583, 153 Sol Jo (No 24) 34 153
Cook v Lewis [1951] SCR 830 (SCC) 72–73
Corr v IBC Vehicles [2008] UKHL 13, [2008] 1 AC 884..49, 54, 57
Creutzfeldt-Jakob Disease Litigation, Groups A and C Plaintiffs 54 BMLR 100
 (QBD).. 135–36
Davies v Taylor [1974] AC 207 (HL)..158
Dickins v O2 plc [2008] EWCA Civ 1144, [2009] IRLR 58 .... 63–64, 67, 94, 101–2, 105–6
Dillon v Twin State Gas & Electric Co 163 A 111 (NH 1932) 105
Dingle v Associated Newspapers [1961] 2 QB 162 (CA) 37, 63, 97
Dixon v Clement Jones [2004] EWCA Civ 1005, (2004) Times, 2 August................. 170–71
Doyle v Wallace [1998] 30 LS Gaz R 25, [1998] PIQR Q 146 (CA) 153

Durham v BAI (Run-off Ltd) [2012] UKSC 14, [2012] 3 All ER 1161 149–51
Elayoubi v Zipser [2008] NSWCA 335 ....85–86, 112
Evans v Queanbeyan City Council (2011) 9 DDCR 541, [2011]
    NSWCA 230 .... 12, 120, 123
Fairchild v Glenhaven Funeral Services [2002] UKHL 22, [2003] 1 AC 32 .... 6, 16, 33,
               35–36, 55, 76, 95, 108, 120, 123–34, 136, 139–51, 166, 169
Fitzgerald v Lane [1989] AC 328 (HL) 103
Flanagan v Greenbanks Ltd (t/a Lazenby Insulation) [2013] EWCA Civ 1702,
    151 Con LR 98 42
Gates v Howard Rotavator Pty Ltd (2000) 20 NSWCCR 7 (NSWCA) 97
Graham v Rourke (1990) 75 OR (2d) 622 (Ont CA) 196
Gregg v Scott [2005] UKHL 2, [2005] 2 AC 176 .... 148, 151–53, 158, 160–61, 164, 170
Hatton v Sutherland [2002] EWCA Civ 76, [2002] 2 All ER 1 100–2
Herskovits v Group Health Cooperative of Puget Sound 664 P 2d 474 (Wash 1983) 118
Hicks v Chief Constable of the South Yorkshire Police [1992] 2 All ER 65 (HL) 151
Home Office v Dorset Yacht Co Ltd [1970] AC 1004 (HL) 46, 48, 56, 78
Hotson v East Berkshire AHA [1987] AC 750 (HL) 97, 158, 160–65
Hughes v Lord Advocate [1963] AC 837 (HL) 54–56
Hughes v McKeown [1985] 3 All ER 284 (QBD) 153
Hymovitz v Eli Lilly & Co 541 NYS 2d 941 (NY 1989) 76
International Energy Group Ltd v Zurich Insurance Plc UK [2013] EWCA Civ 39,
    [2013] 3 All ER 395 132
JD v East Berkshire Community Health NHS Trust [2005] UKHL 23, [2005]
    2 AC 373 143
Jobling v Associated Dairies [1982] AC 794 (HL) 18, 39–41
Jolley v Sutton LBC [2000] 3 All ER 409 (HL) 55–56
Jones v Secretary of State for Energy and Climate Change [2012] EWHC 2936 (QB),
    [2012] All ER (D) 271 (Oct) 117, 130
Kingston v Chicago & Northwest Railway Co 211 NW 913 (Wis 1927) 67, 69–71
Kitchen v RAF Association [1958] 2 All ER 241 (CA) 160
Knightley v Johns [1982] 1 WLR 349 49–50
Laferrière v Lawson [1991] 1 SCR 541 (SCC) 161
Langford v Hebran [2001] EWCA Civ 361 153
Malec v JC Hutton Pty Ltd [1990] HCA 20, (1990) 169 CLR 638 ....106, 157
Mallet v McMonagle [1970] AC 166 (HL) 157
March v Stramare [1991] HCA 12, (1991) 171 CLR 506 59
McGhee v National Coal Board [1972] 3 All ER 1008 (HL) 6, 34, 123–26,
                              128–29, 131, 134, 136, 139–40, 144–45, 162–63
McKew v Holland and Hannen & Cubitts [1969] 3 All ER 1621 (HL) 50
McWilliams v Sir William Arroll Co Ltd [1962] 1 All ER 623 (HL) 154, 166, 171
Menne v Celotex Corp 861 F 2d 1453 (Kan 1988) 72
Merrell Dow Pharmaceuticals v Havner 953 SW 2d 706 (Tex 1997) 121–22
Miller v Minister of Pensions [1947] 2 All ER 372 (KBD) 12
National Insurance Co of New Zealand Ltd v Espagne [1961] HCA 15, (1961)
    105 CLR 569 1
Naxakis v Western General Hospital [1999] HCA 22, (1999) 197 CLR 269 161
Novartis Grimsby Ltd v Cookson [2007] EWCA Civ 1261, [2007] All ER (D)
    465 (Nov) 113–14, 120

Oliver v Miles 110 So 666 (Miss 1926)..72–73
Oropesa, The [1943] P 32 (CA) 48–49
Overseas Tankship (UK) Ltd v Morts Dock and Engineering Co Ltd (The
    Wagon Mound No 1) [1961] AC 388 (PC) 38, 47, 54–56
Paroline v United States et al No 12–8561, April 23, 2014 (USA) 44, 70, 112
Performance Cars Ltd v Abraham [1962] 1 QB 33 (CA)..............................36–37, 39, 94, 97
Phillips v Syndicate 992 Gunner [2003] EWHC 1084 (QB), [2003] All ER (D)
    168 (May)..151
Purkess v Crittenden (1965) 114 CLR 164 123
R v Hughes [2013] UKSC 56, [2013] 4 All ER 613..48
R (on the application of Lumba) v Secretary of State for the Home Department [2011]
    UKSC 12..21
Rahman v Arearose Ltd [2001] QB 351 (CA) 40, 63, 94, 99–102, 110
Rees v Darlington Memorial NHS Trust [2003] UKHL 52, [2004] 1 AC 309 156–57
Reeves v Commissioner of Police for the Metropolis [2000] 1 AC 360 (HL)................49, 78
Resurfice v Hanke [2007] SCC 7, [2007] 1 SCR 333 ....9, 33, 145
Robinson v Post Office [1974] 1 WLR 1176 (CA)..14, 46
Rothwell v Chemical & Insulating Co Ltd [2007] UKHL 39, [2008] AC 281............. 150–51
SAAMCO see South Australia Asset Management Corp v York Montague Ltd
Saunders System Birmingham Co v Adams 117 So 72 (Ala 1928).........46–48, 65–66, 78–86
Schrump v Koot (1977) 18 OR (2d) 337 (Ont CA) 106
Sellars v Adelaide Petroleum NL [1994] HCA 4, (1994) 179 CLR 332 ....153, 155
Seltsam Pty Ltd v McGuiness [2000] NSWCA 29, (2000) 49 NSWLR 262....12, 118–19, 123
Shortell v BICAL Construction, unreported, 16 May 2008 (Liverpool District
    Registry) 114, 120
Sienkiewicz v Greif [2009] EWCA Civ 1159, [2010] QB 370; [2011] UKSC 10,
    [2011] 2 AC 229 ....4, 11–12, 35–36, 94, 98, 115–16, 119, 121,
                                                        127–28, 130–37, 139–41, 144–47, 149, 163–64
Sindell v Abbott Laboratories 26 Cal 3d 588 (Cal 1980)...74–77, 148
Smith v Leech Brain [1962] 2 QB 405 (QBD)..38–39, 41, 54, 116
Smith v Moscovich (1989) 40 BCLR (2d) 49 (BC SC)...149
Snell v Farrell [1990] 2 SCR 311 (SCC)...30
South Australia Asset Management Corp v York Montague Ltd (SAAMCO)
    [1997] AC 191 (HL)..47, 52–53
Summers v Tice 199 P 2d 1 (Cal 1948)..9, 72, 85
Sykes v Midland Bank Executor & Trustee Co [1971] 1 QB 113 (CA) 154
Tabet v Gett [2010] HCA 12, (2010) 240 CLR 537 161
Thompson v Smiths Shiprepairers (North Shields) Ltd [1984] QB 405 (QBD).....37, 96–98
Tubemakers of Australia Ltd v Fernandez (1976) 50 AJLR 720 (HCA).............................109
Wagon Mound No 1, The see Overseas Tankship (UK) Ltd v Morts Dock and
    Engineering Co Ltd
Wieland v Cyril Lord Carpets Ltd [1969] 3 All ER 1006 (QBD)..50
Williams v University of Birmingham [2011] EWCA Civ 1242, [2011] All ER (D)
    25 (Nov) 147, 151
Willmore v Knowsley Borough Council see Sienkiewicz v Greif
Wilsher v Essex Area Health Authority [1988] AC 1074 (HL) 16, 33–34, 36, 48, 149
Wright v Cambridge Medical Group [2011] EWCA Civ 669, [2013] QB 312.....100, 110–12
XYZ v Schering [2002] EWHC 1420 (QB), 70 BMLR 88 115

Yardley v Coombes (1963) 107 Sol Jo 575 (QBD)...160
Ybarra v Spangard 154 P 2d 687 (Cal 1945)...99

TABLE OF LEGISLATION AND OFFICIAL PUBLICATIONS

United Kingdom
Civil Liability (Contribution) Act 1978 100–1
  s 1(1) 100
Consumer Protection Act 1987 115
Compensation Act 2006 139
  s 3 .... 129, 139, 142–43, 146–48
  (1) 146
Grinding of Metals (Miscellaneous Industries) Regulations 1925
  reg 1 95
Law Reform (Contributory Negligence) Act 1945 49
  s 1 131
  s 4 131
Office, Shops and Railway Premises Act 1963 40

United States
Restatement (Third) of Torts: Liability for Physical and Emotional Harm... 67–68, 71–73, 77
  § 26 .... 68, 104
  § 27 .... 68, 71, 77, 113
  § 28 .... 11, 73, 76–77, 122
  § 29 .... 44, 56
Violence against Women Act 1994 112

1

INTRODUCTION

Philosophy and science seek the explanation of phenomena and look to relationships and concurrences. Law is not concerned *rerum cognoscere causas*, but with attributing responsibility to persons.[1]

Causation in negligence has become complicated, convoluted and confused. As a result, it is regarded with trepidation by those forced to engage with it, whether student or Supreme Court Justice. Unlike some other legal minefields, whose conceptual difficulties are the result of academic neglect, causation has suffered from over-analysis or, at the very least, excessive micro-analysis, at the expense of attention paid to the whole. Somewhat unsurprisingly, this has led to a proliferation of theories, tests and approaches, not all of which serve their intended purpose.

The objective of this book is twofold. First, it aims to provide an accessible means of navigating the infamously baffling case law in this area by disentangling the different 'types' of causal problems which arise, and classifying decided cases accordingly. This should make it far easier to establish which authorities apply to which factual situations; something which currently challenges courts across the common law world. For ease of reference, each chapter concerned with a substantive category of causal problem will begin with a list of illustrative cases,[2] as well as a brief account of the distinctive features of that category. The main body of each chapter will then elaborate on why those cases and characteristics belong to the category concerned.

Alongside this objective lies another; one which, to some extent, makes the identification of separate categories of causal problem less significant than they are currently perceived to be: this work also offers a simple analytical formulation which is capable of dealing with all aspects of the causal inquiry in negligence, even those hitherto regarded as difficult. This formulation, referred to as the Necessary Breach Analysis (NBA),[3] eschews detailed philosophical and theoretical handwringing in favour of pragmatic reasoning. After all, whilst it is open to philosophers and abstractionists to assign only secondary importance to

---

[1] *The National Insurance Co of New Zealand Ltd v Espagne* [1961] HCA 15, (1961) 105 CLR 569 at [109] 591 (Windeyer J).
[2] Intended to be a list of the most significant, well known and/or contentious, and not an exhaustive catalogue of all those which belong in either the category or the chapter.
[3] For reasons which will become clear in the following chapter.

practical conclusions, the lawyer has no such luxury.[4] As Professor Stapleton points out:

> [C]onceptual indeterminacy [presents] grave problems for the practice of the Law, where the stakes can be very high, where issues must be authoritatively resolved in the here and now, and where clarity of analysis is a critical goal.[5]

What is more, extended philosophical analysis is simply not beneficial in constructing a causal model which meets the objectives of the tort of negligence. The NBA is based on an acceptance of the fact that there are sound legal justifications for imposing liability on the basis of causal conclusions which both scientists and philosophers would regard as less than satisfactory.[6]

> A good start to sensible doctrinal development would be an acceptance that we should not be hampered by a belief that there is some holy grail of causation, a platonic and pan-disciplinary understanding of that concept which we must struggle to discover and distil. For lawyers, just as 'duty' and 'damage' mean what we decide they should mean in a legal context, so it should be with the concept of causation. What counts as a sufficient causal connection for the purposes of the law is quite properly a decision to be taken with legal considerations in mind.[7]

In terms of the tort of negligence, one such 'legal consideration' is the specificity of the causal question with which negligence is concerned: less promiscuous than an inquiry as to *what* caused a particular event, the tort demands a directed conclusion as to whether an already-established breach of duty on the part of the defendant can be linked to a claimant's damage. Rudimentary tort law this may be, but it is worth emphasising here because it highlights what is distinctive about the causal inquiry in negligence.[8] (It also explains a particular use of language throughout this book, where 'breach' or 'breach of duty' is used in preference to 'tort', since to use the latter in discussing the unresolved outcome of the causal inquiry is to beg the question.)[9] The NBA is fit for purpose precisely because it provides a means of answering this functional question on its own terms, without concerning

---

[4] Whilst Allan Beever claims that the academic lawyer does – see A Beever, *Rediscovering the Tort of Negligence* (Oxford, Hart Publishing, 2009) 34, n 114 – this rather presumes that such lawyers want the doors to their ivory towers to remain locked, or so time consuming to open that practising lawyers will look elsewhere for assistance.

[5] J Stapleton, 'Choosing What We Mean by "Causation" in the Law' (2008) 73 *Missouri Law Review* 433, 441.

[6] See, eg, *Amaca Pty v Ellis* [2010] HCA 5, (2010) 240 CLR 111 at [6] 121–22: 'The courts' response to uncertainty arising from the absence of knowledge must be different from that of the medical practitioner or scientist. The courts cannot respond to a claim that is made by saying that, because science and medicine are not now able to say what caused Mr Cotton's cancer, the claim is neither allowed nor rejected. The courts must decide the claim and either dismiss it or hold the defendant responsible in damages'.

[7] M Hogg, 'Developing Causal Doctrine' in R Goldberg (ed), *Perspectives on Causation* (Oxford, Hart Publishing, 2011) 44.

[8] See D Robertson, 'The Common Sense of Cause-in-Fact' (1997) 75 *Texas Law Review* 69.

[9] But see C Schroeder, 'Corrective Justice and Liability for Increasing Risks' (1990) 37 *UCLA Law Review* 439.

itself with broader or more generic issues of what might count as 'a cause' for other purposes.¹⁰

> There is a distinction between conduct that counts as a cause in an explanatory inquiry, on the one hand, and conduct that is made by the law a basis of liability provided it is causally connected with the harm suffered, on the other.¹¹

That the *scope* of this specific causal inquiry is reduced, however, should in no way suggest that its significance is in any way diminished. On the contrary, the tort of negligence is dependent upon there being a causal link between defendant and claimant. The significance of this link is the reason why the causal inquiry has such resonance for corrective justice:¹²

> A correlatively structured remedy responds to and undoes an injustice only if that injustice is itself correlatively structured. In bringing an action against the defendant the plaintiff is asserting that they are connected as doer and sufferer of the same injustice. As is evidenced by the judgment's simultaneous correction of both sides of the injustice, what the defendant has done and what the plaintiff has suffered are not independent items. Rather, they are the active and passive poles of the same injustice, so that what the defendant has done counts as an injustice only because of what the plaintiff has suffered, and vice versa.¹³

On the same issue, Professor Gardner has written:

> The question of corrective justice is not the question of whether and to what extent and in what form and on what ground [something] should ... be allocated among them full stop, but the question of whether and to what extent and in what form and on what ground it should now be allocated *back* to one party from the other, reversing a transaction that took place between them.¹⁴

There has been to date some (understandable) lament over the apparent lack of judicial attention paid to academic analysis of causation in negligence, and particularly any theory that attempts to take the inquiry beyond its But For rudiments. This is not too surprising, however, given both the number and complexity of the theories which abound, and their subsequent refinements and

---

¹⁰ See also L Hoffmann, 'Causation' (2005) 121 *LQR* 592, 597: 'the question of what should count as a sufficient causal connection is a question of law, just as the question of what makes an act wrongful is a question of law'.
¹¹ A Honoré, 'Necessary and Sufficient Conditions' in D Owen (ed), *Philosophical Foundations of Tort Law* (Oxford, Clarendon Press, 1995) 369.
¹² Used here in the basic Aristotelian sense of being a collection of norms, as distinguished from those of distributive justice. See J Gardner, 'What is Tort Law For? Part 1: The Place of Corrective Justice' (2011) 30 *Law and Philosophy* 1, throughout, but particularly 7–21 and cf Beever, *Rediscovering*, above n 4 at 29, where he defends the idea that corrective justice is itself instantiated by a range of common law decisions, and can be discovered, therefore, by 'reflective equilibrium' on the same.
¹³ E Weinrib, *Corrective Justice* (Oxford, OUP, 2012) 17. See also E Weinrib, 'Correlativity, Personality, and the Emerging Consensus on Corrective Justice' (2001) 2 *Theoretical Inquiries in Law* 107.
¹⁴ Gardner, 'What is Tort Law For?', above n 12 at 9–10. See also J Goldberg and B Zipursky, 'Civil Recourse Revisited' (2011–12) 39 *Florida State Law Review* 341, 348–51. Cf Beever, *Rediscovering*, above n 4 at 11.

modifications.[15] Intellectually brilliant though some of them are,[16] the very depth and detail which makes them academically satisfying can also make them less accessible for forensic purposes. And yet, as a number of recent judicial exertions[17] have shown, there is a real need for a robust and practical means of dealing with causation which will remain consistently effective, regardless of the exigencies of particular factual conditions. This is what the NBA aims to provide. Its form is simple:

Stage 1: Is it more likely than not that *a* defendant's breach of duty changed the normal course of events[18] so that damage (including constituent parts of larger damage) occurred which would not otherwise have done so when it did?

Stage 2, applied to each defendant individually: Was the effect of *this* defendant's breach operative when the damage occurred?

This form of analysis, which will be explored in detail in the next chapter, is offered not as a 'theory of causation' in the way that the more philosophically inclined offerings of NESS[19] and INUS[20] are. It is not in fact offered as a theory of any sort, since it does not attempt to provide a definition of causation, nor an abstract account of what it means to be a cause. Rather, it is intended to provide a framework for analysing any given factual situation, so that the legally relevant causal issues therein can be identified and resolved in a straightforward and effective manner. Whilst it is inevitable, therefore, that some of its conclusions will coincide with those of NESS and INUS, for instance, its objectives are far more localised and finite than either of those means of analysis, because the whole point of the NBA is to provide a forensic device which will be useful on a case-by-case basis. It does not claim, therefore, to tell us what causation is *eo ipso*, but will allow us to know it when we see it. This may look like an overly modest aim. In this very specific context, however, less is more: a universal definition of causation is undoubtedly both intellectually satisfying and a thing worth having, but a practical analytical tool is equally valuable.[21]

Many of the current complications in this area stem from the fact that there have grown up over the last few decades several 'categories' of causation in negligence, each regarded as having its own demands and difficulties, and therefore

---

[15] See, eg, L Hoffmann, 'Causation' in Goldberg, *Perspectives*, above n 7 at 3.

[16] Such as NESS; see Ch 2, under the heading 'Wright'.

[17] *Chester v Afshar* [2004] UKHL 41, [2005] 1 AC 134; *Bailey v Ministry of Defence* [2008] EWCA Civ 883, [2009] 1 WLR 1052; and *Sienkiewicz v Greif* [2011] UKSC 10, [2011] 2 AC 229 to name but a few.

[18] Used herein to describe the course of events which a claimant could expect to experience in the absence of any breaches of duty of care to her. See MS Moore, *Causation and Responsibility* (Oxford, OUP, 2009) 84 and Stapleton, 'Choosing What We Mean by "Causation" in the Law', above n 5 at 453, where she discusses the concept of a claimant's 'normal expectancies'.

[19] Necessary Element of a Sufficient Set. Originally conceived by Hart and Honoré (see HLA Hart and A Honoré, *Causation in the Law*, 2nd edn (Oxford, OUP, 1985) ch V) and popularised by Richard Wright (see Ch 2 under heading 'Wright').

[20] Insufficient but non-redundant parts of a condition that is itself unnecessary but sufficient for the occurrence of the effect. John Mackie's formulation: JL Mackie, *The Cement of the Universe: A Study of Causation* (Oxford, Clarendon Press, 1980) throughout, but particularly ch 3.

[21] See Hoffmann, 'Causation', above n 15.

requiring a specific test. Inevitably, the boundaries and distinctions within this mosaic have become both blurred and, in some cases, completely obscured over time, leaving everyone wondering to which 'category' a particular set of facts belongs, and which 'test' should therefore be applied to it. Whilst the proliferation of tests is to be regretted, not least because of the incoherence it has introduced into the law, the categorisation of factual situations is helpful in clarifying the precise causal issue to be addressed in each case. For this reason, and to make the NBA more immediately accessible to those familiar with navigating material on causation in negligence, the chapters in this book will broadly follow the categories of causal issues which seem to have become an accepted part of the fabric of tort law. One of the effects of this, in addition to demonstrating how a NBA is universally effective, is that the boundary lines of those categories will be redrawn clearly, with an accompanying explanation of what such distinctions achieve.

Structure of the Book

The Necessary Breach Analysis and But For Causation

Chapter 2 explains how the NBA works, what is distinctive about it, and why But For causation is at its core. In doing so, it conducts a review of other current perspectives on But For causation, and outlines the principal points of convergence and divergence between them.

Basic Principles of Causation

Chapter 3 identifies basic causal principles which, when followed in combination with a NBA, will lead to a causal inquiry that is more streamlined and less complex than is often the case when such principles are overlooked. Importantly, this chapter also examines in detail the second stage of the NBA, and deals with the question, pivotal to that analysis, of what 'operative' means for those purposes.

Illustrative Cases: *Wilsher, Performance Cars, Baker, Jobling*

Duplicative (and Potentially Duplicative) Causation

Chapter 4 addresses the issues that form the most fundamental challenge to But For causation, and goes on to explain how an NBA deals with them. It also sets out clearly which factual features indicate that a situation exhibits either duplicative, or potentially duplicative, causation, and shows how each is relevant to that classification. The problems raised by these distinctive factual situations require solutions not only as to where liability should lie, but also as to how and why situations of duplicative causation (over-determination) are conceptually distinct from those of *potentially* duplicative causation (pre-emption). This chapter demonstrates clearly why this matters, and what its implications are for the causal inquiry. Since one such implication is that moral luck could be seen to play a greater role in these causal situations than it does elsewhere, this chapter also considers prominent views on the concept, and what they can contribute to the NBA.

Illustrative Cases: *Summers, Cook, Saunders Systems*

Material Contribution to Injury

The principal aim of Chapter 5 is to distinguish cases to which a material contribution to injury analysis is appropriate from those to which such an analysis has been erroneously applied, or at least attempted. This has occurred most commonly in relation to cases to which basic, orthodox causal principles, as outlined in Chapter 3, should properly be applied. The chapter tackles the question, extant as a result of *Bonnington Castings Ltd v Wardlaw*,[22] of how damages should be quantified as a result of a material contribution to injury analysis. It also includes an examination of the value of epidemiological evidence, and how the 'doubling of the risk' test has been used inappropriately in the past.

Illustrative Cases: *Bonnington, Bailey, O2, Rahman, Fitzgerald*

Material Increase in Risk

Chapter 6 is concerned with the exceptional causal principle derived originally from *McGhee v National Coal Board*,[23] and developed subsequently in *Fairchild v Glenhaven Funeral Services*.[24] Its principal argument is that the exception is a defensible one only in very limited circumstances, and that its application outside of these is highly problematic, and should not be encouraged. It pays particular attention to the 'rock of uncertainty' and 'single agent' aspects of what has come to be known most widely as the 'Fairchild principle'.

Illustrative Cases: *McGhee, Fairchild, Barker, Sienkiewicz*

---

[22] *Bonnington Castings Ltd v Wardlaw* [1956] AC 613 (HL).
[23] *McGhee v National Coal Board* [1972] 3 All ER 1008 (HL).
[24] *Fairchild v Glenhaven Funeral Services* [2002] UKHL 22, [2003] 1 AC 32.

Loss of Chance: Real (and Illusory)

In Chapter 7 'loss of chance' formulations are divided into two types; those in which there is a chance to be lost, and those in which no such chance ever existed as far as the forensic process is concerned. This analysis departs in both form and substance, therefore, from the traditional dichotomy made between lost chances of financial gain and lost chances of positive physical outcomes. In so doing, it offers a more palatable means of recognising some lost chances as actionable, but not others. Chapter 7 also presents a novel interpretation of *Chester v Afshar*,[25] in arguing for it to be understood as a 'loss of autonomy' case.

Illustrative Cases: *Allied Maples, Chester, Gregg, Hotson*

---

[25] *Chester v Afshar* [2004] UKHL 41, [2005] 1 AC 134.

2

The Necessary Breach Analysis and But For Causation

Why But For Causation?

> Why is causation important in tort law? One reason is that to insist on causal connection between conduct and harm ensures that in general we impose liability only on those who, by intervening in the world, have changed the course of events for the worse.[1]

Honoré here highlights the enduring significance of the causal inquiry in tort: the identification of behaviour which has a deleterious effect on the world. In terms of the tort of negligence in particular, it is crucial to add that the only relevant 'conduct' is that which amounts to a breach of a duty of care:[2]

> It is true that tort law already includes a norm of corrective justice, the norm according to which (legally recognized) wrongdoers are required to pay reparative damages in respect of those (legally recognized) losses that they wrongfully occasion, on the ground that they wrongfully occasioned them.[3]

There is little doubt that the causal element of the negligence inquiry is of the utmost significance.[4] Perhaps this explains why it has become so convoluted in recent years, as successive courts struggle to deal with non-standard causal questions. As it happens, however, first principles are far more useful in this exercise than current approaches would suggest.

> The standard legal test for answering the factual causation question is the so-called 'but-for' test: the legally proscribed conduct in question was a factual cause of the outcome in question if (holding everything else constant) the outcome would not have occurred But For that conduct. The but-for test tells us whether the relevant, legally proscribed conduct was a necessary condition for the outcome in question.[5]

---

[1] A Honoré, 'Necessary and Sufficient Conditions in Tort Law' in D Owen (ed), *Philosophical Foundations of Tort Law* (Oxford, Clarendon Press, 1995) 385.

[2] See J Stapleton, 'Occam's Razor Reveals an Orthodox Basis for *Chester v Afshar*' (2006) 122 *LQR* 426, n 7 and accompanying text.

[3] J Gardner, 'What is Tort Law For? Part 1: The Place of Corrective Justice' (2011) 30 *Law and Philosophy* 1, 17.

[4] See MS Moore, *Causation and Responsibility* (Oxford, OUP, 2009) 83.

[5] P Cane, *Responsibility in Law and Morality* (Oxford, Hart Publishing, 2002) 120.

A principal contention of this book is that a But For test remains the best way to approach the causal inquiry in negligence. This is because the essential objective of that inquiry is the identification of claimants whose states at trial are as they are[6] only because *a breach of duty* has intervened in their history, thereby diverting events from their normal course beyond the point of breach.[7] Unless the breach of an established duty of care has had such an effect, the claimant's state is no concern of the tort of negligence.[8] So far, so simple. Such simplicity might, however, be said to come at too high a price. The well-known objection levelled at the traditional But For test is that, in making necessity the touchstone of causation, it does not work in situations of overdetermined cause; that is, where there are multiple causes, each sufficient in its own right to bring about the end result. Take, for instance, the much-used example of the two hunters who both negligently fire into the forest, hitting V and killing him.[9] The But For test, asked of each in turn, tells us that neither caused the death, because but for A's shot, V would still have died, and the same applies to B. In requiring that a factor be individually necessary for the adverse outcome in question, the application of this test to A and to B generates an unsatisfactory answer: two duty-breaching defendants avoid liability,[10] a claimant who has suffered adverse consequences as a result of at least one breach bears that loss himself, and the negligence inquiry thereby defeats its own object.

Whilst it is true that this is essentially a theoretical, rather than a practical, problem since, first, there are relatively few cases of this nature which actually trouble the courts and, second, that judicial pragmatism has usually prevented such a counter-productive result from representing the legal response to such situations,[11] the impact of this theoretical glitch has been profound. The inability of the But For test to cope adequately and helpfully with cases of overdetermined cause has driven some of the finest legal minds of recent times to attempt to formulate an improved alternative.[12] In fact, Professor Wright, one of the leading proponents of perhaps the most effective alternative, popularised by him as the NESS test, admits, as we shall see, that his formulation is really only materially different from, and relevant as a substitute for, the But For test in overdetermined cause situations.[13]

---

[6] Of course, any conclusion as to why things are as they are, made in a forensic context, is one reached on the balance of probabilities. See below, under heading 'The Balance of Probabilities'.
[7] See Moore, *Causation*, above n 4 at 84 and J Stapleton, 'Choosing What we Mean by "Causation"' in the Law' (2008) 73 *Missouri Law Review* 433, 453.
[8] See *Resurfice v Hanke* [2007] SCC 7, [2007] 1 SCR 333 at [22] (per McLachlan CJ).
[9] Based on *Summers v Tice* 199 P 2d 1 (Cal 1948). See Ch 4, under heading 'All Factors Tortious'.
[10] And the same would be true of however many multiple defendants had overdetermined the harm in question.
[11] See Ch 4.
[12] eg Hart and Honoré, Stapleton, Wright, Beever, Porat, Stein, Goldberg and Zipursky, to name but a few.
[13] See R Wright, 'Causation in Tort Law' (1985) 73 *California Law Review* 1735, 1792 and 1803; and C Miller, 'NESS for Beginners' in R Goldberg (ed), *Perspectives on Causation* (Oxford, Hart Publishing, 2011) 336. For more on this, see below, under heading 'Wright'.

He is not alone in asserting that, in the vast majority of cases, the But For test is perfectly fit for purpose.[14]

It is remarkable indeed that the inability of a basic test to accommodate a very specific and relatively unusual set of facts has generated such a wealth of literature, hypotheses and discussion. Hard cases make bad law for the very reason that they favour exceptions over rules. The formulation, therefore, of any rule designed expressly for exceptional circumstances is bound not only to prove an extremely difficult exercise, but also to produce a result which is far more complicated than the vast majority of situations will demand. The form of analysis favoured in this book, by contrast, was formulated with the express purpose of being a forensic model[15] which remains as accessible and effective in difficult cause situations (including the overdetermination scenario outlined above) as it does in the most straightforward circumstances. This formulation, the Necessary Breach Analysis (NBA), has But For causation at its core:

Stage 1: Is it more likely than not that *a* defendant's breach of duty changed the normal course of events so that damage (including constituent parts of larger damage)[16] **occurred which would not otherwise have done so when it did?**[17]

Stage 2, applied to each defendant individually: Was the effect of *this* defendant's breach operative when the damage occurred?

This analysis has three elements of distinctive significance: its initial aggregation of potential causal factors, its specific concept of cause, and its explicit move from the counterfactual to the factual. Before taking a detailed look at these aspects of the analysis, however, it is worth defining more closely one of its elements which might appear at first to be require little elaboration; what is meant by 'more likely than not' and the balance of probabilities standard to which it refers.

The Balance of Probabilities

Although the balance of probabilities standard plays a fundamental role in the forensic treatment of civil claims, interpretations of it are not as uniform as might be expected of such a rudimentary legal device:

---

[14] D Robertson, 'The Common Sense of Cause in Fact' (1997) 75 *Texas Law Review* 1765, 1776. Although Stapleton is of a different opinion, describing the test as 'notoriously inadequate'; J Stapleton, 'Reflections on Common Sense Causation in Australia' in S Degeling, J Edelman and J Goudkamp (eds), *Torts in Commercial Law* (Pyrmont, Thomson Reuters, 2011) 339.

[15] It is intended, as outlined in the Introduction, as a framework for prospective analysis, and not as a comprehensive theory of causation in the way that NESS and INUS are.

[16] This particular element of the NBA is explained in detail in Ch 5, which deals with material contributions to injury.

[17] See Ch 5 for an explanation of why this temporal specificity is important.

There are two main standards of proof of fact in English and American courts. The plaintiff in a civil case must prove on the balance of probabilities, and the prosecutor in a criminal case must prove his conclusion at a level of probability that puts it beyond reasonable doubt. Does the concept of probability involved here conform to the principles of the mathematical calculus of chance? Some philosophers have claimed that it does, some that if such a probability were measurable it would do so, and some that it is not even in principle a mathematical probability. The third of these views is the most defensible.[18]

The third of these views is also the one used in this book. Steve Gold has explained it in simple terms:

> The 'more likely than not' test should be applied only to the requisite strength of the fact-finder's belief, and not to the definition of the factual elements which the parties must prove... Juries should not be instructed that the preponderance standard means that plaintiffs must prove a fact probability greater than 50%. Rather, they should be instructed that plaintiffs must establish certain facts... by evidence which convinces juries that the fact is more likely than not.[19]

Richard Wright is a vociferous proponent of this clear distinction between statistical probabilities and degrees of subjective belief on the part of fact-finders, with the latter being the key to satisfying the civil law standard of persuasion (defined as 'balance of probability' in English law and 'preponderance of the evidence' in the US).[20] This is hardly surprising, given his approach to causation in general, and, in particular, his definition of a causal law:[21]

> A causal law is a law of nature; it describes an empirically based, invariable, nonprobabilistic relation between some minimal set of abstractly described antecedent conditions and some abstractly described consequent condition, such that the concrete instantiation of all the antecedent conditions will always immediately result in the concrete instantiation of the consequent condition.[22]

---

[18] LJ Cohen, *The Probable and the Provable* (Oxford, OUP, 1977) 49. The six chapters which follow this assertion are devoted to showing how the forensic standard of proof is not synonymous with mathematical calculations of probability. Of particular interest is the discussion at 62–64, of whether the balance of probability lies between the plaintiff's and the defendant's contentions. See, for example, at 62, 'it might be argued [that] we should not construe the phrase "the balance of probability" as denoting the balance between the probability of S and the probability of not-S, on Q, where one party to the case asserts S and his opponent asserts not-S, and Q states the facts before the court, but rather as denoting the balance between the probability of S and the probability of, say, R, where one party asserts S, the other asserts R, and S and R, though mutually inconsistent, do not exhaust the domain of possibilities'.

[19] S Gold, 'Causation in Toxic Torts: Burdens of Proof, Standards of Persuasion, and Statistical Evidence' (1986) 96 *Yale Law Journal* 376, 395. Cited with approval in *Sienkiewicz v Greif* [2011] UKSC 10, [2011] 2 AC 229 at [156] 285 (per Lord Rodger), [194] 296 (per Lord Brown) and, most notably, [217]–[222] 302–03 (per Lord Dyson). See also Ch 6, text to n 56.

[20] See also American Law Institute, *Restatement of the Law, Third, Torts: Liability for Physical and Emotional Harm* (St Paul, MN, American Law Institute, 2010) § 28 comment c.

[21] An approach which will be examined in greater detail below, under heading 'Wright'.

[22] R Wright, 'Probability versus Belief' in Goldberg (ed), *Perspectives on Causation*, above n 13 at 205.

There is, therefore, a definitional distinction between probabilistic judgements (which enable us to make predictions or place bets as to what an outcome might be) and 'uniquely instantiating' evidence (which tells us what that outcome actually was).[23] Wright is right: this distinction is crucial. And it is often missed. In substantiating his point, Wright cites the following jury instruction:

> To 'establish by a preponderance of the evidence' means that something is more likely so than not so. In other words, a preponderance of the evidence in the case means such evidence as when considered and compared with that opposed to it, has more convincing force, and produces in your minds belief that what is sought to be proved is more likely true than not true. This rule does not, of course, require proof to an absolute certainty, since proof to an absolute certainty is seldom possible in any case.[24]

With the obvious exception of the references to juries, the point is easily transferable to English law:

> Our civil law does not deal in scientific or logical certainties. The statistical evidence may be so compelling that, to use the terminology of Steve Gold, the court may be able to infer belief probability from fact probability. To permit the drawing of such an inference is not to collapse the distinction between fact probability and belief probability. It merely recognises that, in a particular case, the fact probability may be so strong that the court is satisfied as to belief probability. Whether an inference of belief probability should be drawn in any given case is not a matter of logic. The law does not demand absolute certainty in this context or indeed in any context[25]

and

> 'If the evidence is such that the tribunal can say: "we think it more probable than not", the burden is discharged'.[26]

For the purposes of the NBA, the 'more likely than not' standard is a standard of belief probability.[27] The most significant implications of this will be examined in Chapter 5.

---

[23] ibid, 208.

[24] EH Devitt et al, *3 Federal Jury Practice and Instructions (Civil)*, 4th edn (St Paul, West Publishing Co, 1987) s 72.01 at 32. See also I Hacking, *The Emergence of Probability: A Philosophical Study of Early Ideas about Probability, Induction and Statistical Inference* (Cambridge, CUP, 1975) 89–91.

[25] *Sienkiewicz* at [222] 303 (per Lord Dyson). This is by no means a new controversy, with theoretical and definitional disagreement reaching back to its conception in the decade around 1660: see Hacking, *The Emergence of Probability*, above n 24 at ch 2. Even now, there are 'personalists ... who have said that propensities and statistical frequencies are some sort of 'mysterious pseudo-property', that can be made sense of only through personal probability. There are frequentists who contend that frequency concepts are the only ones that are viable' (at 14). For a highly detailed discussion, see R Weatherford, *Philosophical Foundations of Probability Theory* (London, Routledge and Kegan Paul, 1982).

[26] *Miller v Minister of Pensions* [1947] 2 All ER 372 (KBD), 374 (per Denning J). See also *Seltsam Pty Ltd v McGuiness* [2000] NSWCA 29, (2000) 49 NSWLR 262 at [136] 284–85 (per Spigelman CJ) and *Evans v Queanbeyan City Council* [2011] NSWCA 230, (2011) 9 DDCR 541 at [42] (per Allsop P).

[27] See *Clements v Clements* [2012] SCC 32, [2012] SCR 181 at [9] (per McLachlin CJ): 'The "but for" causation test must be applied in a robust common sense fashion. There is no need for scientific evidence of the precise contribution the defendant's negligence made to the injury'.

Aggregation

The NBA is predicated on But For causation because it is a bespoke means of conducting the causal inquiry in negligence, and that tort is engaged only by breaches of duty which make a difference to a claimant's normal course of events. The necessity with which the NBA is concerned, however, is necessity in the aggregate sense. In other words, as long as *at least one* breach was necessary to divert the claimant from 'the mandated course of events',[28] the first stage of the test is satisfied. Whilst this aggregation will be needed only in cases involving multiple breaches of duty, the formulation loses nothing by being phrased so as to encompass both single and plural breach situations. The crucial objective of this stage is both simple and narrow: to establish whether the claimant's injury is the concern of the law, or whether it is the result of misfortune. At this stage, no consideration of the liability of individual defendants is necessary. So, in the two hunters example outlined above, we know that it is more likely than not that the victim would have avoided being shot were it not for at least one defendant's breach because, on these facts, the death would not have happened but for the two breaches which did in fact occur. This is all that is required to satisfy the first stage of the NBA, and it does not matter that, considered individually, neither breach is a But For cause. For forensic purposes, the only conclusion of relevance *at this point* is that the claimant has been made worse off by at least one breach of duty, and not by her own bad luck.

This aggregation deals directly with the overdetermination issue. In such situations, we know that asking the But For question individually of each breach of duty will give us the intuitively unacceptable answer that none of the breaches caused the claimant's damage. Applied in its conventional form, the But For test denies causal relevance to factors which overdetermine a result because, in attempting to isolate a philosophically defensible conclusion about what caused a particular outcome, it attempts to do too much. This is not what is required of the causal inquiry in negligence, and is not the answer sought by judges in such cases. In maintaining the significance of But For causation, therefore, the NBA emphasises (as its name suggests) the importance of necessity for the causal inquiry in negligence, *but* is at this stage concerned not with physical necessity (referred to by Wright as 'lawful necessity'),[29] but with what could be referred to as tortious necessity: unless at least one breach was necessary to change the claimant's world from the hypothetical no-damage scenario to the real-world damage scenario, that damage falls outside of the tort's remit and so the claimant must bear it. That is not to suggest that the

---

[28] Stapleton, 'Choosing What We Mean by "Causation" in the Law', above n 7 at 460. The important distinction is made herein between 'normality' and 'the mandated course of events' to encompass situations where the claimant's legal entitlement (the denial of which amounts to damage) is not synonymous with 'normality'.

[29] See eg R Wright, 'The NESS Account: Response to Criticisms' in Goldberg (ed), *Perspectives on Causation*, above n 13 at 289.

ultimate substantive aim of the causal inquiry is anything other than, in most cases,[30] to determine whether a claimant should receive damages from a particular defendant and, if so, how much, but merely to state that asking this narrow question *in the first instance and without more* is not the most effective way of furthering this aim.[31]

This is not the end of the process. It would be neither intuitively acceptable, nor in keeping with the aims of the tort of negligence, to hold all defendants liable on the basis that at least one of their breaches altered the claimant's normal course of events for the worse.[32] This is why the second stage of the NBA looks to each defendant in turn, in order to establish whether each breach of duty was operative in terms of the claimant's damage. The concept of 'operative' is clearly a pivotal concept for these purposes and, as such, will be analysed in detail in Chapter 3.

It is also important to note that the aggregation of the first stage prevents factors which would, if considered individually, amount to But For causes from being rendered causally irrelevant by the existence of duplicative breaches. It is only concerned, therefore, with aggregating breaches which would, if considered individually, be necessary for the adverse outcome to occur. Its function is not to inculpate those actions which amount to a breach of duty, but which are not necessary to cause the claimant's damage even in the absence of other factors. This point is illustrated by the facts of *Robinson v Post Office*.[33] In that case, the claimant was injured at work as a result of his employer's breach of duty. In response, his GP administered anti-tetanus medication, but did so without performing the appropriate tests beforehand. Nine days later, the claimant developed encephalitis in reaction to the anti-tetanus serum and suffered brain damage and permanent disability as a consequence. The defendant employers were held liable for the full extent of the damage, notwithstanding the doctor's negligence.[34] In order to understand why this was the correct result, it is necessary to recognise that, although the doctor's actions were undoubtedly a breach of duty, they made no factual difference to the claimant's position because, even had the requisite test been carried out, the claimant's reaction would not have manifested in time for the doctor to decide not to go ahead with the treatment. Even had he acted non-negligently in other words, he would still have taken the same action and the same outcome would have resulted. Since the doctor's action, even had it been considered individually, was not a But For cause of the claimant's injury, it is not one that would satisfy the first stage of the NBA.

---

[30] See Ch 6 for the only genuine exception to this (and its corresponding limits).
[31] For an in-depth discussion of how the NBA deals with overdetermined harm situations, see Ch 4.
[32] Although this may well be the least worst option in exceptional situations, such as cases in which there is a 'rock of uncertainty'. See Ch 6.
[33] *Robinson v Post Office* [1974] 1 WLR 1176 (CA).
[34] The doctor, initially not sued by the claimant, was later added as a co-defendant.

Specific Concept of Cause

As a specialised model of causal reasoning, the NBA explicitly adopts a specific notion of cause. This affects the chronology of its analysis. It is common, for instance, for the question of what caused an outcome to precede the question of whether that outcome counts as legally relevant damage (ie whether that outcome amounts to an alteration of the claimant's normal course of events). As we shall see below, this is the approach favoured by Professors Stapleton, Wright, Stevens and Beever. This is because each of these accounts is concerned not to deny causal relevance to any factor which either 'contributed' to, or was 'involved' in, the claimant's adverse outcome. Such concerns are inevitable in exercises whose object is to avoid a concept of 'cause' which is defined entirely by its legal context, and to use instead the generic, broadly scientific concept of cause with which non-lawyers are familiar. Such approaches could be said to have the advantage of using a more accessible notion of 'cause' because they deem all factors which contribute to an outcome to have causal relevance to it. This, it might be said, is a more authentic causal inquiry than one which sets out self-consciously to recognise as causes only those factors which *alter* a claimant's normal course of events, and to deny (legal) causal relevance to those factors which contribute to an outcome that would have happened even in the absence of those factors. This is not, however, the end of the story: by starting off in a more promiscuous way, those approaches employing a more generic notion of cause have at some stage to yield to the constitutive restrictions of the negligence inquiry. This means that they must distinguish between outcomes which would have occurred but for the breach of duty factor, and those which would not. Addressing the question of cause in this way means that an answer may well take the form:[35]

> X contributed to Y and so was a cause of Y.
> Y would have happened But For X.
> X is not therefore liable for Y in negligence because, although he was a cause of Y, Y is not actionable in the tort of negligence.

By contrast, the first stage of the NBA would give an answer in the form:

> But For X, Y would have happened. X is not therefore a cause in negligence terms and so is not liable for Y.

Both means of conducting this inquiry are predicated on the concept of But For causation. The first, however, answers a general causal question before moving on to address the specific legal question. The second starts and ends with the specific legal question. This explains the assertion made in the Introduction to this book that what the NBA aims to do is explicitly to provide an immediate and accessible means of conducting the causal inquiry in negligence, and it does not purport to

---

[35] For simplicity, I am using X to denote both the defendant and her breach of duty.

say anything about the extra-legal concept of causation, or to provide any universal answers as to what that concept might mean. It is to some extent, therefore, a means of analysis which is incommensurate with the theories of causation referred to here, and analysed in more detail below. Nevertheless, it is worth asking whether, when the objectives of both forms of analysis are the same when applied to the tort of negligence, there is more to be gained from one approach than the other. Is it better, in other words, to build in a priori legal limits to a definition of cause and to use this from the outset, or to be more liberal with this definition, and to select legally relevant causes at a subsequent stage?

There are two principal reasons to favour the NBA form of inquiry as opposed to one involving a more liberal definition of cause: one factual and one practical. In factual terms, the common law itself tells us that judges simply do not deal with causation in any extra-legal sense,[36] but refer instead to a single, specifically legal construct. Extra-judicially, Lord Hoffmann has made the following point:

> On what basis are academic writers entitled to say that judges should take into account a philosophically privileged form of causation which satisfies criteria not required by the law? Of course, anyone is entitled to say that in treating X in some particular context as having caused Y, the courts are stretching the ordinary meaning of 'cause'. But this is engaging in legitimate argument over interpretation and not introducing the concept of 'real' causation as a preliminary test which has to be satisfied before the question of interpretation arises ... In my view, all this wringing of hands is quite unnecessary. It might be easier if, instead of speaking of proof of 'causation', which makes it look as if we are dealing with one monolithic concept ... we spoke of the 'causal requirements' of a legal rule. That would make it clear that causal requirements are creatures of the law and nothing more.[37]

The practical argument in support of the NBA relates to the forensic process. Establishing causation is rarely straightforward, speedy and inexpensive. There is more to be said, therefore, for a means of streamlining the process sooner rather than later, so that resources are not consumed by causal investigations which turn out ultimately to have no legal relevance. In pragmatic terms, proving that something is a physical cause of an outcome, but then going on to conclude that it is not a cause in which the law has any interest, is a potentially wasteful process.[38] It is also the case, however, that, in asking 'is it more likely than not that a defendant's breach of duty changed the normal course of events so that damage occurred which would not otherwise have done so when it did?', the NBA might need to employ the same level of resources as would the broader causal investigation, because sometimes this question cannot be answered without a comprehensive analysis of some, or all, potential factors. It remains the case, however, that in

---

[36] As exemplified by most of the cases discussed in this book, most notably *Chester v Afshar* [2004] UKHL 41, [2005] 1 AC 134 (discussed in Ch 7), *Fairchild v Glenhaven Funeral Services Ltd* [2002] UKHL 22, [2003] 1 AC 32 (discussed in Ch 6), *Wilsher v Essex Area Health Authority* [1988] AC 1074 (HL) (discussed in Ch 3) and *Baker v Willoughby* [1970] AC 467 (HL) (discussed in Ch 3).

[37] L Hoffmann, 'Causation' in Goldberg (ed), *Perspectives on Causation*, above n 13 at 5 and 9.

[38] Whilst it might seem as if the same could be said of the NBA's two stages, there is in fact no analogous argument to be made here since compensation is always awarded in cases which satisfy its first stage: the second stage merely determines the source and extent of that compensation.

expressly limiting the remit of the causal inquiry at the outset, the chronological approach adopted by the NBA will more often lead to a leaner, more focused and immediately appropriate causal inquiry than one which provides no initial filter for factors whose causal influence must be scrutinised by the forensic process.

Counterfactual to Factual

One significant implication of the NBA's aggregative first stage is that its second stage must perform a disaggregation. The nature of this transition is both deliberate and necessary because it is what enables the NBA to deal effectively with situations involving multiple defendants, and particularly with problems of both overdetermination and pre-emption; all of which have proven problematic for conventional But For causation. As we have seen, the first stage asks a counterfactual question of all breaches of duty taken together; something which enables it to be more inclusive than an individualised But For test. This ensures that claimants injured as a result of more than one breach of a duty of care do not fall between two (or more) tortfeasors.[39] Corrective justice, however, is a game of two halves, and ensuring that deserving claimants get compensation does not give us a final score. As Weinrib has made clear:

> what the defendant has done and what the plaintiff has suffered are not independent items. Rather, they are the active and passive poles of the same injustice, so that what the defendant has done counts as an injustice only because of what the plaintiff has suffered, and vice versa ... Because the defendant, if liable, has committed the same injustice that the plaintiff has suffered, the reason the plaintiff wins ought to be the same reason that the defendant loses ... A factor that applies to only one of the parties ... is an inappropriate justification for liability because it is inconsistent with the correlative nature of the liability.[40]

It would not do, then, to hold all defendants liable on the basis that at least one of their breaches of duty made a difference to the claimant's normal course of events. Disaggregation must follow. The NBA's second stage, which asks of each defendant whether her breach was operative at the time the claimant's damage occurred, performs this function. Crucially, whereas the first stage is concerned with the counterfactual, this second stage deals only with the factual. It is this move from the subjunctive to the indicative which enables the NBA to discriminate between factors which *could* have caused an outcome and factors which *did* cause an outcome. The significance of this step is most obvious in situations of pre-emption and, as such will be discussed in more detail in Chapter 4. As an example for present purposes, however, consider the claimant run down and killed by the careless motorcyclist. Had he not been killed by the motorcyclist, he would have drunk the

---

[39] See *Baker v Willoughby* [1970] AC 467 (HL) and Ch 3.
[40] E Weinrib, *Corrective Justice* (Oxford, OUP, 2012) 17.

coffee he was carrying. A few minutes before the accident, a trainee barista had accidentally put a lethal dose of cleaning fluid in the beverage, meaning that, even if the road accident had not killed him, the coffee would have done. On these facts, the first stage of the NBA is easily passed, since the hapless sleepy pedestrian would still have been alive but for *at least one* of the breaches of duty. Rather than being effectively absolved by the presence of the other breach, as would happen under an individualised But For analysis, both defendants are subjected to the second stage of the NBA. The counterfactual nature of the first stage merely tells us that this claimant deserves compensation;[41] it remains to be established who should pay, and for what, and this is the function of the second stage.

The avowedly factual nature of the second stage means that negligence liability will be imposed only on any defendant whose breach of duty *actually* affected the claimant in the real world. At this stage, all hypothetical worlds are ignored. They have served their purpose by showing that the claimant would have been better off without any breaches of duty, and it now falls to establish what actually made the claimant worse off *as it happened*. This allows us clearly to see that, in the example given above, the fact that the motorcyclist's breach *was* operative on the claimant when he died means that the barista's breach *was not* operative, and never could be.[42] On an NBA, therefore, the motorcyclist is liable but the barista is not. It is of course true to say that the barista's breach *would* have been operative in the hypothetical world in which the motorcyclist's breach did not exist, but this is not the world in which any of us lives, and it is not the world with which negligence actions are concerned.[43] The combined effect of stages one and two is to ensure that claimants whose histories have been adversely affected by at least one breach will be compensated, and that that compensation will come from any of those defendants whose breaches of duty have made an actual contribution to the claimant's state at trial.[44] In order for a sole defendant's breach to satisfy the NBA, that breach must have been necessary to change the claimant's course of events for the worse, and it must remain operative until the claimant's damage occurred.[45] Where the breach of a defendant who is one of multiple defendants is analysed in the same way, however, *any* breach must have been necessary to change the claimant's course of events for the worse, but *that particular defendant's* breach must have been operative.[46] A highly significant implication of this is that defendants will not be absolved from liability by the existence of other defendants whose breaches remain operative on the claimant when the damage occurs. Since the combined effect of the separate stages of the NBA is that a defendant's operative breach of duty cannot be

---

[41] Subject to 'remoteness' considerations, dealt with in NBA terms under the concept of 'operative', see Ch 3.

[42] An argument justified in detail in Ch 3.

[43] The implications of this for moral luck are discussed in Ch 4, under the heading 'Moral Luck'.

[44] For detailed discussion of how such awards will be divided up where appropriate, see Chs 5 and 6.

[45] See Ch 3 for a discussion of situations in which the breach ceases to be operative before the ultimate damage manifests itself – eg *Jobling v Associated Dairies* [1982] AC 794 (HL).

[46] And more than one breach can be operative in terms of the same damage.

causally eclipsed by other operative breaches, claimants will not be disadvantaged by being affected by more than one tort.[47]

Current Perspectives on But For Causation

Given that the NBA is predicated on the concept of But For causation, it is worth setting it alongside some of the more prominent current views on the subject.[48] What follows is by no means intended as an exhaustive analysis of those views, to which attention will be given throughout the book, but as an introductory overview of how the various treatments of But For currently compare with one another.

Stapleton

Whilst Professor Stapleton recognises that '[c]ourts throughout the world are agreed that the relation of necessity (that is, "but for") between the breach and outcome is one that the law should designate as causal',[49] she nonetheless believes that the But For test is 'notoriously inadequate'.[50] Her principal reason for reaching this conclusion is that 'the law is also concerned with factors that contribute to an outcome but which are neither necessary nor sufficient for it to exist'[51] and she offers the following illustration of her point:

> Five members of a club's governing committee unanimously, but independently and in breach of duty, vote in favour of a motion to expel member X from the club, where a majority of only three was needed under the club's rules. The vote of Committee Member Number 1 is neither necessary nor sufficient for the motion to pass. This is true of the vote of each member, yet the motion passed. Where there is a liability rule requiring proof that the vote of the individual voter was a factual 'cause' of a motion passing, the law must recognise this relation of one vote to the passing of the motion as 'causal'.[52]

As will become clear in Chapter 4, applying the NBA to these facts would reach an identical conclusion to the one advocated by Stapleton. And yet the NBA adheres to But For necessity, and Stapleton unequivocally rejects it. In common with others,[53] Stapleton discounts the benefits of a test based on necessity because of its perceived failure to deal with these overdetermined cause situations or, to use her term, situations of 'duplicate necessity'.

---

[47] See Ch 4.
[48] All of which will be considered further in the chapters which follow.
[49] Stapleton, 'Reflections on Common Sense Causation in Australia', above n 14 at 338.
[50] ibid, 339.
[51] ibid, 340.
[52] ibid, 340.
[53] Eg A Beever, *Rediscovering the Law of Negligence* (Oxford, Hart Publishing, 2009) ch 12 and Wright, 'Causation in Tort Law', above n 13.

It is only the seductive simplicity of the 'but for' test that distracts courts from enunciating an appropriately wide statement of the relation of 'factual cause': namely, that a factor is a factual cause if it contributes *in any way* to the existence of the phenomenon in issue. Courts should no longer allow the fact that in most cases the contribution in issue will have been one of necessity (but for), to mislead them into regarding necessity as the fundamental form of causal relation recognised by the law: courts should grasp that it is the relation of contribution.[54]

But For causation, however, has more than one seductive quality: not only is it simple, but it is, properly applied, also the *only* means of distinguishing those defendants who have causal relevance to the tort negligence from those who do not. For this very reason, necessity needs to remain the basic form of causal relation in negligence and, by using it to evaluate the contribution of breaches of duty *in aggregate*, the NBA manages to discriminate between those cases which are the concern of negligence, and those which are not, and resolves the potential problem of under-inclusiveness in situations of overdetermined cause. In some of her most recent work on the subject, Stapleton asserts that

> a significant source of confusion has been the failure to distinguish two distinct matters. One is the factual causation issue of whether the breach was a cause of the injury. A quite separate issue is whether this injury is one for which compensatory damages can be claimed.[55]

As outlined above, Stapleton employs an expansive initial causal inquiry, and then truncates the results of that inquiry by reference to a limiting concept of damage:

> It is a defining and well-known principle of the approach the law of torts takes to the assessment of compensatory damages that it should not make a claimant better off than he would have been had he not been the victim of tortious conduct . . . Thus, even if a defendant's breach was a cause of an injury, he is not liable to pay compensatory damages if the same or an equivalent injury would have occurred in the absence of tortious conduct . . . If the innocent contributions alone would have been sufficient to reach the threshold, the injury would still have occurred absent tortious conduct. In such a case, compensatory damages are barred because they would make the claimant better off. Though the defendant was a 'cause' of the injury, the injury did not represent 'damage'.[56]

In relation to this question of 'damage', Stapleton proposes a choice for tort law of two benchmarks:

> '[W]hether the 'no better off' principle . . . is that compensatory damages should not make a claimant better off relative to:
>
> (i) where he would have been But For the individual defendant's tortious contribution; or
>
> (ii) where he would have been But For all the tortious contributions.[57]

---

[54] Stapleton, 'Reflections on Common Sense Causation in Australia', above n 14 at 341.
[55] J Stapleton, 'Unnecessary Causes' (2013) 129 *Law Quarterly Review* 39, 54.
[56] ibid, 55.
[57] ibid, 58–59.

In her view, (ii) would be the preferable option for the law to take and this, of course, is the benchmark employed by the NBA. The difference between the two approaches, however, is that the NBA uses this benchmark at the outset, aggregating breaches in order to determine whether the case in question is one which should attract further legal scrutiny. Stapleton's approach, on the other hand, invites the law to conduct a causal inquiry *independently* of determining whether or not the claimant has suffered any damage in which the law should take an interest. Such an inquiry is at times bound to conclude that, whilst the defendant's breach was a cause of the claimant's injury, there will be no liability since the claimant has suffered no legally recognised damage. Indeed, Stapleton acknowledges this explicitly:

> See also *R (on the application of Lumba) v Secretary of State for the Home Department)* [2011] UKSC 12 ... refusing compensatory damages for false imprisonment because, absent the 'deplorable' conduct of the Home Office, the claimants would have been detained lawfully. The approach advocated in this article would recognize the conduct of the Home Office as a 'cause' of the actual imprisonment being unlawful, while also recognizing that the claimant had, as a result of the unlawfulness of that detention, suffered no 'damage' for the purposes of compensatory damages.[58]

Effectively, this merely postpones the inevitable question, 'critical to the compensatory principle, of whether the injury represents damage to the victim'.[59] If this question is so critical to the central concerns of negligence (and it is), then what is to be achieved by keeping it from centre stage, and assigning it only secondary forensic status? There is something intuitively unwieldy about insisting that the law recognise as causal a factor which, whilst contributing in physical terms to an outcome, does so where that outcome is of no concern to the tort of negligence. This is merely to insist on an extra-legal concept of 'cause', only to reinstate legal criteria at the subsequent stage of evaluating 'damage'. It would seem, however, that this is indeed the chronology of the method formulated by Stapleton.[60] It is unclear exactly what is to be achieved by separating the two issues in this way, other than perhaps to maintain a closer link between what the legal inquiry is doing and what is generally understood by the non-legal definition of cause (although it is not obvious why this should be a valid aim).

As suggested above, there is much to say in support of a process which does not require the court to ask itself potentially difficult causal questions when the answers may well turn out to be unnecessary. To conclude, as the NBA sometimes will, that a defendant's breach was not a cause of the claimant's damage, is not to make any general claim about the physical history of the universe, but only to make a specific judgement about that claimant's negligence claim. Why say 'the defendant's breach was a cause of the claimant's position at trial, but that position is not one for which the claimant can recover in negligence' when the same

---

[58] ibid, n 74.
[59] ibid, 60.
[60] ibid, 55.

conclusion can be reached simply by stating 'the defendant's breach did not cause the claimant's actionable damage'? In negligence terms, the causal inquiry is a very specific one: its aim is to determine whether there is a causal link between a claimant's legally recognised damage and a defendant's breach. The answers to the anterior questions, therefore, of whether the defendants have breached their duty, and whether the claimant has suffered actionable damage, themselves set the parameters of the causal inquiry.

Wright

Professor Wright explains his version of the NESS test, as it stands today, in the following terms:

> When analysing singular instances of causation, an actual condition c was a cause of an actual condition e if and only if c was a part of ... the instantiation of one of the abstract conditions in the completely instantiated antecedent of a causal law, the consequent of which was instantiated by e immediately after the complete instantiation of its antecedent, or (as is more often the case) if c is connected to e through a sequence of such instantiations of causal laws.[61]

In Wright's account, a causal law is:

> an empirically derived statement that describes a successional relation between a set of abstract conditions (properties or features of possible events and states of affairs in our real world) that constitute the antecedent and one or more specified conditions of a distinct abstract event or state of affairs that constitute the consequent such that, regardless of the state of any other conditions, the instantiation of all the conditions in the antecedent entails the immediate instantiation of the consequent, which would not be entailed if less than all of the conditions in the antecedent were instantiated.[62]

Ingenious and intellectually rigorous though this is, it is easy to see why it has not been widely adopted as a forensic tool.[63] Neither does it give the impression of being able to promote the transparency or accessibility of judicial reasoning, albeit that it would no doubt lead to consistency of principle, were it to be applied properly in every case. In actual fact, the idea is quite simple: under NESS, a condition is a cause if and only if it forms part of any sequence of conditions which is *minimally* sufficient for the event in question to occur as soon as that sequence has concluded, regardless of the state of any other conditions. Thus, the NESS test is capable of identifying all factual contributing causes to an event, and its analysis is not derailed by the existence of competing or duplicative factors, but it is in this self-consciously comprehensive remit that it differs most from the NBA:

> Just as Newtonian mechanics serves as an adequate substitute for the more accurate and comprehensive theories of relativity and quantum mechanics in ordinary physical situ-

---

[61] Wright, 'The NESS Account', above n 29 at 291 (footnotes omitted).
[62] ibid, 289 (footnotes omitted).
[63] See Hoffmann, 'Causation', above n 37.

ations, the but-for test serves as an adequate substitute for the NESS test in ordinary causal situations . . . however, the substitute must give way to the more accurate and comprehensive concept when the situation is more subtle and complex.[64]

The best way of explaining how the NESS test and the NBA compare with one another is by describing the former as longitudinal and algorithmic and the latter as latitudinal and explanatory. They relate to one another, therefore, in a perpendicular way, which is to say that they neither conflict with, nor duplicate, the work of the other. Wright characterises his approach as one which 'captures the essence of causation and gives it a comprehensive specification and meaning'.[65] Presuming that this was his aim in developing the NESS test, his project is a broader one than that which is the subject of this book because, as has been made clear, the NBA does not aim to define causation as a generic concept. Rather, its objective is to provide a means of identifying legally relevant causes in particular instances. Its results are therefore unlikely to be at odds with those of the NESS test, but will constitute only a subset of the factors identified as NESS causes. Wright's NESS test focuses on differentiating itself from the *sine qua non* approach on the basis that it is able to deal better with overdetermined cause situations: 'The NESS test collapses into the but-for test if there was only one set of conditions that was actually or hypothetically sufficient for the occurrence of the result on a particular occasion'.[66]

But NESS is a generalistic account; necessarily so since it deals with duplicative causes by disaggregating sets of factors, each of which *would* have been sufficient in the absence of the others. This goes some way towards explaining the use of 'longitudinal' above: the NESS method works by dissecting situations in such a way that duplicative factors are separated from one another, and placed in 'vertical' sets, each a timeline of events which includes a NESS factor and the result in question. Since these sets can, and do, operate in parallel, the analysis is capable of identifying a plurality of causes for the same result. In philosophical terms, there is not much with which to quibble. In terms of the causal inquiry in negligence, however, the method does too much, and therefore achieves too little. What is required instead is more of a particularised (and therefore latitudinal) approach, which has no need for a theoretically comprehensive account of the factual contributions to the claimant's adverse outcome. So, where the NESS approach identifies those factors which are necessary elements of *a* set of factors sufficient to bring about the result in question, the NBA selects those factors which are necessary to the set of factors[67] presented to the court. There will be some overlap in results, but those reached by the NESS method will outnumber those chosen by the NBA.

Wright argues that tests which insist on necessity in a specific and limited set of circumstances in this way are incapable of accommodating instances of

---

[64] Wright, 'Causation in Tort Law', above n 13 at 1792.
[65] Wright, 'The NESS Account', above n 29 at 322.
[66] R Wright, 'Causation, Responsibility, Risk, Probability, Naked Statistics, and Proof: Pruning the Bramble Bush by Clarifying the Concepts' (1988) 73 *Iowa Law Review* 1001, 1021.
[67] Or all of the sets, where applicable.

duplicative causation.[68] The NBA shows that this need not be true. The difference, however, between the NESS and the NBA might not be as significant as the comparison thus far would suggest. Wright is insistently clear at every point of his exposition that NESS is an avowedly factual, empirical inquiry, and is therefore not wholly dispositive of legal causal questions.[69] Like Stapleton, Wright believes that the factual aspect of causation needs to be separated from the normative question of whether, and to what extent, liability should attach to a factual cause. NESS is intended to provide an answer only to the factual inquiry. A direct comparison between the two tests, therefore, needs to be carried out with caution, since the NBA is not intended as a test of avowedly 'factual' causation, but simply as an entire means of identifying legally relevant causes. In this sense, it attributes very little significance to the boundary between the factual and normative inquiries which some hold so dear. The most that needs to be said of the NBA in these terms is that it is a normatively constructed factual inquiry, in the form recognised by Fumerton and Kress:

> [E]ven if we are clear about the distinction between factual and normative issues, another distinction must be made to avoid potential confusion. It may be that determinations of law involve nonnormative matters of fact, but our decision to make a given nonnormative fact relevant to a finding of law is itself grounded in normative considerations. We should distinguish between two levels here. In the first level, we operate as lawyers, and deploy a notion of actual causation that is nonnormatively factual or empirically based. At a second, or meta-level, we justify our deployment of a factual, nonnormative concept of actual causation at the first level, the level of the actual practice of lawyers and courts. This meta-level justification may be fully normative ... Matters become ... complicated ... once we reflect on the fact that disputes arise over causation that do not seem to have much to do with empirical matters. You were carelessly driving down a street while I was crossing against a red light ... Who was causally responsible for my injuries? ... Is it not plausible to suppose that our choice to describe one of the other of these acts as '*the* cause' is guided by our pre-theoretical intuitions about who is responsible? Furthermore, this notion of responsibility goes beyond matters of empirical fact, and reintroduces normative concepts.[70]

Stevens

Professor Stevens, in departing from the more traditional 'loss model' of tort, equates the infringement of a right with the gist of a negligence claim.[71] The 'rights' which form the basis of his analysis are defined essentially as specific claim rights of Hohfeldian derivation.[72] In his analysis, Stevens distinguishes between those situations in which damages should be awarded in recognition of the fact that a

---

[68] Wright, 'Causation in Tort Law', above n 13 at 1791.
[69] Wright, 'The NESS Account', above n 29 at 305.
[70] R Fumerton and K Kress, 'Causation and the Law: Preemption, Lawful Sufficiency, and Causal Sufficiency' (2001) 64 *Law and Contemporary Problems* 83, 86–87 (footnotes omitted).
[71] R Stevens, *Torts and Rights* (Oxford, OUP, 2007) 72.
[72] ibid, 4.

right has been infringed[73] but no consequential damage has ensued, and those in which damages should be awarded for the infringement of the right *as well as* for damage consequential on that infringement:

> In the assessment of damages it is necessary to distinguish between damages awarded as a substitute for the right infringed and consequential damages as compensation for loss to the claimant, or gain to the defendant, consequent upon this infringement. Damages which are substitutive for the right which has been infringed are assessed objectively, save where the infringement is particularly egregious ... Damages are awarded even if there is no loss to the claimant or gain to the defendant consequent upon the infringement of the right ... The distinction between damages which are awarded as a substitute for the right and those awarded to compensate for consequential loss can be obscured because in most cases the value attached to the right is precisely the same as the loss suffered, usually financial, by the claimant ... it is a mistake to think that where no loss is suffered no claim for damages is available.[74]

This conceptualisation of the law of torts (his analysis is not limited to negligence) means that Stevens, whilst not being a critic of the But For test in form, employs it in a very different manner to the way in which it has traditionally been used, by using it to ask a different question or, more accurately, two questions: 'the first question is "has the defendant by his actions infringed the claimant's right ...?" ... The second question is "but for the infringement of the right, what loss would the claimant not have suffered?".'[75]

The malleability of Stevens' concept of a right makes it hard to pinpoint exactly what his understanding of causation in this context is. Take, for instance, the famous desert traveller situation, in which A sets off across the desert with a water bottle. B puts a fatal dose of poison in the bottle and then C, acting independently, empties the water bottle. A consequently dies of thirst.[76] In response to this, Stevens proffers two alternative resolutions: one in which both imposters are liable, and one in which, it appears, only the second imposter is liable. In the first:

> [W]hen B poured the poison into A's water keg, A acquired a right against him that B would take steps to prevent him from drinking. Specifically, B is under an ongoing duty to warn A. Similarly, C is under an ongoing duty to warn A that his keg is empty and to prevent him setting off with no water supply. If either B or C had warned A just before his setting off, A would have discovered that he had no water and would not have died. Therefore it can be said against either B or C independently 'but for your failure to warn A, A would not have died'.[77]

And yet, in the second:

> We must ask, looking back from the perspective of the time of death, whose actions killed A ... At the time of death, the only operative cause of death was the lack of water.

---

[73] Which he terms 'substitutive damages', in contradistinction to consequential damages.
[74] Stevens, *Torts and Rights*, above n 71 at 61.
[75] ibid, 134.
[76] J McLaughlin, 'Proximate Cause' (1925–26) 39 *Harvard Law Review* 149, 155.
[77] Stevens, *Torts and Rights*, above n 71 at 132.

If the poisoner's actions had killed the man he would have been dead already. Judged at the moment of death, the action of the poisoner was a 'shot' which missed.[78]

The second conclusion is also the one that would be reached on an NBA.[79] The NBA would not, however, produce the result reached in the first of Stevens' analyses here, because the poisoner's breach was no longer operative when the damage occurred, meaning that B would be dismissed as a legally irrelevant cause by the second stage of that analysis. Whilst, therefore, Stevens' rights model and the NBA have in common a central role assigned to the But For test, there is much to distinguish between them. First, it is hard to see how useful the But For test can be to any theory within which different results can be reached by employing the same test to the same facts. It arguably detracts from the value, and even the essence, of a right, to make its existence conditional upon a particular interpretation of a set of facts. The other way in which the two approaches differ relates once more to the NBA's aggregative process. (Since Stevens' thesis is presented in terms of the law of torts in general, the following example refers to the tort of nuisance. Its causal structure, however, applies with equal force to negligence.)

> Difficult cases arise where the defendant's actions only amount to an infringement of the claimant's rights when in combination with the actions of others, each of whom is acting independently. If, for example, a public nuisance, would be created by adding over 99 gallons of pollutant into a river, if 10 defendants add 10 gallons each, it can be said against any one individually that but for his actions the nuisance would not have occurred. If, however, there are 11, or more, defendants who add 10 gallons of pollutant it cannot be said as against anyone individually that but for his actions the nuisance would not have occurred. However, this makes no difference to the result. The defendant's actions in combination with the acts of any nine others constitute an infringement of the claimant's right not to have a nuisance committed, regardless of whether there are 10 polluters of the river or 50. *As long as the amount of pollution exceeds 100 gallons as a result of human action, the claimant's right to use the river has been infringed.*[80]

Whilst, on these particular facts, the NBA would reach the same conclusion as Stevens' rights model, it would not do so for the same reason: Stevens' approach requires that the 100-gallon threshold must be reached 'as a result of human action', whereas the NBA would require that the 100-gallon threshold (which was the level of pollution required to change the course of events from a world in which there was no nuisance into a world in which there is a nuisance) must be reached *as a result of at least one breach of duty.* This is, of course, a narrower requirement than the one framed by Stevens but is, as the name suggests, fundamental to an NBA: unless a legally relevant action (ie a breach of a duty where both nuisance and negligence are concerned) has made a difference to the claimant's normal course of events, the case is not the concern of tort. So, courses of events influenced only by human actions which do not amount to legal wrongs will not

---

[78] ibid, 136.
[79] Explained in detail in Ch 4.
[80] Stevens, *Torts and Rights*, above n 71 at 136.

satisfy the first stage of the NBA. This divergence between the two approaches is even clearer in Stevens' later modification of the example:

> Once 99 gallons of pollutant are added to the river, everyone who contributes to the quantum of pollution infringes the rights of others who make use of this public good. Further, it would not matter if 200 gallons are added by an act of God, and only one gallon by the defendant, every gallon over 99 gallons which is added is a fresh infringement of the rights of others to use the river . . . In this example of a nuisance, the harm is divisible, and each defendant should only be liable for the proportion of his contribution.[81]

If 200 gallons of pollutant had been added by an act of God in a situation judged on the basis of the NBA, there would be no liability because no breach was necessary to alter the claimant's normal course of events.[82] Stevens' opposing conclusion on this point is facilitated by his identification and quantification of what constitutes a right and its infringement: a right (here, the right not to have a nuisance committed) exists in the singular, but can be infringed by more than one agent.[83] So, on the basis of the last pollutant example given above, a claimant could sue under Stevens' model, for the less than 0.5 per cent contribution that a defendant made to the infringement of her right. The reason for this position is that Stevens, as we saw above, employs the But For test only at the second stage of his enquiry, to establish whether, but for that infringement, any damage would have occurred. This means necessity is required under his model for the award of compensatory damages for consequential loss, but not for substitutive damages for the infringement of the right. The practical upshot of this is that, as in the pollutant example, claimants would be able to consume a potentially enormous amount of resources, both public and private, to pursue a sum which might well only end up as a tiny fraction, not only of their entire loss, but also of the cost of bringing their claim. This rather prosaic point mounts a considerable challenge to such a rights-based theory, elegant as it is in theory, as a means of deciding cases in practice.

Beever

Beever is another critic of the But For test. He argues that 'causation is centrally concerned with sufficiency rather than with necessity',[84] and explains this conclusion in the following terms:

> Something is *a* cause of an effect if that something was an element of the set that was *the* cause of the event. To spell this out in more detail, we might say that that C is *a* cause of E if and only if C was a member of the set of conditions that were sufficient for E and led to E. Any test inconsistent with this definition must be mistaken. The but for, targeted

---

[81] ibid, 137–38.
[82] And the breach of duty concerned was not severally sufficient – see Ch 4 under sub-heading 'Several Sufficiency'. See also Stapleton, 'Unnecessary Causes', above n 55 at 55.
[83] Stevens, *Torts and Rights*, above n 71 at 133.
[84] A Beever, *Rediscovering the Law of Negligence* (Oxford, Hart Publishing, 2009) 417.

but for[85] and NESS test are examples. This is because they adopt, in different ways, the strategy of using a test of necessity to determine whether an alleged event was a member of the set of events that led to an effect. But there is no reason to think that necessity is an appropriate test. In fact, in the light of the problems that flow from overdetermination, there is every reason to think that it is not. Accordingly, we must conclude not only that the but for, targeted but for and NESS tests are flawed but that any test incorporating necessity must also be.[86]

In common with other critics of the But For test, Beever is here evaluating it as a means of identifying a generic, physical cause, as opposed to a cause of specific legal relevance. As a general proposition, it is obvious that one event need not be necessary for the occurrence of another in order for it to qualify as a cause of that event, but this is too promiscuous a question for the negligence inquiry. Take, for example, a case where the normal vicissitudes of a claimant's life created conditions that were themselves sufficient to bring about the damage of which the claimant complains.[87] The defendant's breach also occurred, and so happened to constitute part of that set. On such facts, although the breach was part of a set sufficient for the deleterious outcome, it was not necessary for the sufficiency of that set because the same course of events would have happened anyway. This is not to deny that the breach had any involvement in the occurrence of the damage, but only to deny that its contribution should have any legal relevance in the negligence context. If it is an incidental, rather than instrumental, factor, in other words, it is not something which requires a legal response.

In any event, despite Beever's hostility towards the But For test, his assertion that causation is 'centrally concerned' with sufficiency rather than with necessity is not incompatible with a *sine qua non* approach, as Wright makes clear:

> Contrary to what many assume, the *sine qua non* analysis relies on an embedded analysis of (lawful rather than causal) sufficiency. To determine if some condition was strongly necessary for the occurrence of some consequence that actually occurred, one must 'rope off' the condition at issue and then, using the relevant causal generalisations, determine whether the remaining existing conditions were lawfully sufficient for the occurrence of the consequence – that is, whether the relevant causal laws would have been fully instantiated in the absence of the condition at issue. If they would have been, the condition at issue was not strongly necessary for the occurrence of the consequence.[88]

Beever's purported rejection of necessity in favour of sufficiency (and his apparent belief that the two are mutually exclusive causal criteria) renders his theory of limited use to a *legal* inquiry. As Wright observes of Beever, he 'rejects any necessity restriction in causal analysis and instead merely requires that a condition be a

---

[85] A formulation offered by Jane Stapleton in J Stapleton, 'Perspectives on Causation' in J Horder (ed), *Oxford Essays in Jurisprudence* (Oxford, OUP, 2000) and J Stapleton, 'Legal Cause: Cause-in-Fact and the Scope of Liability for Consequences' (2001) 54 *Vanderbilt Law Review* 941.
[86] Beever, *Rediscovering the Law of Negligence*, above n 84 at 425–26.
[87] At the time when it did. For the importance of the temporal dimension, see Ch 5, in discussion of the first stage of the NBA.
[88] Wright, 'The NESS Account', above n 29 at 291–92.

member of a sufficient set. This "weak sufficiency" analysis opens the door to treating every condition as a cause'.[89] What is more, Beever's reference to overdetermination suggests that he, like others, has made the mistake of judging a foundational element of negligence on the basis of a single, and relatively unusual, set of facts. As has been suggested here, and as Chapter 4 on Duplicative Causation will show, this is unnecessary.

It is also generally accepted, even by those searching for an alternative to But For causation, that necessity is both effective and important outside of overdetermined cause situations, meaning that it is only problematic, even from their point of view, within a very narrow fact range.[90] Beever, on the other hand, is of the opinion that the But For test is never effective, claiming that 'the law adopts for the most part an inappropriate test for factual causation, but, when that test generates absurd results, relies on an alternative that is effectively meaningless. This is not good'.[91]

It is difficult to see, however, how a test for causation which completely eschews any necessity requirement (or claims to) can be compatible with the corrective justice which is integral to Beever's 'principled approach' to the tort of negligence. Causation is, according to Beever, 'an essential element of corrective justice'[92] for the following reason:

> The law of negligence seeks to remedy wrongs committed by one person against another. I have expressed this earlier by saying that the defendant will be liable if and only if he created an unreasonable risk of the actual injury that the claimant suffered. But if the defendant did not cause the claimant's injury, then the claimant did not suffer an injury *as the result of the defendant's wrongdoing*. Hence, the law of negligence is uninterested in activities that do not result in injury.[93] (emphasis supplied)

The italicised words highlight a point crucial to corrective justice principles.[94] Yet, in order to ensure that only claimants whose injuries *do* result from a defendant's wrongdoing recover in tort, the role of causation must do more than merely identify a defendant's actions as being part of a set of conditions that would have been sufficient for the damage to occur in any event. Recovery in negligence by any claimant whose injury would have occurred as a result of tortiously irrelevant factors would amount to a windfall for that claimant, effectively rendering her better off as a result of the tort than she would have been in the hypothetical world in

---

[89] ibid, 291, n 40.
[90] See, for instance, ibid, 291.
[91] Beever, *Rediscovering the Law of Negligence*, above n 84 at 417.
[92] ibid, 413.
[93] ibid, 414.
[94] As identified above, text to n 40, and in contrast to other principles, such as distributive justice, first party insurance or state compensation schemes. See, inter alia, P Cane, 'Distributive Justice in Tort Law' [2001] *New Zealand Law Review* 401; J Stapleton, 'Torts, Insurance and Ideology' (1995) 58 *MLR* 820; R Mullender, 'Corrective Justice, Distributive Justice, and the Law of Negligence' (2001) 17 *Professional Negligence* 35; J Coleman, 'The Practice of Corrective Justice' in Owen (ed), *Philosophical Foundations of Tort Law*, above n 1.

which the tort did not occur, but in which she was injured anyway.[95] The corollary of this overcompensation of the claimant is, of course, excessive liability for the defendant, who is being asked to pay for damage which she did not cause. The key to avoiding such a situation, and so preserving the link between wrongdoing and injury, is the criterion of necessity. Without this, any factor is, *in legal terms*, not causative but merely an also-ran, because the damage in question was already inevitable when it occurred. Beever acknowledges that the common law recognises this, meaning that on the following facts, as stated by him, D would not be found liable: 'D and B injured the claimant. Had either D or B not acted, the claimant would have been injured nevertheless'.[96] It is in Beever's treatment of this situation that we first see how he accommodates his corrective justice commitment within a theory of causation which ostensibly has no room for necessity. His approach involves two queries: 'Should the defendant be liable and, if so, what is the quantum of damages that the defendant must pay?'[97] His answer to the first question is that the defendant should be liable because he 'was prima facie liable and a factual cause of the claimant's personal injury'.[98] In response to the second, he concludes that 'the claimant would have suffered exactly the same injury had the defendant not acted negligently. Hence, the quantum of damages is zero. In other words, the claimant is entitled only to nominal damages'.[99] This distinction is impossible to make without reference to necessity. Rather than using it, however, as a conventional *sine qua non* approach does, to distinguish between liability and no liability, Beever uses it to distinguish between liability for nominal damages and liability for substantial damages. The difference between Beever's 'principled approach' and the NBA lies once more in what they understand the gist of a negligence claim to be. Beever believes that a non-necessary but sufficient factor should lead to liability because the defendant responsible for that factor 'violated the claimant's rights causing factual injury [but] in doing so the defendant left the claimant no worse off than the claimant would have been had the defendant not acted negligently'.[100] The principled approach, therefore, in common with Stevens' rights model, regards the gist of a negligence claim as the infringement of a right *eo ipso*.[101] The NBA, on the other hand, would discourage such activity by refusing to recognise liability even where a claimant's right has been infringed, unless it has resulted in a departure from the claimant's normal course of events.[102]

It would seem, therefore, that the criterion of necessity is not as alien to the principled approach as Beever makes out: although it is certainly true that it plays

---

[95] See *Snell v Farrell* [1990] 2 SCR 311 (SCC) at 327 (per Sopinka J).
[96] Beever, *Rediscovering the Law of Negligence*, above n 84 at 435. In a footnote to this example, however, he also describes D and B as being independently sufficient (which they are) and jointly necessary (which they are not).
[97] ibid, 435.
[98] ibid, 435.
[99] ibid, 435.
[100] ibid, 436.
[101] See above, text to n 71.
[102] See Ch 7.

no role in distinguishing between liability and no liability, it still performs the highly significant task of discriminating between those liable for substantial damages and those who, in Beever's view, should have to pay something, but not very much. Since this last position is not currently the one adopted by the common law, Beever's theory of the tort of negligence does not, in its purported rejection of necessity, accurately explain or reflect that tort as it is. There is a gap between reality and the principled approach, and it is a necessity-shaped hole.

The purpose of this comparative exercise was to put the objectives of the NBA in context by providing an overview of how they relate to current perspectives on But For causation. The chapters which follow aim to demonstrate how that analysis functions in practice, and what it can contribute to the causal inquiry in negligence.

3

Basic Principles

Illustrative Cases: *Wilsher, Performance Cars, Baker, Jobling*

Basic Causal Principles

Many of the perceived difficulties with causation in negligence can be either minimised or eliminated through a proper application of basic causal principles. Confusion has arisen where these principles have been overlooked in favour of unnecessarily nuanced analysis. At their most elemental, these principles are as follows:

- A defendant will only be liable where she has, on the balance of probabilities, made a difference to the claimant's normal course of events.
- A defendant will only be liable for that difference which, on the balance of probabilities, she can be determined to have made to the claimant's course of events.[1]
- A defendant is entitled to take her victim as she finds her at the time of her breach of duty[2] (and this can work to both the advantage and disadvantage of the defendant).

These principles, adhered to in combination with an NBA, will enable the causal inquiry to reach answers which are both consistent and predictable, regardless of the fact pattern to which they are applied. Chapter 2 has already discussed the first of these, and outlined why a breach of duty has to have made a difference to the claimant's normal course of events if it is to be considered by the law to be causally relevant. This is a criterion which, in spite of (or perhaps because of) its simplicity, is prone to be discounted too readily as a useful guide for decision making. More specifically, it has recently been challenged by repeated attempts to misapply the material contribution to injury analysis (discussed in Chapter 5) and/or to extend

---

[1] This might seem like overly delicate wording for a basic principle, but it accommodates those defendants who, by materially contributing to an injury which is indivisible (at least in practice), are held liable for the whole of that injury. See Ch 5 for a more detailed discussion.

[2] This is to accommodate the idea that only pre-existing characteristics can be taken into account in assessing consequential damages, and not anything which develops independently subsequent to the tort. See HLA Hart and A Honoré, *Causation in the Law*, 2nd edn (Oxford, OUP, 1985) 343.

the exceptional material contribution to risk analysis (discussed in Chapter 6). Neither is appropriate.[3]

Defendant Only Liable Where Her Breach Has, on the Balance of Probabilities, Made a Difference to the Claimant's Normal Course of Events

An example of a decision made on the basis of this fundamental principle is *Wilsher v Essex Area Health Authority*.[4] The claimant in this case had been born almost three months early. As a result of his premature birth, he had suffered in his first few days from a number of conditions which often affect premature babies. In addition to these unfortunate, but naturally arising, conditions, the defendant's hospital also breached its duty of care to the claimant during his treatment in its special care baby unit. As a result of the negligent insertion of a catheter into an umbilical vein instead of an artery, too much oxygen was administered to baby Wilsher. Shortly after his birth, the claimant developed a condition called Retrolental Fibroplasia, or RLF, which left him blind in one eye and with severely impaired vision in the other. The specific problem for the causal inquiry was that RLF could have been caused by *any* of the naturally arising conditions from which the claimant suffered, or by the excess oxygen to which he was exposed by the defendant's breach of duty. Whilst both the trial judge and the Court of Appeal were satisfied that a sufficient causal link existed between the defendant's breach and the RLF, the House of Lords, with good reason, disagreed.[5] Appellate courts have rightly adhered to this decision ever since, despite the various challenges which orthodox causal principles have faced during the years since it was decided.[6] This might be due in part to the fact that the reasoning in *Wilsher* itself makes it very clear both why it was a case in which the claimant had failed to establish causation, and why neither a material contribution to injury nor material contribution to risk analysis was appropriate on the facts, despite both being raised by counsel therein.

The crucial element in *Wilsher*, which sets it apart from either of these material contribution analyses, is the fact that there was no specific causal relationship established between the agency of the defendant's breach of duty and the claimant's injury. It was, therefore, truly a case of indeterminate *cause*, for which there

---

[3] For an object lesson in this, see the judgment of the Supreme Court of Canada in *Resurfice v Hanke* [2007] SCC 7, [2007] 1 SCR 333.
[4] *Wilsher v Essex Area Health Authority* [1988] AC 1074 (HL).
[5] The House of Lords did not dismiss the case against the defendant, but rather ordered a retrial of the causation issue in front of a different judge. It concluded that the trial judge had misdirected himself on the correct test to apply to the evidence, but that it would be inappropriate, given the complexity of the evidence and the extent of the disagreements between the relevant experts, for an appellate court, with access only to a transcript, to come to any final decision on liability based on causation.
[6] Such as the *Fairchild* (*Fairchild v Glenhaven Funeral Services Ltd* [2002] UKHL 22, [2003] 1 AC 32) line of cases, for discussion of which see Ch 6.

can be no liability on any But For basis. In *Bonnington Castings v Wardlaw*,[7] however, and the line of cases which followed it along 'material contribution to injury' lines, the indeterminacy did not relate to the *causal agent* of the relevant damage. In *Wardlaw* itself, for example, the causal agent which brought about the pneumoconiosis was accepted as being silica dust, and it was only the *proportion* of that dust which resulted from the breach of duty that was indeterminate. Similarly, in the material contribution to risk cases, such as *McGhee v National Coal Board*,[8] the causal agent linked to the claimant's dermatitis was brick dust, and the indeterminacy on these facts, although more extensive, did not relate to the causal agent itself, but to the source/s from which that substance came. There is, therefore, on both material contribution analyses, a specific causal link established between a particular agent and the injury of which the claimant complains. This is what was lacking on the facts of *Wilsher*.[9] In the House of Lords, Lord Bridge recognised this in stating that he was 'quite unable to find any fault' with the following dissenting judgment of Sir Nicholas Browne-Wilkinson VC in the court below:

> To apply the principle in *McGhee v National Coal Board* [1973] 1 WLR 1 to the present case would constitute an extension of that principle. In the *McGhee case* there was no doubt that the pursuer's dermatitis was physically caused by brick dust: the only question was whether the continued presence of such brick dust on the pursuer's skin after the time when he should have been provided with a shower caused or materially contributed to the dermatitis which he contracted. There was only one possible agent which could have caused the dermatitis, viz., brick dust, and there was no doubt that the dermatitis from which he suffered was caused by that brick dust.
>
> In the present case the question is different. There are a number of different agents which could have caused the RLF. Excess oxygen was one of them. The defendants failed to take reasonable precautions to prevent one of the possible causative agents (e.g. excess oxygen) from causing RLF. But no one can tell in this case whether excess oxygen did or did not cause or contribute to the RLF suffered by the plaintiff. The plaintiff's RLF may have been caused by some completely different agent or agents, e.g. hypercarbia, intraventricular haemorrhage, apnoea or patent ductus arteriosus. In addition to oxygen, each of those conditions has been implicated as a possible cause of RLF. This baby suffered from each of those conditions at various times in the first two months of his life. There is no satisfactory evidence that excess oxygen is more likely than any of those other four candidates to have caused RLF in this baby. To my mind, the occurrence of RLF following a failure to take a necessary precaution to prevent excess oxygen causing RLF provides no evidence and raises no presumption that it was excess oxygen rather than one or more of the four other possible agents which caused or contributed to RLF in this case . . . A failure to take preventative measures against one out of five possible causes is no evidence as to which of those five caused the injury.[10]

---

[7] *Bonnington Castings v Wardlaw* [1956] AC 613 (HL). This line of cases is discussed in detail in Ch 5.
[8] *McGhee v National Coal Board* [1972] 3 All ER 1008 (HL). This line of cases is discussed in Ch 6.
[9] [1988] AC 1074 (HL) at 1090.
[10] [1987] QB 730 (CA) at 779 (per Sir Nicholas Browne-Wilkinson VC).

This principle was vindicated in *AB & Ors v Ministry of Defence*,[11] a group action brought by ex-servicemen who had been involved in nuclear tests carried out by the Ministry of Defence in the 1950s. Their contention was that the radiation to which they had been exposed during those tests had caused or contributed to the various illnesses, injuries and disabilities from which they had suffered in recent years, some resulting in death. There were two principal issues involved in the appeal – one relating to limitation and the other to causation. Whilst the former of these took up by far the greater part of the courts' time, the causation issue was dealt with in a way which was as convincing as it was concise. In essence, the claimants' case on causation was that, despite the fact that they had not managed to establish a link between the radiation and their injuries, and that there were other idiopathic and/or environmental factors[12] which could have brought about their conditions, they could nevertheless satisfy the causation requirement in negligence on the basis of either material contribution to injury, or material contribution to risk. They conceded that, in order to succeed on the latter point, the exceptional principle in *Fairchild v Glenhaven Funeral Services Ltd*[13] would have to be extended. Quite correctly, both arguments were given short shrift by both the Court of Appeal and the Supreme Court. The Court of Appeal considered the causation issue to a somewhat greater extent, and its reasoning was both clear and succinct where the applicability of this basic principle was concerned:[14]

> So, we conclude that there is no foreseeable possibility that the Supreme Court would be willing to extend the *Fairchild* exception so as to cover conditions such as we are here concerned with, which have multiple potential causes some of which have not even been identified. We reject as highly unlikely the suggestion that the Supreme Court might be prepared, on policy grounds, to extend the exception well beyond that which was contemplated at the time of *Fairchild* or *Barker*. We say that because, to effect such a change would be to upset completely the long established principle on which proof of causation is based. It is true that *Fairchild* itself made a small inroad into that principle. The inroad is slight and there were strong policy reasons for it. But the inroad applies only to cases where the cause of the condition is known. It does not apply where the cause is unknown. Here the causes of the Claimants' conditions are not known. All that can be said in these cases is that radiation exposure is one of several possible causes.[15]

By the time the appeal reached the Supreme Court, it had already handed down its own judgment in *Sienkiewicz v Greif*,[16] which emphatically limited the reach of

---

[11] *AB & Ors v Ministry of Defence* [2012] UKSC 9, [2013] 1 AC 78.
[12] For instance, many of the servicemen were suffering from various types of cancer which are relatively common amongst men of comparable ages who were not involved in the nuclear tests.
[13] *Fairchild v Glenhaven Funeral Services Ltd* [2002] UKHL 22, [2003] 1 AC 32.
[14] Unfortunately, several other issues were confused in the dicta of Smith LJ: [2010] EWCA Civ 1317, (2011) 117 BMLR 101 at [134] 140–41, which describe the material contribution to injury analysis, as applied in *Bonnington Castings v Wardlaw* [1956] AC 613 (HL), as being a modification of the But For test; and the apparent acceptance of the 'doubling of the risk' (DTR) test at [140] 142, [146] 143 and [151] 144. For a criticism of both approaches, see Ch 5.
[15] [2010] EWCA Civ 1317, (2011) 117 BMLR 101 at [154] 145.
[16] *Sienkiewicz v Greif* [2011] UKSC 10, [2011] 2 AC 229.

36 Basic Principles

the *Fairchild* exception. This may well explain why its dismissal of the claimant's causation point took up only a tiny fraction of the overall judgment:

> The most that the veterans as a group are currently in a position to establish is that there is a possibility that some of them were exposed to a raised, albeit low level, of fallout radiation and that this may have increased the risk of contracting some at least of the injuries in respect of which they claim. This falls well short of establishing causation according to established principles of English law. Foskett J was prepared to contemplate the possibility that the Supreme Court would extend the principle in *Fairchild v Glenhaven Funeral Services Ltd* [2003] 1 AC 32 so as to equate causing an increase of risk with causing injury. The Court of Appeal at para 154 held that there was no foreseeable possibility of this. In the light of the observations of this court in *Sienkiewicz v Greif* [2011] 2 AC 229, the Court of Appeal was plainly correct.[17]

This is the principle enshrined in the first stage of the NBA. In *Wilsher* and in *AB v MOD*, the claimants failed to prove on the balance of probabilities that, but for at least one defendant's breach of duty, their damage would not have happened. Fortunately, then, the result in *AB v MOD* does no harm to basic causation principles, despite the twofold opportunity for it to have done so. Almost as crucial as the outcome itself, however, is the fact that both appellate courts recognised how easy it was to answer the causal question they were being asked, and neither decision muddies the water by entertaining unnecessary causal complications.

Defendant Only Liable to the Extent of the Difference Her Breach Has Made

The second basic principle of causation in negligence is that a defendant will only be liable for the measure of difference which, on the balance of probabilities, she can be determined to have made to the claimant's normal course of events.[18] There are several clear articulations of this principle in well-known cases dealing with causation. In *Performance Cars Ltd v Abraham*,[19] for instance, the defendant negligently collided with the claimant's Rolls Royce, thereby causing damage necessitating a respray. The car had, however, already been damaged in an earlier collision with a third party (against whom the claimants had an unsatisfied judgment). Evershed MR regarded the question of the extent of the defendant's liability as 'interesting and novel',[20] and dealt with it thus:

> I have in the end felt compelled to the conclusion that the necessity for respraying was not the result of the defendant's wrongdoing because that necessity already existed. The

---

[17] [2012] UKSC 9, [2013] 1 AC 78 at [157] 137 (per Lord Phillips). This paragraph is the most substantial evaluation of the issue in the entire judgment. Given the importance of the limitation issue, and the obvious answer to the causal question, this is unsurprising.

[18] But see the rule where a defendant's breach has made a material contribution to an indivisible injury (see Ch 5) or made a material contribution to the risk of the claimant suffering an injury (see Ch 6).

[19] *Performance Cars Ltd v Abraham* [1962] 1 QB 33 (CA).

[20] [1962] 1 QB 33 (CA) at 37.

Rolls Royce, when the defendant struck it, was in a condition which already required that it should be resprayed in any event . . . In my judgment in the present case the defendant should be taken to have injured a motor-car that was already in certain respects (that is, in respect of the need for respraying) injured; with the result that to the extent of that need or injury the damage claimed did not flow from the defendant's wrongdoing.[21]

In *Dingle v Associated Newspapers*,[22] a libel case, but one in which Lord Devlin also discussed the principles of damages for personal injury, his Lordship stated that where

> there is not one indivisible injury but two separate injuries, the second wrongdoer – at any rate in cases where the legal consequences of the first injury are complete and ascertained before the second injury is done – is liable only for the excess of damage done by the second injury. That excess is not necessarily smaller than the damage done by the first injury. The earning capacity of a man who has lost an eye may be diminished by perhaps 10 per cent. If he loses his other eye, his earning capacity will probably disappear altogether. A defendant who is responsible for the loss of the second eye will have to pay for much more than half of the consequences of total blindness. In the example I have taken, he will have to pay for 90 per cent of the loss of earnings; but he will not have to pay the full 100 per cent because he can properly plead that the plaintiff's earning capacity was already damaged.[23]

The claimants in *Thompson v Smiths Shiprepairers (North Shields) Ltd*[24] suffered from deafness as a result of industrial exposure by several different employers. Mustill J (as he then was) offered the following succinct account of this basic point:

> It would be an injustice to employer B to make him liable for damage already done before he had any connection with the plaintiff. His liability, first principles suggest, should be limited to compensation for (a) the perpetuation and amplification of the handicaps already being suffered at the moment when the employment changed hands, and (b) the bringing to fruit in the shape of current hardship those symptoms which had previously been no more than potential.[25]

This has been recognised as an aspect of corrective (in)justice by Weinrib, through what he terms his 'limitation thesis':

> [B]ecause the relationship between the parties is one of equal freedom, the plaintiff's freedom does not entitle the court to coerce the defendant into providing the plaintiff with a windfall over and above the restored right, for that, in turn, would be inconsistent with the defendant's freedom.[26]

---

[21] [1962] 1 QB 33 (CA) at 39–40. See also see M Green, 'The Intersection of Factual Causation and Damages' (2005–06) 55 *DePaul Law Review* 671, 680.
[22] *Dingle v Associated Newspapers* [1961] 2 QB 162 (CA).
[23] [1961] 2 QB 162 (CA) at 194.
[24] *Thompson v Smiths Shiprepairers (North Shields) Ltd* [1984] QB 405 (QBD).
[25] [1984] QB 405 (QBD) at 438.
[26] E Weinrib, *Corrective Justice* (Oxford, OUP, 2012) 92.

Taking a Victim as Found

It is axiomatic that a defendant in a negligence action takes her victim as she finds her.

> It has always been the law of this country that a tortfeasor takes his victim as he finds him. It is unnecessary to do more than refer to the short passage in the decision of Kennedy J in *Dulieu v White & Sons*, where he said: 'If a man is negligently run over or otherwise negligently injured in his body, it is no answer to the sufferer's claim for damages that he would have suffered less injury, or no injury at all, if he had not had an unusually thin skull or an unusually weak heart.'.[27]

In *Smith v Leech Brain*,[28] the case in which Lord Parker CJ made this assertion, the claimant's husband had died from cancer after contracting a burn on his lip as a result of the defendant's breach of duty as his employer. The fact that he was found to have an existing predisposition to cancer[29] was regarded as the 'initiating process' of his illness, but the subsequent burn as a 'promoting process',[30] rendering the defendants liable for the full extent of his ultimate injury:

> [I]t seems to me that this is plainly a case which comes within the old principle. The test is not whether these employers could reasonably have foreseen that a burn would cause cancer and that he would die. The question is whether these employers could reasonably foresee the type of injury he suffered, namely, the burn. What, in the particular case, is the amount of damage which he suffers as a result of that burn, depends upon the characteristics and constitution of the victim.[31]

This rule, variously referred to as the eggshell or thin skull rule, puts a claimant's pre-existing vulnerabilities on the defendant's account.[32] As such, its pedigree is not purely causal, but is rather a normatively determined,[33] a priori risk allocation device. In a situation in which the question amounted to 'Was the claimant's injury caused by her pre-existing vulnerabilities, or by the effect of the defendant's

---

[27] *Smith v Leech Brain* [1962] 2 QB 405 (QBD) at 414 (per Lord Parker CJ) (footnotes omitted).
[28] [1962] 2 QB 405 (QBD).
[29] Which was probably because of his history of working in the gas industry but, in any event, was unrelated to the defendant's breach.
[30] [1962] 2 QB 405 at 414.
[31] [1962] 2 QB 405 (QBD) at 415 (per Lord Parker CJ). His Lordship here interpreted *Overseas Tankship (UK) Ltd v Morts Dock and Engineering Co Ltd (The Wagon Mound No 1)* [1961] AC 388 (PC) as restricting reasonable foreseeability to the *type*, rather than to the *extent*, of damage: an interpretation which turned out to be immanent. See Hart and Honoré, *Causation*, above n 2 at 274.
[32] See also J King Jr, 'Causation, Valuation and Chance in Personal Injury Torts involving Pre-Existing Conditions and Future Consequences' (1981) 90 *Yale Law Journal* 1353, 1361.
[33] See Ch 2, n 69 and R Fumerton and K Kress, 'Causation and the Law: Preemption, Lawful Sufficiency, and Causal Sufficiency' (2001) 64 *Law and Contemporary Problems* 83, 86.

breach of duty?', a conclusion in favour of either factor to the exclusion of the other could not be described as physically causal because both factors played a part in the damage occurring when it did. If Mr Smith's lip had not been burned at work, he would not have developed cancer when he did, and, if he did not have a pre-existing susceptibility, his burn would have healed and led to no further damage. Holding the defendant liable, therefore, for the consequent cancer and death is not a conclusion based on physical causation,[34] but the result of an a priori decision to put such inherent vulnerability at the risk of defendants as a class, as opposed to claimants as a class. The thin skull rule, therefore, has clear implications for the causal inquiry in negligence. Dealing as it has to in the medium of risk, it operates both to any given party's potential benefit, as well as to its detriment.[35] In *Smith*, it clearly disadvantaged the defendant to have injured a particularly vulnerable individual but, in other cases, the rule has worked to the defendant's advantage. In *Performance Cars Ltd v Abraham*,[36] for instance, the second defendant 'found' a victim whose property had already been damaged. So, whilst the same collision with an undamaged car would have rendered him liable for the cost of making a damaged car pristine once more, the thin skull rule on this occasion worked to his advantage by making him liable only for the *additional* damage his collision caused to an already imperfect car. Such is the nature of a risk allocation tool, and, applied in a context in which we are all potential claimants as well as potential defendants, there is no systemic injustice whichever way the risk is allocated.

By making this a priori decision about where such risks should lie, the law simplifies its own causal inquiry in what would otherwise be particularly challenging circumstances. The thin skull rule, when combined with an NBA, is able to explain cases which have generally been regarded as anomalous or hard to reconcile; the two most obvious examples being *Baker v Willoughby*[37] and *Jobling v Associated Dairies*.[38] In the former case, the claimant's left leg had been injured in a collision caused by the defendant's careless driving. Later, but before his claim came to court, he was shot in the same leg during an armed robbery at the scrap metal yard at which he worked, which led to his left leg having to be amputated. The robbers, who were the second tortfeasors, were, unsurprisingly, not before the court. The defendant argued that he should not be liable for damages beyond the point at which the leg was amputated, but the House of Lords disagreed, and he remained liable for all of the consequences of the claimant's original injury, as if the intervening event had never occurred. Had the defendant's liability ended at the point at which the leg was removed, the claimant would have been worse off as a result

---

[34] What Richard Wright would refer to as 'lawful' causation; the instantiation of a causal law. See eg R Wright, 'The NESS Account: Response to Criticisms' in R Goldberg (ed), *Perspectives on Causation* (Oxford, Hart Publishing, 2009) 289.
[35] See P Cane, *Responsibility in Law and Morality* (Oxford, Hart Publishing, 2002) 121–22.
[36] See above n 19.
[37] *Baker v Willoughby* [1970] AC 467 (HL).
[38] *Jobling v Associated Dairies* [1982] AC 794 (HL).

of suffering from two torts than he would have been had he just been a victim of the defendant's carelessness, since then there would have been no question of his receiving damages for being deprived of the use of a good leg for the rest of his life. This respects basic principles of causation; the defendant pays for depriving the claimant of the use of a good leg for ever and, were the robbers to have been sued, they would have been liable for the *additional* damage they caused for an already injured leg.[39] Consequently, they, as second tortfeasors, would have benefited from the a priori risk allocation effect of the thin skull rule. It might be reasonable to ask, therefore, why it has been seen as a contentious decision. There are two discernible reasons for this: first, that the decision is hard to reconcile with the later case of *Jobling*, and second, the question of whether the shooting should be regarded by the law as an intervening act sufficient to 'eclipse' the effect of the first (not least because the first tort was negligent whereas the second was intentional). In *Jobling*, the claimant was injured at work as a result of his employer's breach of duty.[40] His back was damaged and his earning capacity thereby reduced. Three years later, and completely independent of his accident, Jobling developed, as a natural occurrence, a back disease which rendered him completely unfit for work. The House of Lords held that the defendants should not be liable for the claimant's consequential losses after the point at which the disease manifested itself. The decision in *Baker* came in for some criticism along the way, but remained unaffected as a matter of precedent.[41] According to *Clerk & Lindsell*, '*Jobling v Associated Dairies* cannot be satisfactorily reconciled with *Baker v Willoughby*'.[42] Presumably this is because, in the former case, the intervening event served to truncate the defendant's liability whilst in the latter, it did not. To conclude on this basis, however, that the two cannot be reconciled is to miss the legal import of the difference between the two sets of facts.

It has generally come to be accepted that, as suggested by Lords Russell and Keith in *Jobling*,[43] the *Baker* approach is applicable where the two events are successive torts and the *Jobling* approach appropriate where the subsequent event is not tortious. On an NBA, it is easy to see why this is correct, and why the two cases are really nothing more than instances in which basic and orthodox causal principles have been applied. The facts of *Jobling* do not satisfy the first stage of the NBA, since, but for at least one breach, the claimant would still have been in the position he was at trial, owing to his naturally occurring illness. Since the claimant's state of affairs beyond the development of the natural illness has not been made worse by a breach of duty, therefore, the claimant's ultimate injury falls

---

[39] On the question of whether they, as intentional tortfeasors, should be able to take advantage of this rule as against a merely negligent first tortfeasor, see below, text to n 50.

[40] A statutory one under the Offices, Shops and Railway Premises Act 1963.

[41] See, for instance, [1982] AC 794 at 821 (per Lord Bridge).

[42] WHB Lindsell, AM Dugdale and MA Jones, *Clerk & Lindsell on Torts*, 20th edn (London, Sweet & Maxwell, 2010) 2–99, although the two were regarded as being consistent with one another in *Rahman v Arearose Ltd* [2001] QB 351 (CA) at [32] (per Laws LJ).

[43] [1982] AC 794 (CA) at 810 and 815, although this was not universally accepted – see 819–21 (per Lord Bridge).

outside of the remit of negligence.[44] The facts of *Baker*, on the other hand, do satisfy the first stage of the NBA, because, but for at least one breach of duty, the claimant would not have been in the condition he was at trial. On the basis of the causal inquiry as formulated for the purposes of negligence, the ultimate injury in *Baker* had a legally relevant cause, whereas the ultimate injury in *Jobling* did not. The two cases are on this basis simple to reconcile and to understand.

In the course of the *Jobling* decision, some thought was given to the implications of the thin skull rule for the specific facts of that case.[45] After all, so the argument goes, if a defendant takes her victim as found, should she not remain liable for the consequences of her breach notwithstanding the future development of any natural condition of the claimant which subsumes the effects of her actions? In other words, why did the claimant's dormant myelopathy in *Jobling* operate so as to reduce the defendant's liability when the deceased's pre-malignancy in *Smith* did no such thing? The key to this distinction is independence. In *Smith*, the predisposition and the negligently caused injury were linked and *inter*dependent: as outlined above, the malignancy was triggered by the burn and the burn exacerbated by the malignancy. In *Jobling*, by contrast, the effects of the breach neither affected, nor were affected by, the physiological event. It is crucial to note that the purpose of the thin skull rule is not to dispense with the need for any causal link whatsoever between the breach and the characteristics of the claimant, but only to dispense with the need to disentangle the two factors when and where they do interact. It was clear on the facts of *Jobling* that there was no interaction between the breach and the naturally occurring event,[46] and so the thin skull rule was simply not applicable.[47]

The decision in *Jobling* was justified further by reference to the 'vicissitudes of life' principle. Since damages for personal injury would normally be discounted to recognise the possibility that the claimant would not have remained unimpeded for the rest of his life in any event, all the decision in *Jobling* did was to substitute concrete knowledge of such a vicissitude having happened for the speculation it would normally make.[48] This might lead to the question of why the second tort in *Baker* did not count as the manifestation of such a vicissitude with similar effect. The answer to this is simply that others' breaches of duty are not 'vicissitudes of life' for this purpose and do not, therefore, count as part of any claimant's 'normal course of events'. Although there is no obvious judicial consensus on this point, as *Jobling* demonstrates,[49] it is easily explicable on the basis of an NBA. The 'vicissitudes of life' principle exists so as to provide a means of discounting a claimant's

---

[44] This does not prevent, however, a claimant recovering for the earlier period during which he was temporarily worse off as a result of at least one tort – ie after the breach of duty but before his illness materialised. This would be covered in the orthodox way by an assessment of special damages.

[45] See, for example, [1982] AC 794 (CA) at 810 (per Lord Russell) and 813 (per Lord Keith).

[46] [1982] AC 794 (CA) at 816 (per Lord Bridge).

[47] As recognised by [1982] AC 794 (CA) at 810 (per Lord Russell), describing it as 'wholly irrelevant' on the facts.

[48] See [1982] AC 794 (CA) at 810–11 (per Lord Russell).

[49] See [1982] AC 794 (CA) at 811 (per Lord Russell), but note 815 (per Lord Keith).

damages for future contingencies and, where more than one breach satisfies both stages of the NBA, there is nothing to justify reducing the liability of any one of those on the basis that at least one other exists. To do this would be both to make a claimant's post-judgment position worse than it would have been had she suffered from only one tort, and to grant a defendant a windfall on the basis that hers was not the only breach of duty to affect the claimant adversely: hardly a convincing, or even coherent, justification.

Despite the fact, therefore, that *Baker v Willoughby* is almost universally given negative academic and judicial treatment,[50] it is in fact a perfectly good example of basic causation principles being applied to good effect. As Lord Pearson recognised in that case,

> The supervening event has not made the plaintiff less lame nor less disabled nor less deprived of amenities. It has not shortened the period over which he will be suffering. It has made him more lame, more disabled, more deprived of amenities. He should not have less damages through being worse off than might have been expected.[51]

In other words, the effects of both breaches remained operative on the claimant: there was nothing about the second event which neutralised or improved the injury caused by the first, nor that reduced the duration of the claimant's suffering. The shooting served only to exacerbate the claimant's condition. On this point, rights-based reasoning is helpful: the infringement of the claimant's right (in this case to bodily integrity) did not stop once the second event occurred, and so losses consequent on it did not disappear:

> For the defendant driver, the subject matter of the right infringed was a good leg. The defendant driver must pay for all of the loss consequent upon making a good leg stiff, and infringements of the right to the bodily safety by others . . . do not reduce this loss. The actions of other wrongdoers, both actual and hypothetical, do not reduce the loss consequent upon the infringement of the right.[52]

Where this is the case, no defendant[53] who has made a difference to a claimant's normal course of events, and whose breach is operative when the ultimate injury occurred, should be relieved of liability by the existence of other defendants in the same position.[54] Such a principle could be defended on the grounds of corrective justice (as conceived by Weinrib):

> The plaintiff remains linked to the defendant through a right that pertains to the object as an undamaged thing. Although the defendant's wrong has modified the physical condition of the object embodying the plaintiff's right, the right remains intact as the normative marker of the relationship between them with respect to that object.[55]

---

[50] But see R Stevens, *Torts and Rights* (Oxford, OUP, 2007) 139.
[51] [1970] AC 467 (HL) at 495.
[52] Stevens, *Torts and Rights*, above n 50 at 140–41.
[53] Of course, only one was before the court in *Barker*, but the principle should nevertheless apply.
[54] See also *Borealis AB v Geogas Trading SA* [2010] EWHC 2789 (Comm), [2011] 1 Lloyd's Rep 482 and *Flanagan v Greenbanks Ltd (t/a Lazenby Insulation)* [2013] EWCA Civ 1702, 151 Con LR 98.
[55] Weinrib, *Corrective Justice*, above n 26 at 90.

The application of this principle ensures that claimants are not worse off, post judgment, as a result of suffering multiple torts than they would be had they suffered fewer. That is not to say that they should be better off either; they should not recover for more than the value of the infringement of their right plus any losses consequent on it,[56] but, as *Baker* demonstrates, this will not happen if each defendant pays only for the losses she caused. What it does mean is that the only recovery risk faced by claimants is the universal one of the disappearing defendant (the robbers in *Baker*). Where this principle is applied, at least the absence of a defendant whose contribution came later in the physical sequence of events will not detract from a claimant's ability to recover for the ongoing effects of earlier torts (as would have been the case had the defendant's argument in *Baker* been accepted by the House of Lords). The gap in recovery will then be the difference between the claimant's earlier injured state and her ultimate injured state; not the entire difference between a whole claimant and her final injuries. This is clearly preferable to the Court of Appeal's approach to the facts of *Baker*, which saw it prepared to hold the second defendants liable for the whole of the loss on the basis that they had deprived the claimant of his ability to establish a case against the first defendant.[57] The unusual facts of this case illustrate exactly why this result is not consonant with the objectives of the law of negligence: by restricting the possibility of recovery for damage resulting from multiple torts to only one party, the law would thereby increase the wronged claimant's risk of non-recovery at the same time as granting a reprieve to the initial, wrongdoing defendant. As the Court of Appeal's treatment of *Baker* shows, if a defendant to whom all liability has been transferred is judgment-proof, the final outcome performs no compensatory or corrective functions at all. The House of Lords' approach, on the other hand, made the best of a notoriously difficult situation without under-compensating the claimant, and without either over-burdening or arbitrarily absolving the defendant.

'Operative': the Second Stage of the NBA

Stage 2, applied to each defendant individually: Was the effect of *this* defendant's breach operative when the damage occurred?

It is a common theme in recent academic discourse for the question of remoteness to be sharply and clearly distinguished from the anterior question of 'factual' causation. As Stapleton has long insisted:

> ... the two substantive underlying arguments relevant to the dispute about responsibility: the question of fact about whether the tortious conduct of the defendant, D, was

---

[56] See Stevens, *Torts and Rights*, above n 50 at 133–34.
[57] For a closely related argument, see Ariel Porat and Alex Stein's case for imposing liability on the basis of 'evidential damage' – A Porat and A Stein, *Tort Liability Under Uncertainty* (Oxford, OUP, 2001) particularly ch vi.

historically involved with the claimant, C, suffering actionable damage; and the normative inquiry into whether the consequence of the tort that C is complaining about should be judged to be within the appropriate scope of liability for consequences of the tortious conduct ... are best kept quite separate ... The Reporters of the American Law Institute's Restatement (Third) of Torts: Liability for Physical Harm (Basic Principles) have recognised the confused state of the structure of past case law in this area and that of the past Restatements. Accordingly, their current 2003 draft completely abandons the text of the former Restatements. The draft separates into two different chapters the factual issue of cause-in-fact (i.e. historical involvement) from the scope-of-liability-for-consequences chapter. Moreover the draft eliminates all causal terminology from the scope chapter, taking the bold but necessary step of abandoning long-standing terminology such as 'legal cause'.[58]

Moore explains the difference between the two questions thus:

> The conventional wisdom about the causation requirement in ... torts is that in reality it consists of two very different requirements. The first requirement is that of 'cause-in-fact'. This is said to be the truly *causal* component of the law's two requirements framed in causal terms, because this doctrine adopts what is thought of as the 'scientific' notion of causation. Whether cigarette smoking causes cancer, or whether the presence of hydrogen or helium caused an explosion, are factual questions to be resolved by the best science the courts can muster. By contrast, the second requirement that of 'proximate' or 'legal' cause, is said to be an evaluative issue, to be resolved by arguments of policy and not arguments of scientific fact. Suppose a defendant knifes his victim who then dies because her religious convictions are such that she refuses medical treatment. Has such a defendant (legally) caused her death? The answer to such questions, it is said, depends on the policies behind liability, not on any factual issues; factually, it is thought, the knifing surely caused her death.[59]

At the stage of linking a defendant's breach with a claimant's damage, it is clear that there are two types of question being asked – one indicative and one evaluative. There are few grounds, and no reasons, for denying this. To insist, however, that the two inquiries be carried out as two distinct exercises is both to overstate the point, and to underestimate judicial proficiency. It is understandable that more extensive guidance is necessary where juries require direction (as in the US),[60] but the same considerations simply do not apply in a system in which facts are assessed by those with extensive legal training and expertise. There is, therefore, no reason to suppose that the second stage of the NBA, dealing with the

---

[58] J Stapleton, 'Cause-in-Fact and Scope of Liability for Consequences' (2003) 119 *LQR* 388, 388–89. See also J Stapleton, 'Unpacking Causation' in HLA Hart and A Honoré (eds), *Relating to Responsibility* (Oxford, Hart Publishing, 2001) 167–68.

[59] MS Moore, *Causation and Responsibility* (Oxford, OUP, 2009) 83–84. 'Proximate cause' is the conceptual device employed by US courts, and is something of a protean idea – see eg *Paroline v United States et al* No 12-8561, April 23, 2014 (USA) at II.

[60] See, for instance, American Law Institute, *Restatement of the Law, Third, Torts: Liability for Physical and Emotional Harm* (St Paul, MN, American Law Institute, 2010) § 29: 'Jury instructions that separate these two components of the case facilitate focus on the appropriate matter. Even in cases in which both issues are disputed, separate instructions and separate consideration of each issue should clarify the requisite inquiries for the jury'.

question of whether the effects of a defendant's breach are operative, is unable to accommodate both types of consideration. It is, after all, a relatively straightforward exercise for a judge to ensure that any breach ultimately deemed to satisfy that second stage has had a physical effect on the claimant's position *as well as* having a connection with that position which the law regards as culpable. In semantic terms, it is hardly stretching any definition of 'operative' to apply it to both tasks simultaneously. Just as the law means something very specific by the terms 'property', 'consent', 'intention' and 'consideration', for instance, so it can define precisely what it means in this context by 'operative':

> The kind of causal connection required to create liability, whether it is the standard criteria or some special variant on those criteria, is just as much a legal concept as the kind of conduct required to constitute negligence or murder.[61]

Whilst this might not always be a substantively easy inquiry to conduct, it would be made no easier by being formally divided into two separate stages. Whilst it is true that the factual determination must precede the evaluative process (for a factor that is not a factual cause cannot, in negligence terms, be a culpable one),[62] there is no compelling reason to manufacture an additional process in order to accommodate this. In essence, the forensic objective at this stage of the negligence inquiry is to establish whether there is a causal link between the defendant's breach of duty and the claimant's damage sufficient to impose liability, and whilst this is a question of both fact and degree, it remains appropriate to treat it as a single question with two dimensions. As Lord Hoffmann has said, 'causal requirements are creatures of the law and nothing more'.[63]

There are several practical reasons for streamlining the inquiry in this way, in addition to the theoretical point that the concept of 'operative' is easily able to accommodate both factual and normative considerations. First, since the establishment of a causal link is itself a constituent part of a wider inquiry, there is an argument for making it as lean a process as possible. Secondly, the law, and therefore the judiciary, are no strangers to such two-dimensional assessments, and have long applied them with minimal fuss elsewhere.[64] Whilst it is easy to see, therefore, how the two aspects of the question lend themselves very well to academic division, this has no necessary bearing on the way in which they are employed in practice.

> Forensic necessity may well make bedfellows of 'causal' and 'scope of rule' limitations. When it does so the judge or jury, if they are to limit responsibility at all, must purport

---

[61] L Hoffmann, 'Causation' (2005) 121 *LQR* 592, 598.
[62] Although see the discussion of *Chester v Afshar* [2004] UKHL 41, [2005] 1 AC 134 in Ch 7, text to n 60.
[63] L Hoffmann, 'Causation' in Goldberg (ed), *Perspectives on Causation*, above n 34 at 7–9.
[64] Consider, for instance, the questions asked by the court when trying to establish whether a contract was concluded under duress or undue influence: was the threat/inducement illegitimate or undue, *and* did it make a difference to the claimant's position as regards the contract. In both cases, these questions, whilst being of different types, are asked together, in a single inquisitorial stage. See E Peel, *Treitel on the Law of Contract*, 13th edn (London, Sweet & Maxwell, 2011), 10-006 and 10-012–10-016.

to do so on the ground that the harm is 'too remote' or 'not proximate', whether the reason which really weighs with them is that the harm was not caused by the wrongful conduct of the defendant or that the scope of the rule he violated does not extend to compensating for that type of harm. But the theorist, who is concerned to understand rather than to manipulate the principle of legal responsibility, must keep them separate.[65]

The NBA, which owes its content to theory, but its form to forensic necessity, accommodates both as bedfellows in its second stage. Since 'operative' is a pivotal concept in this second stage, it is one which requires robust definition. Taking this into account, therefore, the NBA adopts the following criteria for determining when breaches of duty will be operative for the purposes of the *second* stage of its analysis. It must be borne in mind that this analysis is only applied to breaches of duty where the first stage of the NBA has been passed (and it has thereby been established that at least one breach of duty has affected the claimant's normal course of events).

When will a Breach be 'Operative'?

A breach of duty will be deemed operative under the second stage of the NBA where the risk created by that breach has actually eventuated in at least part of the claimant's damage[66] (as opposed to being a counterfactual alternative).[67] Whilst this is a simple enough formulation, however, it is not always straightforward to apply. The following criteria, therefore, which are neither additive nor mutually exclusive, provide some elaboration. The first three will be needed only in cases involving causal issues which have historically proven difficult to resolve, whilst the final one is of universal application.

A breach will be operative:[68]

1. where any intervening event is not coincidental upon the breach* – *Home Office v Dorset Yacht Co Ltd*;[69]
2. notwithstanding any coincidental intervening event, unless that intervening event has brought the infringement to an end or improved the claimant's position – *Baker v Willoughby*;[70]

---

[65] Hart and Honoré, *Causation*, above n 2 at 132.
[66] This point is what justifies the 'single agent' limitation on the exceptional material contribution to risk principle – see Ch 6, under heading 'Single Agent'.
[67] As in *Saunders System Birmingham Co v Adams* 117 So 72 (Ala 1928), discussed in detail in Ch 4, under the heading 'Pre-emption'.
[68] A case illustrating the function and significance of each condition has been provided, and elaboration follows below.
[69] *Home Office v Dorset Yacht Co Ltd* [1970] AC 1004 (HL).
[70] [1970] AC 467 (HL). *Robinson v Post Office* [1974] 2 All ER 737 (CA) will also be discussed under this heading, below.

3. where it affects a claimant's ability to manage her own risk with respect to the damage which results – *South Australia Asset Management Corp v York Montague Ltd (SAAMCO)*;[71]
4. in relation to damage which is of a reasonably foreseeable type – *Wagon Mound (No 1)*[72] as long as it occurs within a time frame considered by the court to be reasonable in the circumstances of the case.

* Where any such intervening event is a breach of duty, liability should be joint and several in the case of defendants, and liable to reduction on a contributory negligence basis in the case of the claimant.

The Significance of a Risk Which Has Actually Eventuated

One aspect of this distinction is discussed in Chapter 4,[73] as part of the analysis of duplicative causation. To summarise briefly for present purposes, this condition is the means of discriminating between those breaches of duty which create *a risk* of damage to the claimant and those which create *the risk* which actually eventuated.[74] Where necessary, the second stage of the NBA disaggregates the factors included by the first stage.[75] The transition from the irrealis mood of the first stage to the realis mood of the second stage is what is distinctive about the NBA; the objective of the former being to identify where a remedy is in principle due, and the objective of the latter being to identify its source in practice. The second stage is the point, therefore, at which inclusion yields to discrimination, and, as we will see, this is highly significant in situations of duplicative and potentially duplicative causation.

In *Saunders*,[76] for example, in which the claimant was injured by the negligence of a driver, who, although driving a car with faulty brakes, failed even to attempt to brake, it is easy to see why the first stage is satisfied; but for at least one breach of duty, the claimant would not have suffered the injuries for which he was claiming.[77] Once the first stage of the NBA has established, however, that the claimant is in his current position because of at least one breach of duty, it is clear that, rather

---

[71] *South Australia Asset Management Corp v York Montague Ltd (SAAMCO)* [1997] AC 191 (HL). This is consonant with the autonomy point, discussed in Ch 7.
[72] [1961] AC 388 (PC).
[73] Under heading 'Pre-emption'.
[74] This is where the significance of the single agent limitation to the *Fairchild* exception is evident – see Ch 6, under heading 'Single Agent'.
[75] Such aggregation being necessary to prevent duplicative causal factors being erroneously classified as irrelevant on But For grounds – see Ch 2.
[76] 117 So 72 (Ala 1928).
[77] Were this aggregate question not to be asked, however, and a traditional But For test performed, neither the faulty brakes nor the careless driving would have satisfied the test because, but for either one, the other would have functioned to cause the injuries in question.

than excluding both factors, as conventional But For analysis would do, each needs to be submitted to further scrutiny. At this point, the NBA is not interested in the counterfactual, but only in the factual, and this explains why the condition discussed here applies in order for a breach to be 'operative'. So, on the facts of *Saunders*, the only factor to have generated the risk which eventuated in the claimant's damage was the negligent conduct of the driver. That driving created a risk and 'the injury . . . fell "squarely" within that risk'.[78] The faulty brakes were merely a counterfactual 'backup' risk which *would* have had an effect in the absence of negligent driving; that is, if the driver had tried to use them to avoid hitting the claimant.[79] On the other hand, in cases of genuinely duplicative causation (overdetermination), such as the case of the two careless hunters and the claimant injured by a gunshot,[80] both hunters' breaches generated a risk of a gunshot wound, which was what eventuated. Both, therefore, would be liable on an NBA. This condition of risk eventuation, therefore, is one which must be satisfied if a breach is to be considered 'operative', because a breach of duty which had no real-world effect on the claimant cannot ground a claim in negligence.

A breach of duty will be operative where any intervening event is not coincidental upon it.*

* Where any such intervening event is a breach of duty, liability should be joint and several in the case of defendants, and liable to reduction on a contributory negligence basis in the case of the claimant.

This is hardly a contentious condition, and it is illustrated clearly by the facts of several well-known decisions. One such case is *Home Office v Dorset Yacht Co Ltd*,[81] in which Home Office employees, in breach of their duty of care, left young offenders unsupervised on an island. When the latter damaged the claimant's yacht, the Home Office was held liable. Much was made in the judgments of the degree to which the acts of the boys were a foreseeable, probable or likely consequence of the officers' inadequate supervision. In *The Oropesa*,[82] another case which illustrates the effect of this condition, Lord Wright said:

> There are some propositions which are beyond question in connexion with this class of case. One is that human action does not per se sever the connected sequence of acts. The mere fact that human action intervenes does not prevent the sufferer from saying that injury which is due to that human action as one of the elements in the sequence is recoverable from the original wrongdoer.[83]

---

[78] *Wilsher v Essex AHA* [1987] QB 730 (CA) at 780 (per Sir Nicholas Browne-Wilkinson).
[79] See also *R v Hughes* [2013] UKSC 56, [2013] 4 All ER 613, in which the defendant's uninsured, unlicensed, but physically faultless driving was found not to have caused the death of another driver who, driving erratically and under the influence of heroin, had swerved into his path. By analogy, Hughes' breach was not operative in terms of the other driver's death because it did not *eo ipso* create a risk of that particular outcome.
[80] See Ch 4.
[81] [1970] AC 1004 (HL).
[82] *The Oropesa* [1943] P 32 (CA).
[83] [1943] P 32 (CA) at 37 (per Lord Wright).

In that case, a collision at sea had been caused by the negligent navigation of *The Oropesa*. In an effort to manage the situation, the captain of the other vessel took a boat containing several of his crew to *The Oropesa*. Owing to poor weather conditions, the boat overturned and nine crew members were lost. The Court of Appeal found that the initial careless navigation remained legally relevant in terms of the loss of life because the intervening act did not independently arise, but was something which became more likely to happen as a result of the earlier carelessness. More recently, in *Corr v IBC Vehicles*[84] and *Reeves v Commissioner of Police for the Metropolis*,[85] in which the victims ultimately committed suicide, each defendant was held liable for the deaths on the basis that neither claimant's final act was coincidental upon the breach. In *Corr*, the defendant was the deceased's employer when he suffered grave head injuries at work as a result of malfunctioning machinery. His injuries led to his suffering from post-traumatic stress disorder and depression. Despite the fact, therefore, that the defendant argued that his suicide constituted a *novus actus interveniens*, which should have truncated its liability, the House of Lords disagreed on the basis that the victim's act was a consequence of their breach and not independent of it. In *Reeves*, in which a prisoner deemed to be a suicide risk hanged himself in a police cell, a similar point was taken. In this case, however, the claimant's damages were reduced by half on contributory negligence grounds because, in contrast to *Corr*, the deceased in *Reeves* had been of sound mind when he took his own life.[86] This was the correct result, since the defining characteristic of modern contributory negligence[87] is that the existence of carelessness on the part of the claimant should not necessarily[88] prevent a defendant's breach from being operative, but will function instead as a factor with partial causal relevance. In this sense, the claimant's negligent behaviour is treated by the second stage of the NBA in the same way as any other breach: if it remains operative on the claimant at the time the ultimate damage occurred, it will be deemed a partial cause of that damage.

A different result was reached in *Knightley v Johns*.[89] The first defendant in that case had negligently overturned his car in a busy traffic tunnel. The fourth defendant, the police inspector who arrived on the scene, failed immediately to seal the tunnel to traffic in breach of his duty of care. When, some time later, in order to do this, two police constables drove their motorcycles, against the flow of traffic, one of them was injured in a collision with the second defendant. The Court of Appeal found that the first defendant was not liable for the constable's injury because the intervening act of the police inspector broke the chain of causation between them. So, in this case, despite the fact that no injury would have occurred but for the first defendant's negligent driving, his breach was deemed to be no

---

[84] *Corr v IBC Vehicles* [2008] UKHL 13, [2008] 1 AC 884.
[85] *Reeves v Commissioner of Police for the Metropolis* [2000] 1 AC 360 (AC).
[86] See [2000] 1 AC 360 (AC) at 394 (per Lord Jauncey).
[87] And the reason for the enactment of the Law Reform (Contributory Negligence) Act 1945.
[88] Unless it is so great as to prevent the defendant's breach being operative at all, and so constitute a *novus actus interveniens* – see below, text to nn 92–93.
[89] *Knightley v Johns* [1982] 1 WLR 349.

longer operative (in NBA language) by the time of the claimant's injury. The concept of coincidence is crucial in justifying the difference between these outcomes, and a 'loss is coincidental if, as things turned out, its risk was not increased *for this claimant* by the wrong which occurred'.[90]

The existence of a traffic incident did not, for instance, increase the likelihood of the police officer getting injured by driving into oncoming traffic. Whilst the initial incident presented the *opportunity* for the inspector's negligence to occur, this is, without more, insufficient for a breach to remain operative. The same point is evident from the following example of Hart and Honoré's:

> the case of the householder whose prudential storing of firewood in the cellar gave a pyromaniac his opportunity to burn it down would be distinguished from that of the careless friend who left the house unlocked: the fire would not naturally be described as a consequence of the storing of the wood though the loss of the spoons was a consequence of leaving the house unlocked.[91]

The difference between these two situations lies in the specificity of the relationship between the act and the subsequent event. Any action could be said to provide opportunities for other actions or events to follow, but only some will affect the likelihood of certain consequences. Where the initial action[92] is a breach of a duty of care, and it increases the likelihood of a subsequent event which, in itself, leads to actionable damage, that subsequent event will not be regarded by the law as coincidental on the original breach.

This can also be seen in cases in which the intervening event is an act of the claimant. The obvious two cases for comparison are *McKew v Holland and Hannen & Cubitts*[93] and *Wieland v Cyril Lord Carpets Ltd*.[94] In both of these cases, the claimants, having been injured by the defendants, fell down stairs sustaining further injury. In the first case, the claimant, whose knee had been weakened as a result of the initial injury, attempted to descend a steep staircase which had no handrail. During his descent, his knee started to go numb and, on jumping to the floor in response, he injured himself further. The defendant was not held liable for the injury subsequent to the fall. In the second case, the claimant had to wear a neck brace as a result of her initial injury, which meant that she could not use her bifocal spectacles as she needed to. This led to her falling down a flight of stairs. The defendant was held liable for the claimant's ultimate damage. The distinction between the two results is that injuring someone, and thereby creating a weakness, does not increase the likelihood of their acting unreasonably in relation to that weakness, but it does increase the likelihood of that weakness making even prudent behaviour result in further injury.

---

[90] See Ch 7, text to n 65, Stevens, *Torts and Rights*, above n 50 at 165. See also Hart and Honoré, *Causation*, above n 2 at 169.
[91] Hart and Honoré, *Causation*, above n 2 at 60.
[92] Or, exceptionally, omission.
[93] *McKew v Holland and Hannen & Cubitts* [1969] 3 All ER 1621 (HL).
[94] *Wieland v Cyril Lord Carpets Ltd* [1969] 3 All ER 1006 (QBD).

Whilst much is made in all of these cases of concepts such as foreseeability, voluntariness, directness and what is probable and natural,[95] an appropriately defined concept of coincidence does the same work in simpler terms.

A breach of duty will be operative, notwithstanding any coincidental intervening event, unless that intervening event has brought the infringement to an end or improved the claimant's position.

This is simply a formalised statement of the issues dealt with in the *Baker v Willoughby* discussion, above.[96] The basic principle behind it is that no claimant should be worse off as a result of suffering from multiple torts than she would have been had she suffered fewer. Further, no defendant should be relieved of liability by the existence of another tortfeasor.[97] The one exception to this latter principle (which is not an exception to the first) is that where a subsequent breach of duty has ameliorated the claimant's position by, for instance, bringing the adverse effects of the first breach to an end, that initial breach will no longer be deemed operative, and the later breach will be deemed to have superseded the first. This is unlikely to occur very often, but it is conceivable. Take, for example, a building, negligently set on fire by defendant A. After only a few seconds, the sprinkler system, negligently installed by defendant B and therefore defective, floods the burning building. In so doing, it prevents the fire from destroying the property, but causes extensive water damage. The second breach here has effectively improved the claimant's position as far as the first breach was concerned because it has left her with a damaged building rather than no building at all. In this situation, the claimant loses nothing by a court holding that the second breach functioned to truncate the liability of the first defendant.

This is, of course, a counter-example to the facts of *Baker v Willoughby*, where the second breach was held not to truncate the liability of the first defendant. The distinction is not necessarily straightforward because one of the arguments in *Baker* was that, by depriving the claimant of his leg, the second tortfeasors had thereby relieved him of the pain and suffering he would otherwise have felt for the rest of his life, had he had to live with an injured leg, as opposed to an absent one. This makes reference to a very specific head of damages, and this technicality aside, it is very difficult to argue that someone without a leg is better off than someone with a stiff leg. This is therefore distinct from the case in the fire/flood example given above, in which the operation of the first breach had effectively ceased.[98] Lord Pearson explained this concisely in *Baker*:

> I think a solution of the theoretical problem can be found in cases such as this by taking a comprehensive and unitary view of the damage caused by the original accident.

---

[95] See also Hart and Honoré, *Causation*, above n 2 at 164–72.
[96] Text to n 37.
[97] See D Fischer, 'Successive Causes and the Enigma of Duplicated Harm' (1999) 66 *Tennessee Law Review* 1127, 1129–30.
[98] See also [1970] AC 467 (HL) at 495 (per Lord Pearson) for the counter-example exemplified by the facts of *Baker*.

Itemisation of the damages by dividing them into heads and sub-heads is often convenient, but is not essential. In the end judgment is given for a single lump sum of damages and not for a total of items set out under heads and sub-heads. The original accident caused what may be called a 'devaluation' of the plaintiff, in the sense that it produced a general reduction of his capacity to do things, to earn money and to enjoy life. For that devaluation the original tortfeasor should be and remain responsible to the full extent, unless before the assessment of the damages something has happened which either diminishes the devaluation (e.g. if there is an unexpected recovery from some of the adverse effects of the accident) or by shortening the expectation of life diminishes the period over which the plaintiff will suffer from the devaluation. If the supervening event is a tort, the second tortfeasor should be responsible for the additional devaluation caused by him.[99]

A breach of duty will be operative where it affects a claimant's ability to manage her own risk with respect to the damage which results.

This is a condition whose formulation is better represented by the Court of Appeal judgment in its illustrative case, which was subsequently reversed. The relevant question in *South Australia Asset Management Corp v York Montague Ltd* (*SAAMCO*)[100] was whether the defendants, who had negligently overvalued properties which the claimants had then used as securities for mortgages, should be liable for the difference between the negligent valuations and the true valuations, or whether they should also be liable for the extra reductions in value caused by an intervening market fall. Lord Hoffmann summarised the Court of Appeal approach as follows:

> The Court of Appeal (*Banque Bruxelles Lambert SA v Eagle Star Insurance Co Ltd* [1995] QB 375) decided that in a case in which the lender would not otherwise have lent (which they called a 'no-transaction' case), he is entitled to recover the difference between the sum which he lent, together with a reasonable rate of interest, and the net sum which he actually got back. The valuer bears the whole risk of a transaction which, but for his negligence, would not have happened. He is therefore liable for all the loss attributable to a fall in the market. They distinguished what they called a 'successful transaction' case, in which the evidence shows that if the lender had been correctly advised, he would still have lent a lesser sum on the same security. In such a case, the lender can recover only the difference between what he has actually lost and what he would have lost if he had lent the lesser amount. Since the fall in the property market is a common element in both the actual and the hypothetical calculations, it does not increase the valuer's liability.[101]

His Lordship, with whom the rest of the court agreed, thought that this was the wrong approach:[102]

---

[99] [1970] AC 467 (HL) at 496.
[100] [1997] AC 191 (HL).
[101] [1997] AC 191 (HL) at 210 (per Lord Hoffmann).
[102] Although the practical outcome of some of the cases do not align fully with his reasoning: see J Stapleton, 'Negligent Valuers and Falls in the Property Market' (1997) 113 *LQR* 1, 6–7.

Rules which make the wrongdoer liable for all the consequences of his wrongful conduct are exceptional and need to be justified by some special policy. Normally the law limits liability to those consequences which are attributable to that which made the act wrongful. In the case of liability in negligence for providing inaccurate information, this would mean liability for the consequences of the information being inaccurate.

... A mountaineer about to undertake a difficult climb is concerned about the fitness of his knee. He goes to a doctor who negligently makes a superficial examination and pronounces the knee fit. The climber goes on the expedition, which he would not have undertaken if the doctor had told him the true state of his knee. He suffers an injury which is an entirely foreseeable consequence of mountaineering but has nothing to do with his knee.

On the Court of Appeal's principle, the doctor is responsible for the injury suffered by the mountaineer because it is damage which would not have occurred if he had been given correct information about his knee. He would not have gone on the expedition and would have suffered no injury. On what I have suggested is the more usual principle, the doctor is not liable. The injury has not been caused by the doctor's bad advice because it would have occurred even if the advice had been correct.[103]

Whilst it is true, however, that the doctor's breach should not have been deemed operative in relation to the mountaineer's injury, this is not a true analogy with the facts of *SAAMCO*, and the valuers' breaches therein *did* remain operative. The risks of a non-knee-injury-related mountaineering accident are not intrinsic to the risk of mountaineering with a knee injury, and are in fact completely discrete. A climber's ability to manage those risks, therefore, is in no way compromised by climbing with a weak knee. (Obviously, the risks of a knee-injury-related accident are very much compromised by the doctor's breach, but this was not *the* risk which eventuated on the facts.) The risks of a fall in the property market, however, are intrinsic to the risk of lending money on the basis of an overvalued security, and are not separable from that risk. As Lord Hoffmann himself points out:

> There is again agreement on the purpose for which the information was provided. It was to form part of the material on which the lender was to decide whether, and if so how much, he would lend. The valuation tells the lender how much, at current values, he is likely to recover if he has to resort to his security. This enables him to decide what margin, if any, an advance of a given amount will allow for a fall in the market ... The valuer will know that if he overestimates the value of the property, the lender's margin for all these purposes will be correspondingly less.[104]

The Court of Appeal recognised this point and, in doing so, reached an outcome which is far closer an approximation of corrective justice than the result achieved by the House of Lords. Whilst it is true that the fall in property prices was not a consequence of the negligent valuation, the lenders' *ability to manage their risk in relation to that fall* was just such a consequence; a fact that was known to both parties and which was not unrelated to the valuers' duties of care in the first place.

---

[103] [1997] AC 191 (HL) at 213 (per Lord Hoffmann).
[104] [1997] AC 191 (HL) at 211 (per Lord Hoffmann).

Where, therefore, a breach of duty affects a claimant's capacity to manage his risk in relation to an intervening event, that breach remains operative.

A breach of duty will remain operative in relation to damage which is of a reasonably foreseeable type as long as it occurs within a timeframe considered by the court to be reasonable in the circumstances of the case.

In its own right, this is by no means a novel criterion. The first part is derived from *The Wagon Mound (No 1)*[105] or at least from its interpretation in the line of cases which followed it.[106] The second element, relating to time frame, is really just a verbalisation of a principle which is perhaps so obvious that it is rarely stated.[107] The important point to note is that foreseeability is limited to type and time frame, but not to extent, and this fits with the basic axioms of negligence law discussed earlier in this chapter.[108] Stevens, however, has levelled an objection to this approach on the basis that

> there is no necessarily right answer to the question of what counts as the same type of damage, and no criterion by which it can be determined. The 'type' is wholly dependent upon the level of generality with which the damage is described.[109]

This is true, as recent cases have shown,[110] but his examples of why he regards the lack of a 'right answer' as being problematic actually make the opposite point. The absence of any a priori conclusion about what a right answer looks like means that courts can distinguish between cases of varying factual complexity. To use the examples given by Stevens:

> X negligently drops Y's vase. Unknown to X the vase contained a powerful Mills bomb which explodes killing Y's racehorse Dobbin. Dobbin's death could not have been foreseen. Ought X to be held liable for Dobbin's death, although in relation to it he was not negligent?[111]

And then, alternatively: 'Let us assume that the explosion does not kill Dobbin, but destroys X's antique crockery collection'.[112] According to Stevens, on the *Wagon Mound* approach, it is not obvious whether Dobbin's death should be classified as

---

[105] [1961] AC 388 (PC), in which damage by oil was deemed to be recoverable, but damage by fire was not. See Lindsell, Dugdale and Jones, *Clerk & Lindsell on Torts*, above n 42 at 2-144–2-151.

[106] See Stevens, *Torts and Rights*, above n 50 at 154–58. Also, *Hughes v Lord Advocate* [1963] AC 837 (HL), in which it was decided that the exact manner of the damage occurring did not have to be foreseeable, as long as it led to damage of a foreseeable type, and *Smith v Leech Brain* [1962] 2 QB 405 (QBD), which illustrates the thin skull rule – see above, text to n 28.

[107] Although see J Stapleton, 'Negligent Valuers', above n 102 at 6–7.

[108] That is, the thin skull rule and the axiom that a defendant takes her victim as she finds him. See above, text to n 28. See also the discussion of moral luck in relation to this in Ch 4.

[109] Stevens, *Torts and Rights*, above n 50 at 155.

[110] For example, *Corr v IBC Vehicles* [2008] UKHL 13, [2008] 1 AC 884 in which the courts appear to give a broader meaning to 'type' where personal injury is caused, as compared to the way in which they treat property damage. In that case, the foreseeability of psychological harm was regarded as a concomitant of foreseeable physical harm: see 912 (per Lord Walker).

[111] Stevens, *Torts and Rights*, above n 50 at 156.

[112] ibid, 156.

the same type of damage as the vase (even though both are the subject matter of Y's property rights), but it is very likely that the crockery collection and the vase would be classified as the same type of damage. On the basis of his rights model, Stevens goes on to argue that neither should be recoverable because both are examples of unforeseeable consequential losses, to be distinguished from the substitutive damages for the destruction of the vase. So, Stevens' method would rule out, a priori, recovery for the death of the horse and for the loss of the crockery, and he uses the difference between these two hypotheticals to illustrate the 'instability' of the *Wagon Mound* test.[113] It is not obvious, however, that the *Wagon Mound* approach would conclude that either the loss of the crockery or the racehorse is recoverable: it would certainly be possible for a court applying that test to conclude that damage caused by dropping was of a reasonably foreseeable type, but that damage caused by an explosive (the presence of which was unknown to the defendant) was not. In *Jolley v Sutton LBC*,[114] for example, when discussing the combined effect of *The Wagon Mound* and *Hughes v Lord Advocate*, Lord Steyn said:

> The scope of the two modifiers – the precise manner in which the injury came about and its extent – is not definitively answered by either *The Wagon Mound (No 1)* or *Hughes's case*. It requires determination in the context of an intense focus on the circumstances of each case.[115]

In the same case, Lord Hoffmann made the following point:

> It is also agreed that what must have been foreseen is not the precise injury which occurred but injury of a given description . . . And the description is formulated by reference to the nature of the risk which ought to have been foreseen. So, in *Hughes v Lord Advocate* . . . the foreseeable risk was that a child would be injured by falling in the hole or being burned by a lamp or by a combination of both.[116]

It is likely, therefore, that, despite Stevens' assertion to the contrary, neither consequence of his hypothetical Mills bomb exploding would be deemed recoverable under the *Wagon Mound* test. He is correct, however, that this conclusion could only be reached after an examination of a particular set of facts. Stevens calls this characteristic 'instability', but one person's 'instability' is another's flexibility.

It is true that this is the criterion of a breach being operative that relies most heavily, compared to the others listed above, on the court's discretion and, correspondingly, on the facts of the particular case to which it is applied. Rather than being a problem, however, such flexibility is a necessary element at this stage of the negligence inquiry because, dealing as it does with issues which are inevitably and intrinsically case-specific, its definition must not be overly prescriptive if it is to be of any real forensic use.[117] A framework which insisted, for instance, that such

---

[113] ibid, 156.
[114] *Jolley v Sutton LBC* [2000] 3 All ER 409 (HL).
[115] [2000] 3 All ER 409 (HL) at 417.
[116] [2000] 3 All ER 409 (HL) at 418.
[117] See, for instance, in *Fairchild v Glenhaven Funeral Services Ltd* [2002] UKHL 22, [2003] 1 AC 32 at [40] 69 (per Lord Nicholls).

considerations could ever be subject to a formulaic test, might look conceptually elegant and intellectually neat, but would have a limited practical value. It might seem disingenuous to use Professor Stapleton's words here, since she is perhaps the principal advocate of the need always to separate purely causal questions from those relating to scope of duty.[118] She would not, therefore, agree with the NBA's amalgamated approach to establishing whether a given breach of duty is 'operative'. Nonetheless, the point she makes remains pertinent, whether the question is ultimately asked discretely, or as part of a more generic inquiry:

> [T]here is no mechanical test which can be applied and ... the court has to make a judgment about the extent of legal responsibility, a judgment which one hopes will explicitly evaluate the often complex concerns which go to produce the boundaries placed on civil obligations. For example, in a situation like that in *Dorset Yacht Co Ltd v Home Office* [1970] AC 1004 the Home Office will not be held responsible for all the foreseeable criminal damage that the escapers do, including that done in the escape itself, later that day, later that week and later that year, even though, ... all such losses are foreseeable and would not have happened but for the defendant's breach. A line must be drawn between types of foreseeable 'but for' consequences ... Typically, there is no consensus where it should be drawn nor even which factors should weigh in its determination, but it must be drawn and should be drawn with explicit evaluation of the factors taken into account.[119]

There is nothing to be gained from pretending that questions of this nature are amenable to a mechanical analysis. The common law is at its best when it achieves the correct balance between flexibility and predictability, and it does this most effectively when its methods are clear and coherent. The approach derived from the *Wagon Mound/Hughes/Jolley* line of cases fits both of these descriptions, as is suggested by the fact that it has been largely uncontentious in terms of its substantive application for over half a century. What has been more contentious has been the form in which it has been applied (or should be applied). As alluded to above, there has been a strong push for separating questions of remoteness from those of 'factual' causation, and for considering the former as 'scope of duty' considerations, or the question of 'whether the harm that occurred is among the harms the risks of which made the actor's conduct tortious'.[120] Such an approach, championed principally by Professor Stapleton,[121] clearly has much to contribute to an assessment of whether a breach remains operative. Whilst such a question might well aid an evaluation of this type, however, it is not the most comprehensive or direct means of identifying operative breaches of duty in its own right. Take, for

---

[118] See eg Stapleton, 'Unpacking Causation' above n 58; Stapleton, 'Cause-in-Fact and Scope of Liability', above n 58; and J Stapleton, 'Choosing What We Mean by "Causation" in the Law' (2008) 73 *Missouri Law Review* 433.

[119] Stapleton, 'Negligent Valuers', above n 102 at 6–7. See also Hoffmann, 'Causation', above n 61 at 603.

[120] American Law Institute, *Restatement of the Law*, above n 60, § 29, comment h. See also G Williams, 'The Risk Principle' (1961) 77 *LQR* 179 and Moore, above n 59 at ch 7.

[121] Stapleton, 'Cause-in-Fact and Scope of Liability', above n 58.

example, the case of *Corr v IBC Vehicles*,[122] discussed above. The defendant employer in that case was held liable for the ultimate death, by suicide, of its former employee. It is easier to argue that death by suicide is damage of a reasonably foreseeable type, given the circumstances of that breach of duty, than it is to argue that suicide is one of the risks which makes an employer's duty to maintain safe machinery tortious. The latter is, at best, a tenuous proposition. The former inquiry, therefore, functioning as a criterion to identify operative breaches, is better able to encompass the broad range of situations in which damage might appropriately be deemed relevant for the tort of negligence.

---

[122] [2008] UKHL 13, [2008] 1 AC 884.

4

Duplicative Causation (Real and Potential): Overdetermination and Pre-emption

Illustrative Cases: (Overdetermination/Real Duplicative Causation) *Summers v Tice, Cook v Lewis*; (Pre-emption/Potential Duplicative Causation) *Saunders Systems*

Factual Basis

These analyses are applied where:

- (**overdetermination**) an event has multiple causal factors, all of which:
  - are severally sufficient;
  - are independent of one another;
  - lead to indivisible injury;
  - lead to the same type of damage;
  - affect the claimant simultaneously;
- (**pre-emption**) an event has multiple causal factors, all of which:
  - have a non-duplicative effect on the claimant (ie the risks generated by at least one factor must never eventuate in damage to the claimant);
  - are independent of one another.

'Duplicative causation' refers here not to all situations in which there are multiple potential causal factors, but only to those in which one factor is causally *substitutive* for another. In other words, it refers to those situations in which the traditional But For test would give the wrong answer; all the time in cases of overdetermination and half the time in cases of pre-emption. This is because, where there exists a causal substitute for every factor (real or potential), the conventional But For question will always be answered in the affirmative for each of those factors (but for X, Y would have caused the result; but for Y, X would have caused the result, and so on). Duplicative causation cases, therefore, are those for which the 'notoriously inadequate'[1] traditional But For test will conclude that the

---

[1] J Stapleton, 'Reflections on Common Sense Causation in Australia' in S Degeling, J Edelman and J Goudkamp (eds), *Torts in Commercial Law* (Pyrmont, Thomson Reuters, 2011) 339.

outcome was effectively un-caused. The aim of this chapter will be to show how the NBA is able to deal with such situations in the same way as it deals with the simplest of causal issues.

First, however, the relevant terminology needs some clarification. The terms 'overdetermination' and 'pre-emption' in the literature dealing with duplicative causation are not used in any universally consistent way, which makes the analysis of such situations potentially confusing. A crucial effect of the second stage of the NBA, however, is that it distinguishes between factors which either determine or overdetermine an outcome (which are inculpated) and those factors which have been pre-empted (which are exculpated). In fact, according to this analysis, true pre-emption cases are not even strictly instances of duplicative causation, but only of *potentially* duplicative causation, since true duplicative causation presupposes that at least one factor is operating in a way which duplicates the function of another. By definition, however, where one factor has pre-empted another, that pre-empted factor is no longer operative and so cannot duplicate the work of the other.[2] In NBA terms, then, situations of overdetermined harm are those in which *more than one* breach of duty satisfies both stages of the analysis.[3] These are those cases in which there are multiple sources of risk, more than one of which *ultimately materialises* in the form of the claimant's damage. Pre-empted factors, on the other hand, are those which do not satisfy the second stage of the NBA because the risk generated by them never eventuates in any harm.

Overdetermination (Real Duplicative Causation)

[T]he preferable course is to use the *causa sine qua non* test as the exclusive test of causation. One obvious exception to this rule must be the unusual case where the damage is the result of the simultaneous operation of two or more separate and independent events each of which was sufficient to cause the damage. None of the various tests of causation suggested by courts and writers, however, is satisfactory in dealing with this exceptional case.[4]

Factual situations in which there are multiple factors, each sufficient to bring about a particular result, are commonly referred to as instances of overdetermination. The overdetermination problem has a long history in the literature on causation, partly because it exposes in vivid detail the shortcomings of the traditional But For test. Take, for example,

---

[2] Nonetheless, such cases will be analysed in this chapter because of the instructive nature of their contrast with cases of real duplicative causation.
[3] If the outcome were overdetermined in the sense that one factor was a natural event and the other a breach of duty, the conclusion of an NBA would be that there should be no liability. This is determined by the first stage of the test and is discussed below, under heading 'Combination of Tortious and Non-Tortious Factors'.
[4] *March v Stramare* [1991] HCA 12, (1991) 171 CLR 506 at 533 (per McHugh J).

a case of perfectly symmetrical overdetermination: Suzy and Billy both throw rocks at a window; the rocks strike at the same time, with exactly the same force; the window shatters. Furthermore, each rock strikes with sufficient force to shatter the window all by itself. There is some intuition here that both Suzy's and Billy's throws are causes of the shattering. But this intuition is far from firm.[5]

There is also something intuitively unsatisfying about concluding, as a counterfactual But For analysis is almost bound to,[6] that neither throw caused the window to shatter. So, we instinctively question the polarity of solutions which ascribe causal relevance either to both throws or to neither throw. Our intuition is vindicated here by logical evaluation of the implications of both: holding neither throw to be causative is tantamount to accepting that it was more likely that the window spontaneously shattered, whilst holding both to be causative ignores a fundamental and definitional feature of the situation – that either impact was sufficient to bring about the ultimate result.[7] Unsurprisingly, this is a problem that continues to be vexatious to philosophers, and probably always will, since anyone seeking to establish a perfectly theoretically defensible account of causation on the basis of such facts is facing a monumental challenge. The lawyer, on the other hand, has no such objective, and is saved from having to strive for such logical perfection by the law's need for a pragmatic solution. What is to the philosopher (and perhaps to academics more broadly) a fiercely difficult puzzle is to those practising and making use of the law a more pressing, and yet less searching, question. When lawyers are conducting a causal inquiry in practice, they are performing a specific exercise, distinct from discussions about what it means to say in physical terms that an event was the cause of a result. Consequently, the overdetermination issue is less of a problem in practice than its place in the literature would suggest. Since, as we have seen,[8] the causal inquiry in negligence operates to match defendants who have breached a duty of care with claimants who have suffered corresponding damage, its conclusion does not have to be philosophically unimpeachable in order fully to serve this purpose. The lawyer's task is easier than the philosopher's: 'Under conditions of uncertainty, the choice is not between rejecting a hypothesis and accepting it, but between reject and not-reject'.[9]

In overdetermined harm situations, findings of liability reached on the basis of the NBA will be synonymous with conclusions of not-reject. Often, in such circumstances, a court will have sufficient information to reach a resolution which does not conflict with the objectives of the tort of negligence. First, there needs to be a claimant whose normal course of events has been changed for the worse by at

---

[5] J Collins, N Hall and LA Paul, 'Causation and Counterfactuals: History, Problems and Prospects' in J Collins, N Hall and LA Paul (eds), *Causation and Counterfactuals* (Cambridge, MA, The MIT Press, 2004) 32–33.

[6] Of course, the counterfactual inquiry with which lawyers are familiar – the But For test – is bound to generate this answer. It is not, however, the only counterfactual approach available: see ibid for a number of variations.

[7] See below, under sub-heading 'Several Sufficiency'.

[8] See Ch 2.

[9] PL Bernstein, *Against the Gods: The Remarkable Story of Risk* (New York, NY, Wiley, 1996) 207.

least one breach of duty. Once this has been done, less offence is caused to corrective justice principles by 'not-rejecting' a claimant's case against any defendant whose breach of duty generated the risk of that very outcome occurring[10] than would be by a resolution which rejected such a case because the traditional But For test allowed multiple sufficient factors to absolve one another of causal relevance.

What Constitutes an Overdetermined Event?

Several Sufficiency

Before proceeding any further, however, it is necessary to establish exactly what marks out an event as being overdetermined. Such situations are usually described as those in which there are multiple factors, each sufficient in its own right, to bring about the end result.[11] It is not necessarily true, however, that each factor needs to be sufficient in its own right in order to overdetermine the result, and a more helpful description would be 'severally sufficient'. The significance of this distinction is most obvious in threshold situations[12] in which each factor's input is both identical and simultaneous in its qualitative effect.[13] To use Stapleton's example once more:

> Five members of a club's governing committee unanimously, but independently and in breach of duty, vote in favour of a motion to expel member X from the club, where a majority of only three was needed under the club's rules. The vote of Committee Member Number 1 is neither necessary nor sufficient for the motion to pass. This is true of the vote of each member, yet the motion passed. Where there is a liability rule requiring proof that the vote of the individual voter was a factual 'cause' of a motion passing, the law must recognise this relation of one vote to the passing of the motion as 'causal'.[14]

In such a case, each factor is not sufficient in its own right to reach that threshold and trigger the harm. The resultant harm is nonetheless overdetermined because there exists within that set of factors a subset of factors sufficient to cause the harm. That subset is one which can exist independently of each individual factor because its members are fungible in terms of every factor which has an identical

---

[10] ie any breach which remains 'operative' when the damage occurred – see Ch 3.
[11] eg A Beever, *Rediscovering Negligence* (Oxford, Hart Publishing, 2007) 432; R Stevens, *Torts and Rights* (Oxford, OUP, 2007) 132; HLA Hart and A Honoré, *Causation in the Law*, 2nd edn (Oxford, OUP, 1985) 122–25.
[12] A situation in which there are several factors contributing at the same time to a result which is instantiated when a certain level of contribution is reached – referred to by Stapleton as a 'threshold' – see, eg, J Stapleton, 'Unnecessary Causes' (2013) 129 *LQR* 39, 41.
[13] To be distinguished from quantitative effects, which do not have to be identical in order for over-determined harm to occur, such as where several different defendants each add different quantities of the same pollutant to a body of water. See below, at text to n 50.
[14] Stapleton, 'Reflections on Common Sense Causation in Australia', above n 1 at 340.

and simultaneous effect. If the court is seeking to evaluate whether a particular input (a duty-breaching defendant) was a 'cause' of the claimant's injury on such facts, its traditional inquiry would be 'but for the defendant's breach, would the claimant's injury have occurred?' Since there is sufficiency here, independent of that one breach (ie *any* three of the other inputs), the answer to the orthodox But For question is yes and the defendant avoids liability. The problem with this approach is obvious: asked of any of the five inputs, the question will always elicit an affirmative response. This will generate the bizarre fiction that the result was un-caused, when we know as a matter of external logic that it was actually caused by at least three of those inputs. The adverse outcome on these facts was, therefore, overdetermined, despite the fact that none of the factors was sufficient in *its own right* to bring about the end result.

This suggests that what is important in identifying cases of overdetermined harm is the existence within the entire set of potential causal factors of a particular subset of factors.[15] That subset of factors, which can of course consist of only one factor as it would in the two hunters case, must:

- be sufficient to cause the adverse outcome;
- exist independently of any individual factor;
- be a subset whose membership is fungible within the set of factors which have simultaneous and identical qualitative effects.

The existence of such a subset will be referred to in what follows as 'several sufficiency'. In the *absence* of such a subset, the traditional But For question remains fit for purpose because it would find that the threshold would not have been reached without each factor it examines, thereby correctly inculpating all such factors. In situations in which several sufficiency features, however, the NBA is far more effective.

This concept of several sufficiency does not, however, tell us the whole story of the overdetermined event. In order to be truly overdetermined in legal terms, an event must be one that has multiple causal factors, all of which:

- are severally sufficient;
- are independent of one another;
- lead to indivisible injury;
- lead to the same type of damage;
- affect the claimant simultaneously.

---

[15] See also R Wright, 'Causation, Responsibility, Risk, Probability, Naked Statistics, and Proof: Pruning the Bramble Bush by Clarifying the Concepts' (1988) 73 *Iowa Law Review* 1001, 1035; R Wright, 'Acts and Omissions as Positive and Negative Causes' in J Neyers, E Chamberlain and S Pitel (eds), *Emerging Issues in Tort Law* (Oxford, Hart Publishing, 2007) 289–298; and J Stapleton, 'Choosing What We Mean by "Causation" in the Law' (2008) 73 *Missouri Law Review* 433, 476 for their respective means of dealing with what is here termed 'several sufficiency' through disaggregation of factors. Wright limits his approach to *permitting* disaggregation of factors within NESS, whilst Stapleton favours a more mandatory approach which *ensures* that all contributing factors are inculpated. See also R Bagshaw, 'Causing the Behaviour of Others and Other Causal Mixtures' in R Goldberg (ed), *Perspectives on Tort Law* (Oxford, Hart Publishing, 2011) 370–73.

Multiple Factors Independent of One Another

Unless multiple factors are independent of one another, they lack the capacity to act discretely, and function instead as several contributions to an injury. Dependence also precludes several sufficiency. The full significance of this criterion, therefore, will be clear once both categories have been discussed.[16]

Indivisible Injury as an Outcome

By definition, divisible injuries cannot be overdetermined because they can be broken down into constituent parts which can then be attributed to different causative, or at least potentially causative, factors. Where a divisible injury has been brought about by multiple factors, causative potency can be addressed in relation to each part of the injury on the orthodox basis of *Dingle v Associated Newspapers*,[17] or on the basis of material contribution to injury on the basis of *Bonnington Castings v Wardlaw*.[18] Of course, it is possible that a divided part of a divisible injury might in itself be overdetermined, in which case the court must proceed in relation to this part as it would in relation to an entire injury which is overdetermined. To use Devlin LJ's famous indivisible injury example from *Dingle v Associated Newspapers*:[19]

> If four men, acting severally and not in concert, strike the plaintiff one after another and as a result of his injuries he suffers shock and is detained in hospital and loses a month's wages, each wrongdoer is liable to compensate for the whole loss of earnings. If there were four distinct physical injuries, each man would be liable only for the consequences peculiar to the injury he inflicted . . .[20]

Only a minor modification is needed in order to make this a situation in which the victim suffers from several distinct physical injuries *and* one resultant psychological injury. In this latter case, each defendant could easily be connected to the divided physical effects of his own breach. The whole of the psychological injury resulting from the breaches in aggregate, on the other hand, would constitute an overdetermined injury, for which each defendant's (additional) liability would need to be worked out accordingly.[21]

---

[16] See also Ch 5.
[17] *Dingle v Associated Newspapers* [1961] 2 QB 162 (CA).
[18] *Bonnington Castings v Wardlaw* [1956] AC 613. For a detailed examination of this causal analysis, see Ch 5.
[19] [1961] 2 QB 162 (CA).
[20] [1961] 2 QB 162 (CA) at 189.
[21] Despite the treatment of this issue in *Rahman v Arearose Ltd* [2001] QB 351(CA) such psychological trauma is now likely to be regarded as inherently indivisible. See *Dickins v O2 plc* [2008] EWCA

Must Lead to the Same Type of Injury

This condition is a necessary characteristic of an overdetermined event because, if this were not the case, then the type of injury itself would determine which of the factors was in fact responsible. Where, however, the precise type of injury is not discoverable, it will, for forensic purposes, be overdetermined. Take, as an example, the indivisible injury that is death. Death by decapitation, for instance, is not the same type of injury as death by asphyxiation. So, in the case of a particularly unfortunate and unpopular individual who has been both smothered and decapitated at the same time,[22] if medical evidence can pinpoint a cause of death as being either the smothering or the decapitation, the result is not overdetermined because one cause has pre-empted the other.[23] Where, however, such an identification is not possible, *and the other conditions outlined here for overdetermination are present*, that death will be, for forensic purposes, overdetermined.

Affect the Claimant Simultaneously

The final condition necessary for an event to be classed accurately as overdetermined is that all factors must affect the claimant simultaneously. Here, the verb 'affect' is pivotal: all overdetermining factors must *actually* affect the claimant in the sense that her damage must be an eventuation of the precise risk generated by each.[24] This characteristic is essential in distinguishing events which are overdetermined from those where at least one cause has pre-empted another. Such a distinction is highly significant for the purposes of the NBA because it works so as to exculpate pre-empted factors, whilst inculpating all factors which either determine or overdetermine an event. This will be discussed in more detail below. Suffice it to say for present purposes that the distinction is made at the second stage of the NBA, which discriminates between breaches whose effects remain operative at the time of the claimant's injury and those whose effects have been eclipsed by another factor.[25]

The desert traveller scenario, with suitably amended facts, illustrates this significant distinction.[26] In the classic account, imposter A puts poison in the

---

1144, [2009] IRLR 58 at [45]–[46] 64 (per Smith LJ) and [53] 65 (Sedley LJ). For the implications of this indivisibility to the material contribution point, and a full analysis of this case, see Ch 5.

[22] See following section for the importance of simultaneous effect.

[23] A concept which will be discussed in detail below, under heading 'Pre-emption'.

[24] This phenomenon is dealt with by the NBA using the criterion of breaches which remain 'operative' – see Ch 3. On one level, this might seem an obvious point but, unless it is recognised, criticisms can be levelled against the analysis on the grounds that it is overly promiscuous because it inculpates simultaneous actions, wherever they occur – see MD Green, 'The Intersection of Factual Causation and Damages' (2006) 55 *DePaul Law Review* 671, 682.

[25] See Ch 3.

[26] See Ch 2 and Hart and Honoré, *Causation in the Law*, above n 11 at 238.

traveller's water flask, whilst imposter B later empties that flask with the result that the traveller dies of thirst. On these facts, the death of the traveller is not overdetermined because, whilst there is indeed an indivisible injury (death), it is death by dehydration rather than poisoning, and so the injury is not of the same type as would have been brought about by factor A. Moreover, the effects of both factors did not affect the claimant at the same time: the poisoning did not, in fact, operate to affect the claimant at all because the risk created by putting water in the flask was prevented from ever eventuating by the subsequent emptying of the flask. On slightly modified facts, however, the importance of simultaneity becomes clearer because the other conditions of overdetermination would be met.

The amended facts call for both factors to be causes of the same 'type' of death. Let this be dehydration. So, imposter A, rather than poisoning the water in the flask, empties it. A short time later, imposter B independently makes a large hole in the (now empty) flask. Had the flask not been empty, the hole made by imposter B would have caused all of the water to drain out within seconds. The traveller dies of thirst. Here, we have independent factors all leading to the same indivisible injury (death), an injury which amounts to the eventuation of the risk created by both imposters (death by dehydration), and several sufficiency in that either A or B's act would have brought about the death on its own. It is not, however, an overdetermined situation because of the lack of simultaneity *in the effects* of the two actions. The fact of imposter A's emptying of the flask meant that the traveller had no water as soon as that action had occurred. In other words, the traveller was doomed at this point. As a result of there being no water in the flask when imposter B made the hole, B's action was *never* operative because A's act had pre-empted its effects. In order to make the desert traveller situation one where the traveller's death was overdetermined, the facts would need to be altered so that the effects of factors A and B were simultaneous. For example, if both A and B made similarly effective holes in the flask at the same time, such facts would constitute a situation of overdetermined harm. Significantly, it is not the acts themselves which must be simultaneous; what is important is the simultaneity of their effects.

Double Omissions

There is in tort often a distinction to be made between positive acts and omissions. In terms of defining overdetermined events, however, the distinction is of no relevance. That this is not immediately obvious might be because, in the case law in which there have been multiple potential causal factors, each of which has been an omission as opposed to a positive act, there has also been an interval of time between those omissions occurring. It is, however, the lack of simultaneity, and not the fact that the factors are both omissions, which prevents the outcome from being overdetermined. In the well-known case of *Saunders System Birmingham Co v Adams*,²⁷ for instance (discussed in detail below), the mechanic's failure to repair the car's brakes clearly preceded the hirer's failure to use them. Also, in situations common to product liability, such as those in *Anderson v Hedstrom Corp*,²⁸ in which the manufacturer failed to provide an adequate warning in circumstances in which the claimant would not have read it in any event, the failure to warn clearly pre-dates the failure to read. As we will see, such situations are not overdetermined but are instead examples of pre-emption because the effects of one factor prevent the other factor from ever becoming operative. Consider as an example a claimant with a back injury caused by heavy lifting at work. His physiotherapist had failed to warn him to wear a back support whenever he engages in any lifting activity. His employer had failed to train him how to lift correctly. Consequently, he lifts some furniture incorrectly and whilst not wearing a back support, thereby injuring his back. Each omission was severally sufficient to cause the injury: he undertook the lift in such a way that he would have been injured even had he not had a weak back which required a support and, even had he performed the lift correctly, he would still have injured his back whilst not wearing his support. Thus, such a situation meets all of the criteria identified above as being necessary for results to be overdetermined.

Combination of Tortious and Non-Tortious Factors

One of the implications of overdetermined fact situations being as they are is that, where a defendant's breach is one of two overdetermining factors and the other factor is non-tortious, that defendant will be absolved of liability by stage one of the NBA because the claimant will be unable to prove, on the balance of probabilities, that her damage was caused by the breach rather than by the non-breach factor:

Stage 1: Is it more likely than not that *a* defendant's breach of duty changed the normal course of events so that damage occurred which would not otherwise have done so when it did?

In fact, in NBA terms, this will be the case wherever the number of tortious overdetermining factors equals the number of those that are non-tortious, because it will never be possible on such facts for a claimant to establish causation on the balance of probabilities.²⁹ Whilst, according to Beever, the same no-liability result would be reached on current common law principles,³⁰ he cites no authority to corroborate his suggestion. As it happens, the no-liability approach is one which

---

²⁷ *Saunders System Birmingham Co v Adams* 117 So 72 (Ala 1928). For a detailed discussion of this, see below under the heading 'Pre-emption'.
²⁸ *Anderson v Hedstrom Corp* 76 F Supp 2d 422 (NY 1999).
²⁹ See also discussion of belief probabilities in Ch 2, under heading 'The Balance of Probabilities'.
³⁰ Beever, *Rediscovering Negligence*, above n 11 at 432.

is out of line with the position suggested by the (albeit scarce) case law available in the common law world. There would appear to be no English authority on the point,[31] but the oft-cited US case of *Anderson v Minneapolis, St Paul & Sault-Ste Marie Railway*,[32] in which the defendant was found liable when a fire started by his railway most probably combined with naturally occurring fire[33] to cause property damage to the plaintiff, has been interpreted, notably by Stevens,[34] as standing for the contrary point:

> A difficult case is where the defendant's actions infringe the right of the claimant and at the same moment, the subject matter of the right is destroyed by a natural event. A simple example is where two fires, each of which is independently sufficient to burn down the claimant's house, join together and destroy the house. If one fire is started by the negligence of the defendant and another by a bolt of lightning, can the claimant successfully claim for the value of his property? Although there is no English authority, cases in the United States and the Third American Restatement of Torts conclude that the answer is yes.[35] This seems correct. The defendant has infringed the claimant's right to the house, and a claim for the full value of the right should be capable of being maintained.[36]

This seems incorrect. To say that the 'defendant has infringed the claimant's right to the house' is to beg the question. The very issue with overdetermined factual situations is that we do not know whether factor A or factor B caused a particular result,[37] and, consequently, cannot conclude that the defendant infringed the claimant's right to the house. An individual will only be liable in negligence for making a difference to another's course of events, and whether a defendant made such a difference is precisely what the claimant cannot prove in relation to any one factor in an overdetermined harm case. Where there are an equal number of tortious and non-tortious factors, a claimant will be unable to prove, even by aggregating the breaches of duty, that a breach was more likely than not to have made a difference to his course of events. In his analysis above, Stevens seems to be declaring that the defendant on such facts has infringed the claimant's right because the defendant's right has been infringed: his argument is missing its

---

[31] The cases of *Bailey v Ministry of Defence* [2007] EWHC 2913 (QB), [2009] 1 WLR 1052 and *Dickins v O2 plc* [2008] EWCA 1144, [2009] IRLR 58, despite having been referred to on occasion as being examples of overdetermination, are actually material contribution to injury cases and, as such, are dealt with in Ch 5.

[32] *Anderson v Minneapolis, St Paul & Sault-Ste Marie Railway* 179 NW 45 (Minn 1920).

[33] The reporting of the case is somewhat obfuscatory on this factual point. The later case of *Kingston v Chicago & Northwestern Railway Co* 211 NW 913 (Wis 1927) (discussed below), however, interprets it as having been a case in which '[t]here was nothing to indicate that the fire of unknown origin was not set by some human agency. The evidence in the case merely failed to identify the agency' and '[t]he conclusion of the court exempting the railroad company from liability seems to be based upon the single fact that one fire had no responsible origin, or no known responsible origin' at [3]–[4] 914.

[34] Stevens, *Torts and Rights*, above n 11 at 137.

[35] Stevens refers here to *Anderson v Minneapolis, St Paul and Sault-Ste Marie Railway* 179 NW 45 (Minn 1920) and *Kingston v Chicago & Northwestern Railway* 211 NW 913 (Wis 1927).

[36] Stevens, *Torts and Rights*, above n 11 at 137.

[37] See Hart and Honoré, *Causation in the Law*, above n 11 at 236.

crucial justificatory causal link. He is correct, however, in asserting that the US *Restatement (Third) of Torts* adopts the same approach.

Section 27 of the *Restatement* deals with what it refers to as 'Multiple Sufficient Causes'.[38] Its rule is as follows:

> If multiple acts occur, each of which under § 26[39] alone would have been a factual cause of the physical harm at the same time in the absence of the other act(s), each act is regarded as a factual cause of the harm.

Comment c contains the rationale for this rule:

> A defendant whose tortious act was fully capable of causing the plaintiff's harm should not escape liability merely because of the fortuity of another sufficient cause. That justification is not entirely satisfactory. *Tortious acts occur, with some frequency, that fortuitously do not cause harm. Nevertheless, the actors committing these acts are not held liable in tort.* When two tortious multiple sufficient causes exist, to deny liability would make the plaintiff worse off due to multiple tortfeasors than would have been the case if only one of the tortfeasors had existed. Perhaps most significant is the recognition that, while the but-for standard provided in § 26 is a helpful method for identifying causes, it is not the exclusive means for determining a factual cause. Multiple sufficient causes are also factual causes because we recognize them as such in our common understanding of causation, even if the but-for standard does not. Thus, the standard for causation in this Section comports with deep-seated intuitions about causation and fairness in attributing responsibility.

The italicised words (emphasised for current purposes) contain an argument which is clearly outweighed by the phenomenon discussed immediately below it: a claimant who has suffered the effects of more than one tort.[40] Comment d, however, deals specifically with those situations in which one cause is tortious and the other innocent:

> This section applies in a case of multiple sufficient causes, regardless of whether the competing cause involves tortious conduct or consists only of innocent conduct. So long as each of these competing causes was sufficient to produce the same harm as the defendant's tortious conduct, this section is applicable. Conduct is a factual cause of harm regardless of whether it is tortious or innocent and regardless of any other cause with which it concurs to produce overdetermined harm.[41]

Within the specific parameters of the causal inquiry in negligence, however, arguments about what amounts to a 'factual cause' lose much of their potency,

---

[38] Essentially, this covers the same phenomena referred to herein as 'overdetermining factors'. As outlined above, this is preferable as a label to 'multiple sufficient causes' because there exist real overdetermined situations in which not every factor is sufficient in itself, but where sufficiency exists within the equation independent of the factor being tested.

[39] The rule under § 26 is: 'Tortious conduct must be a factual cause of harm for liability to be imposed. Conduct is a factual cause of harm when the harm would not have occurred absent the conduct. Tortious conduct may also be a factual cause of the harm under § 27.'

[40] This is subject, according to NBA, to the proviso that there were more tortious than non-tortious factors in total.

[41] See also Green, 'The Intersection of Factual Causation and Damages', above n 24 at 683–88.

particularly when considered alongside the burden of proof rule in tort. The nature of the overdetermination problem is such that, as outlined above, those conducting the causal inquiry in negligence are in no position to favour factor A over factor B.[42] Under such conditions of uncertainty, the principle of indifference tells us that it is simply not possible to establish, on the balance of probabilities, that one factor, rather than another, was the relevant cause. This principle, widely attributed in its original form to James Bernoulli,[43] was restated by Keynes as:

> [I]f there is no *known* reason for predicating of our subject one rather than another of several alternatives, then relatively to such knowledge the assertions of each of these alternatives have an *equal* probability. Thus *equal* probabilities must be assigned to each of several arguments, if there is an absence of positive ground for assigning *unequal* ones.[44]

Of course, there is nothing mandatory about this principle, and Keynes himself devoted an entire chapter of his treatise to its criticism, but it is hard to escape from its conclusion, given the particular form of the causal inquiry in cases of overdetermined harm. In fact, the form of this inquiry, with its specific parameters, would put it within the limited situations in which even Keynes would concede its worth. It fits, for example, the following conditions outlined by him as being necessary for the principle to be applicable and effective:

> There must be no *relevant* evidence relating to one alternative, unless there is *corresponding* evidence relating to the other; our relevant evidence, that is to say, must be symmetrical with regard to the alternatives, and must be applicable to each in the same manner.[45]

Weatherford corroborates the point:

> [T]he Principle works best in a situation of ignorance: it is not just that the universe conforms itself to our expectations, it is just that our probability of success in a random guess is a fixed and equal-valued function of the alternatives we judge to be equipossible, *whether or not those alternatives are equipossible in some objective sense*.[46]

This describes overdetermined harm situations perfectly: the several sufficiency and simultaneous effect aspects which characterise them mean that we have no more information relative to any one factor which would allow us to distinguish it in causal terms from any other. Take, for example, the overdetermined fire case, *Kingston v Chicago & Northwest Railway Co*,[47] in which there were two fires, both

---

[42] For current purposes, the example will involve just two factors, one tortious and one non-tortious, although the argument applies equally to any situation in which the number of torts in consideration = the number of non-tortious acts.
[43] And referred to as the principle of non-sufficient reason, until JM Keynes effectively renamed it in chapter IV of his *Treatise on Probability* (Cambridge, CUP, reprinted 1996) 44.
[44] ibid, 44.
[45] ibid, 60 [15]. For those who prefer symbolic representations of such arguments, this too is provided in [16].
[46] R Weatherford, *Philosophical Foundations of Probability Theory* (London, Routledge and Kegan Paul, 1982) 247.
[47] 211 NW 913 (Wis 1927).

of which originated from non-natural causes.[48] Only one defendant was before the court (the precise originator of the other fire being unidentified). The Supreme Court of Wisconsin held the defendant liable on the now-familiar policy basis that to do otherwise would be to allow those in breach of duty to use each other's wrongs as exculpatory excuses at the expense of the injured claimant.[49] On these facts, there are no grounds on which we might prefer fire A to fire B as the legally relevant cause of burning down C's house. We know they were both sufficient, they both happened, and that C's house was burned down by both.

The same argument applies even to overdetermined threshold situations, although it looks at first glance less apposite. In such a case, each factor might be not be equal in size and influence, which might suggest that a 'smaller' factor should have a different causal valence to a 'bigger' factor. Take, for instance, a river that requires eight units of pollutant to become toxic to the crops which draw water from it. Simultaneously, and in breach of duty, defendant A adds three units, defendant B adds three units, defendant C adds two units and defendant D adds four units. Despite the fact that some defendants add more than others, the simultaneity of their action, and the fact that there exists within the equation several sufficiency (ie there is a sufficient subset of factors independent of any one factor being tested), means that there is nothing to choose between them in terms of causal significance for the purposes of negligence.[50] Both types of overdetermined event are, therefore, ideal subjects for the principle of indifference, even within the conceptual limits on which Keynes insisted.

The principle of indifference has been used forensically to deal with the uncertainty relating to future outcomes but, as Eggleston points out,

> if we are asked to guess the result of a toss, after the coin has been tossed, but before its face has been exposed, we are in the same position as if we were asked to guess before the toss was made. As far as the lawyer is concerned, the proposition that – from the point of view of assessing the probabilities – facts that are unknown are in the same position as facts that have not yet happened is of some significance.[51]

Since the law's primary means of dealing with the inevitable uncertainty it faces in terms of past facts[52] is the balance of probabilities standard, and since the defining principle of this standard is 'that it strikes a balance between the interests of the

---

[48] Again, the evidence was not incontrovertible on the point, but this was assumed to be the position for the purposes of the judgment: 'There being no attempt on the part of the defendant to prove that the northwest fire was due to an irresponsible origin – that is, an origin not attributable to a human being – and the evidence in the case affording no reason to believe that it had an origin not attributable to a human being, and it appearing that the northeast fire, for the origin of which the defendant is responsible, was a proximate cause of the plaintiff's loss the defendant is responsible for the entire amount of that loss' 211 N W 913 (Wis 1927) at 915 (per Owen J).

[49] Any Anglo-American comparative analysis of causation is bound to be complicated somewhat by the use in the United States of the concept of 'proximate cause', which is a term that has never been adopted in England. See eg *Paroline v United States et al* No 12-8561, April 23, 2014 (USA) at II.

[50] See also above, n 12.

[51] R Eggleston, *Evidence, Proof and Probability*, 2nd edn (London, Weidenfeld and Nicolson, 1983) 25–26.

[52] See Ch 7 for a discussion of the difference in the law's treatment of past and future uncertainty.

doers and sufferers: provided the victim can prove that the tortious conduct was more probably than not a factual cause of the harm, damages can be recovered for that harm',[53] the principle of indifference is ideally suited to the task. In overdetermined harm cases, in which the number of breach factors is equal to the number of non-breach factors, the principle of indifference will prevent a claimant from satisfying the first stage of the NBA, meaning that her claim would be unsuccessful:

> [W]hether D ... is liable to P for the loss ... depends upon whether the evil that befell P was a result of a civil wrong by another person or not. If it was, D is liable, if it was not, D is not liable. It may seem strange that D's liability should depend on this extraneous factor, but the conclusion appears to be unavoidable.[54]

There is very little case law concerning overdetermined events, and even less involving a combination of breach factors and non-breach factors. This may be because of the intrinsically unusual nature of the fact combinations necessary to make up such a case, or maybe because, as suggested in the *Restatement*,[55] legal professionals often fail to identify such cases correctly, and treat them accordingly. Applying an NBA means there is no need for any prior identification of 'type' in order for the correct answer to be reached.

All Factors Tortious

The situation is far more straightforward where there are more overdetermining breach factors than non-breach factors. The very nature of overdetermining factors means, as explained above, that the principle of indifference allows us to allocate to each an equal causal valence. For instance, were there to be four overdetermining factors, each would have a causal valence of ¼. If two are breaches and two are non-breaches, it is easy to see that the aggregate causal valence of the breach factors (½) is equal to the aggregate causal valence of the non-breach factors (½).[56] Where, on the other hand, three are breaches and only one is a non-breach, it is clear that it is more likely (¾) than not (¼) that that a breach had legal causal significance. On the latter facts, but not the former, the first stage of the NBA will be passed because the claimant will be able to prove, on the balance of probabilities, that her course of events was altered by at least one defendant's breach of duty.[57] Consequently, all defendants whose breaches remain operative at

---

[53] P Cane, *Atiyah's Accidents, Compensation and the Law*, 7th edn (Cambridge, CUP, 2006) 114. See also K Barker, P Cane, M Lunney and F Trinidade, *The Law of Torts in Australia*, 5th edn (Oxford, OUP, 2012) 538. See Ch 7.
[54] G Williams, 'Causation in the Law' (1961) 19 *CLJ* 62, 76. See also *Kingston v Chicago & Northwest Railway Co* 211 NW 913 (Wis 1927) at 914–15: '[W]e are not disposed to criticise the doctrine which exempts from liability a wrongdoer who sets a fire which unites with a fire originating from natural causes, such as lightning, not attributable to any human agency, resulting in damage'.
[55] § 27, Reporter's Notes to comment d.
[56] Where the question is 'What is the likelihood of the cause being either breach A or breach B?', the probabilities are added together: ¼ + ¼ = ½.
[57] See also M Geistfeld, 'The Doctrinal Unity of Alternative Liability and Market Share Liability' (2006) 155 *University of Pennsylvania Law Review* 447, 472 for his 'evidential grouping' theory; a

the time the claimant's injury was sustained, and so satisfy the second stage as well, will be subject to liability. A defendant cannot, therefore, escape liability on the basis that there exists at least one other duplicative breach. In its aggregative approach, the NBA avoids what would otherwise be a potential problem for claimants affected by more than two defendants, as identified by Wright:

> [I]f there are more than two defendants, and the standard for persuasion is satisfied by a mere 50+ per cent mathematical probability, each defendant ordinarily would easily be able to prove that she was not the cause of the injury, even though it is certain that one of them caused the injury. For example, if there were three defendants, each equally likely to have been the cause of the plaintiff's injury, each defendant can 'prove' that she was not the cause, since there is a 67 per cent probability that she was not the cause, which leads to the paradoxical result that it can be 'proven' that none of the defendants was the cause, even though we know that one of them was the cause.[58]

One obvious justification for the law not wanting to subject claimants to such a difficulty has been outlined above, as adopted by the US *Restatement*.[59] On a similar justificatory note, according to the NBA, all defendants should be liable on such facts because it is more likely than not that *a* breach caused the harm, and defendants should not benefit, any more than claimants should lose out, from the fact that there have been multiple breaches of duty:

> If two hunters are negligent and liable as independent tortfeasors for firing in the direction of a third hunter who is injured thereby, the innocent wronged hunter should not be deprived of his right to redress where the matter of apportionment of damages is incapable of proof. The wrongdoers should be left to work out between themselves any apportionment.[60]

This excerpt is from the infamous case of *Summers v Tice*,[61] a case decided by the Supreme Court of California, and remarkably similar on its facts to two other cases: one decided a few years later by the Supreme Court of Canada, *Cook v Lewis*;[62] and the other a much earlier judgment of the Supreme Court of Mississippi, *Oliver v Miles*.[63] All three cases concerned the now classic scenario in which multiple defendants each discharged their firearms in breach of duty, and a claimant was consequently injured by a shot or shots. The problem in each case was that it

---

method which achieves much the same as the first stage of the NBA. The second stage of the NBA, unlike Geistfeld's evidential grouping, *does* discriminate between defendants whose breach is still operative and those whose breach is not. Owing to its second limb, the NBA could not be subject to a rejection on the same grounds – see, eg, *Menne v Celotex Corp* 861 F 2d 1453, 1459–61 (Kan 1988), and is also able, unlike evidential grouping, to distinguish between overdetermined and pre-empted cause situations – see Geistfeld at 465, text to n 52.

[58] R Wright, 'Proving Causation: Probability v Belief' in Goldberg (ed), above n 15 at 212. See also M Geistfeld, 'The Doctrinal Unity of Alternative Liability and Market Share Liability' (2006) 155 *University of Pennsylvania Law Review* 447, 455.

[59] Above, text to n 38.

[60] *Summers v Tice* 199 P 2d 1 (Cal 1948) at headnote (9).

[61] 199 P 2d 1 (Cal 1948).

[62] *Cook v Lewis* [1951] SCR 830 (SCC).

[63] *Oliver v Miles* 110 So 666 (Miss 1926).

was not possible for the court to ascertain which of the defendants caused the claimant's injuries, although it is of course clear that, but for *a* breach, the claimant would not have been injured at all. In the first two cases, as the excerpt above suggests, the courts dealt with the overdetermination problem by reversing the burden of proof, and holding both defendants jointly liable until, and to the extent that, each could establish he did *not* cause the claimant's injuries. This is known in the US as 'alternative liability'.[64] In *Oliver v Miles*, the same result was reached by means of an apparently different method, viewing both defendants as having been engaged in a joint enterprise:

> We think that they were jointly engaged in the unlawful enterprise of shooting at birds flying over the highway; that they were in pursuit of a common purpose; that each did an unlawful act, in the pursuit thereof; and that each is liable for the resulting injury to the boy, although no one can say definitely who actually shot him. To hold otherwise would be to exonerate both from liability, although each was negligent, and the injury resulted from such negligence.[65]

The essence of this particular forensic quandary – that some negligence apparently caused the injury, but that it is impossible to link the injury to one breach rather than another – is negated by the first stage of the NBA. In being satisfied by aggregated breaches of duty, the initial step in this analysis avoids the obviously erroneous outcome of exonerating all negligent defendants, and inculpates them instead. The only breaches then to avoid being designated causes would be those that are no longer operative at the time the injury occurs,[66] and 'operative' refers only to those breaches which create the risk which actually eventuated in the claimant's harm. Since this is true of each of the defendants' negligent actions in the three shooting cases (because each claimant suffered from gunshot wounds), the result of the NBA would be the same as the outcomes in the cases themselves: joint liability of all duty-breaching defendants. It avoids, however, both the subversion of orthodox rules of proof and the adoption of mercurial and thorny concepts such as joint enterprise.[67] It is important to emphasise, however, that since the NBA employs aggregation only in the first of its two steps, it does not *allocate liability* in any aggregative sense: individual defendants are, through the second stage of the analysis, identified as legally relevant factors only when the risk generated by their individual breaches remains operative on the claimant. This sets the NBA apart from other aggregative approaches and therefore from the objections that can be raised in response to them. It does not do, for instance, what Geistfeld's 'evidential

---

[64] American Law Institute, *Restatement of the Law, Third, Torts: Liability for Physical and Emotional Harm* (St Paul, MN, American Law Institute, 2010) § 28.

[65] 110 So 666 (Miss 1926) at [3] 668. The reference herein to joint engagement in an unlawful enterprise was presumably the means chosen by the court to ensure that the defendants were jointly, as opposed to severally, liable to the claimant.

[66] On which, see below under heading 'Pre-emption', and Ch 3.

[67] The latter of which is both ill-suited to non-intentional torts (as the court in *Cook v Lewis* [1951] SCR 830 (SCC) recognised) and, even in the criminal law, in such a state of flux as to make it unhelpful. See also PS Davies, *Accessory Liability* (Oxford, Hart Publishing, 2015) ch 3.

grouping' approach does, which is to make 'each defendant responsible for the way in which his tortious conduct interacted with the tortious conduct of the other defendant'.[68] Although there might well be interaction between the risks created by individual breaches, this is incidental to, rather than determinative of, liability under the NBA.[69]

A set of facts which presents a potential problem for this analysis is *Sindell v Abbott Laboratories*,[70] a case in which the Supreme Court of California developed a famously innovative response to overdetermined harm, where only a small proportion of those who created the relevant risk were defendants before the court. The claimant in that class action had developed adenocarcinoma, in common with many other daughters of women who had taken Diethystillbestrol (DES) during their pregnancies. This drug, marketed as being capable of preventing miscarriages, was eventually produced by approximately 200 different companies, but was then supplied to wholesalers and prescribed to patients (as is standard practice) under its generic, as opposed to its brand name. This situation, in combination with the long latency period before the effects of DES became known (essentially a whole generation, between the claimant's development in utero and her early adulthood when the disease manifested itself)[71] meant that it was virtually impossible for either claimant or defendant to establish which manufacturer was responsible for supplying the drugs actually taken by the claimant's mother. So, every defendant had breached its duty of care and created a risk of adenocarcinoma in the daughters of those who took its product during pregnancy, but only five out of the 200 manufacturers were before the court, meaning that the producers of the drugs actually taken by the claimants' mothers could well not be a party to the action.

This is a case, therefore, which fulfils all of the criteria of an overdetermined harm situation, but differs from those already identified in an important respect. Whereas holding both defendants jointly liable (or applying the US approach of alternative liability) in the hunter cases would mean that liability would be imposed on the defendant actually responsible for the claimant's injury, this would not necessarily have been the result in *Sindell*, because that particular manufacturer might well not have been one of those before the court.[72] On these facts, the first stage of the NBA is clearly passed, because all of the sources of DES were duty-breaching defendants, so it was more likely than not that the claimant's normal course of events was changed for the worse by at least one breach of duty. The difficulty arises at the second stage because it was not possible to say whether the effects of any of the defendants' breaches of duty were *ever* effective on the claim-

---

[68] Geistfeld, 'Doctrinal Unity', above n 57, and D Fischer, 'Insufficient Causes' (2006) *Kentucky Law Journal* 278.
[69] Which basis could instead be characterised as a breach whose effects are operative on the claimant.
[70] *Sindell v Abbott Laboratories* 26 Cal 3d 588 (Cal 1980).
[71] 12–25 years, according to R Goldberg, *Causation and Risk in the Law of Torts* (Oxford, Hart Publishing, 1999) 56.
[72] Hart and Honoré, *Causation in the Law*, above n 11 at 424.

ant. The second stage could not, therefore, be satisfied.[73] The same problem arose on the conventional analysis applied in that case:

> As defendants candidly admit, there is little likelihood that all the manufacturers who made DES at the time in question are still in business or that they are subject to the jurisdiction of the California courts. There are, however, forceful arguments in favor of holding that plaintiff has a cause of action. In our contemporary complex industrialized society, advances in science and technology create fungible goods which may harm consumers and which cannot be traced to any specific producer. The response of the courts can be either to adhere rigidly to prior doctrine, denying recovery to those injured by such products, or to fashion remedies to meet these changing needs . . . The most persuasive reason for finding plaintiff states a cause of action is that . . . as between an innocent plaintiff and negligent defendants, the latter should bear the cost of the injury.[74]

The solution proposed by the court in that case was to apportion liability according to each defendant's market share, on the basis that this should approximate to their relative contributions to the injuries caused by the drug in aggregate, (the court estimating that

> the number of women who took the drug during pregnancy range from 1½ million to 3 million. Hundreds, perhaps thousands, of the daughters of these women suffer from adenocarcinoma, and the incidence of vaginal adenosis among them is 30 to 90 percent.[75]):
>
> . . .
>
> [It is] reasonable in the present context to measure the likelihood that any of the defendants supplied the product which allegedly injured the plaintiff by the percentage which the DES sold by each of them. . .bears to the entire production of the drug sold by all.[76]

---

[73] This is not a problem for Geistfeld in applying his evidential grouping approach, since all that is required for that to be relevant is a violation of the principle of evidential grouping: 'According to that principle, each defendant cannot avoid liability by relying upon the tortious conduct of other defendants when that form of exculpatory proof would enable all of the defendants to avoid liability, despite the plaintiff's proof that each defendant may have tortiously caused the harm, and at least one of them did' (M Geistfeld, 'The Doctrinal Unity of Alternative Liability and Market Share Liability' (2006) 155 *University of Pennsylvania Law Review* 447, 472). Another means of addressing this (and other) evidential problems has been suggested by Porat and Stein in the form of their 'evidential damage doctrine'. In A Porat and A Stein, 'Liability for Uncertainty: Making Evidential Damage Actionable' (1997) 18 *Cardozo Law Review* 1891 and A Porat and A Stein, *Tort Liability under Uncertainty* (New York, OUP, 2001) chapters VI and VII (and particularly 186–88), the authors suggest replacing liability *under* uncertainty with liability *for* uncertainty, and so holding defendants liable for the damage they do to a claimant's ability to establish her case. According to their thesis, this would take the form either of reversing the burden of persuasion or of awarding damages for the value of the evidence lost. On the facts of *Sindell*, as it happens, they suggest that the result reached would be similar to that reached through market share liability. This is more a question of gist than it is of causation, since it requires reformulation of the damage requirement in negligence rather than any reconsideration of causal principles per se. See also Ch 6, text to n 109.
[74] 26 Cal 3d 588 (Cal 1980) at 610–11.
[75] 26 Cal 3d 588 (Cal 1980) at 597.
[76] 26 Cal 3d 588 (Cal 1980) at 611–12.

In essence, what the Supreme Court of California did was to aggregate defendants and potential claimants in an attempt to achieve something approximating corrective justice as between two *classes* of individuals.[77] In a later case, *Hymovitz v Eli Lilly & Co*,[78] this was taken a step further, as market share liability was imposed even where a producer could establish that his product did not cause a particular plaintiff's injury because it had not been used.[79]

Where it is clear that all of the claimants' damage was caused by at least one breach of duty, but it is also known that several duty-breaching defendants responsible for creating the risk of the injuries which actually occurred are not before the court, the imposition of joint and several liability[80] would be difficult to justify. On the other hand, allowing all duty-breaching defendants to escape liability when it is clear that they all created the same risk of injury, and it is highly probable that many of those risks eventuated in the claimants' injuries, would be no more defensible. A strict application of the NBA would militate against the imposition of liability because the effects of no defendant's breach could be proved to be operative on any particular claimant when the damage occurred.[81] In the light of the inherent and specific difficulties of the case, therefore, market share liability would seem to achieve a reasonable compromise between requiring either claimants as a class or defendants as a class to bear the whole of the loss.[82] Indeed, Teff describes the approach as a 'practical solution to the problems created by harmful fungible goods which cannot be traced to any specific producer'.[83] Hart and Honoré claim that, in imposing market share liability,

> in effect the court dispenses with the need to prove fault and causal connection and instead treats the manufacturers of the drug as collectively insuring, in proportion to the market share of each, those who suffer harm after using the drug. As Richardson J.

---

[77] Although the Court did add the proviso that, in order for market share liability to be imposed, those before the court must represent a substantial share of the market in question. For a trenchant criticism of this, see Goldberg, *Causation and Risk*, above n 71 at ch 2.

[78] *Hymovitz v Eli Lilly & Co* 541 NYS 2d 941 (NY 1989).

[79] Goldberg, *Causation and Risk*, above n 71 at 66.

[80] Market share liability, by contrast, is several but not joint.

[81] In contrast to *Fairchild v Glenhaven Funeral Services Ltd* [2002] UKHL 22, [2003] 1 AC 32 (for a full discussion of which, see Ch 6), in which all employers' breaches were operative in the sense that they had physically exposed each claimant to the risk of developing mesothelioma. In *Sindell*, however, it was impossible to say whether the products of any of the five defendants ever even came into contact with the claimant, regardless of their causal effect.

[82] See also R Wright, 'Liability for Possible Wrongs: Causation, Statistical Probability and the Burden of Proof' (2008) 41 *Loyola Los Angeles Law Review* 1295, 1326–30; Geistfeld, 'The Doctrinal Unity of Alternative Liability and Market-Share Liability', above n 73 at 451; and Porat and Stein, *Tort Liability*, above n 73 at 135, who see this situation as apt for an application of their conception of collective liability: 'This outcome is perfectly aligned with corrective justice'.

[83] H Teff, 'Market Share' Liability – A Novel Approach to Causation' (1982) 31 *ICLQ* 840, 842. Although he also points out that a preferable practical approach to such inherent problems would be an administrative no-fault scheme funded by manufacturers of generic products. See also American Law Institute, *Restatement of the Law, Third, Torts: Liability for Physical and Emotional Harm*, above n 64 at § 28, comment p.

points out in his dissent, this is a radical departure from traditional conceptions of tort law, with their emphasis on the matching of claimants and defendants.[84]

Despite the interesting nature of the particular causal problem in these cases, however, market share liability has never been adopted in English law[85] and, even in the US,

> [m]arket-share liability, while an important and controversial issue two decades ago, has become a modest doctrine without widespread implications for tort law. It has been limited to DES and a handful of smaller case congregations in which the products pose equivalent risks to each other.[86]

Owing to the particularly unusual facts of *Sindell*,[87] market share liability appears to have had almost a niche relevance to that specific situation. As now recognised by the US *Restatement (Third) of Torts*,[88] and suggested by the continued absence of any analogous commonwealth decision, the doctrine itself is of diminishing practical significance.[89]

Pre-emption

As outlined above, situations involving pre-empted causes are not strictly cases of duplicative causation, but only of *potentially* duplicative causation.[90] Pre-empted factors can easily be distinguished from overdetermining factors (genuinely duplicative causes) by the fact that the former are not operative when the claimant's injury occurs, whilst overdetermining factors are, by definition, all operative when that damage materialises.[91] Furthermore, it is important to make clear that pre-empted factors are those that *never* become operative, so as to distinguish them

---

[84] Hart and Honoré, *Causation in the Law*, above n 11 at 424 (footnotes omitted).
[85] Nor given much attention by academics – see K Oliphant, 'Causal Uncertainty and Proportional Liability in England and Wales' in I Gilead, M Green and BA Koch (eds), *Proportional Liability: Analytical and Comparative Perspectives* (Berlin, De Gruyter, 2013) n 34.
[86] MD Green, 'Causal Uncertainty and Proportional Liability in the US' in I Gilead, M Green and BA Koch (eds), *Proportional Liability: Analytical and Comparative Perspectives* (De Gruyter, Berlin, 2013) 357.
[87] In particular the fact that the drug was not patented and so could be sold by many different producers, and the fact that all of the producers involved had the same marketing and production processes, meaning the risks generated by each were equivalent.
[88] See also American Law Institute, *Restatement of the Law, Third, Torts: Liability for Physical and Emotional Harm*, above n 64 at § 28, comment p.
[89] The broader concept of proportional liability in some form remains of some significance, but it is of far greater academic than practical import – see MD Green, 'The Future of Proportional Responsibility' in S Madden (ed), *Exploring Tort Law* (New York, CUP, 2005) throughout, but particularly 396–99. See also Chs 7 and 8.
[90] See R Wright, 'Causation in Tort Law' (1985) 73 *California Law Review* 1735, 1775.
[91] See also American Law Institute, *Restatement of the Law, Third, Torts: Liability for Physical and Emotional Harm*, above n 64 at § 27, comment e: the difference between alternative causes, where one factor operates instead of another, and overdetermining causes, where both factors operate contemporaneously.

from those factors which have had an effect on the claimant, but have been superseded by the effect of another factor.[92] Pre-empted factors are those which expose the claimant to a *risk* of injury, but do so in situations in which that risk never eventuates in harm to the claimant, because the injury which actually occurs is brought about by a different and independent factor.

Significantly, however, it is not the case that a factor must chronologically precede another for it to amount to pre-emption, as demonstrated by the well-known facts of *Saunders System Birmingham Co v Adams*.[93] In that case, a car hire company had allowed one of its vehicles to be leased with faulty brakes. The car was then involved in a collision which injured the claimant, but the claimant was struck following the driver's failure to use the brakes in a timely fashion[94] (without knowledge that they were faulty). This is a situation of pre-empted cause: the driver's failure to brake pre-empted the faulty brakes from becoming an operative factor, even though, chronologically, the brakes were faulty before the driver failed to use them. The claimant was at no point affected by the faultiness of the brakes because, although they created a risk of injury to pedestrians, that risk never came to fruition. Instead, it was the wholly separate and independent risk[95] created by the driver's breach of duty which actually culminated in the injury. This simple but crucial conclusion is one that we can reach on the basis of 'what we know about the natural processes that can plausibly be described as involving "one thing bringing about another"'.[96]

This is another situation in which the NBA is an obvious improvement on the But For test. In *Saunders*-type situations, the But For test is unable to distinguish between pre-empting and pre-empted factors because, again, asked of each factor in turn, the answer will be an affirmative one, which will deny all factors causal relevance. This is because, if the pre-empting cause is removed from the equation (in this example, the failure to brake), then in the resulting hypothetical, the driver must have attempted to brake. Once the driver attempts to brake, the effect of the mechanic's breach in failing to fix the brakes becomes operative, meaning that the damage would have occurred even without the driver's breach. Treating both breaches as causes, however, is not the correct approach because it does not deal

---

[92] See Ch 3.
[93] 117 So 72 (Ala 1928).
[94] Certainly relative to the speed at which she was said to have been travelling.
[95] Whilst the risk created by faulty brakes and the risk of not using them are similar in terms of the type of harm they are likely to bring about, they remain in this context independent because the behaviour giving rise to each risk was neither affected by, nor did it affect, the behaviour which brought about the other. Where two or more instances of risk-creating behaviour are not independent in this way, the situation is not one of pre-emption but becomes a question of supersession (see Ch 3, text to n 98), since the lack of independence makes the question of which breaches' effects are operative harder to answer. For instance, where one breach of duty consists in the failure to prevent behaviour of a particular type (as in *Home Office v Dorset Yacht Co Ltd* [1970] AC 1004 (HL) and *Reeves v Metropolitan Police Commissioner* [2001] 1 AC 360 (HL)), the risk created by the latter behaviour cannot necessarily be disassociated from the former. See also E Weinrib, *Corrective Justice* (Oxford, OUP, 2012) 44.
[96] According to Bagshaw, the legal concept of cause should not be inconsistent with conclusions reached on such bases – see Bagshaw, 'Causing the Behaviour of Others', above n 15 at 366 and 372.

with past *fact* but with past *possibilities*. In so doing, it arbitrarily inculpates some potentially causative factors whilst ignoring (and so exculpating) others.

For instance, on the facts of *Saunders*, the mechanic's negligent failure to repair the brakes could have caused an accident but, in the event, it did not.[97] Were the causal inquiry in negligence to ascribe liability to all such *potential* causative factors, there would often be a long, and increasingly irrelevant, list of 'causes' of a particular event. This is why even an inquiry which is as concerned with facts as that of causation in negligence must have normative parameters imposed at its outset.[98] Deeming to be causative those potential (but, in the event, pre-empted) factors which happen, by either temporal or heuristic accident, to be before the court, is arbitrary, inconsistent and therefore legally objectionable. Most, if not all, factors which are ultimately found (correctly) to have caused a claimant's injury will have pre-empted other potentially causative factors in doing so: there is no justification for ascribing liability to some of these (those before the court) and not to others (those not before the court). If the only reason such pre-empted potential factors are inculpated by the causal inquiry is that they are But For causes, this is insufficient, because it fails to distinguish them from other pre-empted potential factors. As Nagel points out, it is actual as opposed to hypothetical, consequences on which the law bases its decisions:

> [I]f certain surrounding circumstances had been different, then no unfortunate consequences would have followed ... and no seriously culpable act would have been performed; but since the circumstances were *not* different, and the agent *in fact* succeeded in perpetrating a particularly cruel murder, *that* is what he did, and that is what he is responsible for.[99]

Stevens defends the same position on the basis of his rights model:

> The mechanic never infringed any right of the claimant by his careless repair of the brakes. The mechanic owes the driver a contractual duty carefully to repair the car, a duty he has breached, but the claimant is not privy to this right of the driver. If, however, the driver had relied upon the brakes having been repaired by hitting them to avoid injuring the claimant, then the mechanic would have infringed the claimant's right to bodily safety. However, the risk he ran which potentially made his careless conduct a violation of the claimant's right to bodily safety, was the potential failure of the brakes. This risk never eventuated. It is as if he has shot the claimant but, before his bullet has struck, the driver's bullet had already hit home. You cannot kill a corpse.

---

[97] It created a risk which never, as it happened, eventuated in any harm. See JL Mackie, *The Cement of the Universe: A Study of Causation* (Oxford, Clarendon Press, 1980) 46.

[98] R Fumerton and K Kress, 'Causation and the Law: Preemption, Lawful Sufficiency, and Causal Sufficiency' (2001) 64 *Law and Contemporary Problems* 83, 86.

[99] T Nagel, 'Moral Luck' in T Nagel (ed), *Mortal Questions* (Cambridge, CUP, 1979) 35. Whilst this example clearly uses a criminal law example, it applies a fortiori to tort since, as Perry points out (S Perry, 'Honoré on Responsibility for Outcomes' in P Cane and J Gardner (eds), *Relating to Responsibility: Essays in Honour of Tony Honoré on his 80th Birthday* (Oxford, Hart Publishing, 2011) 73) outcome responsibility is more universally applicable to the latter than it is to the former (for which action responsibility is often more relevant). See also P Cane, *Responsibility in Law and Morality* (Oxford, Hart Publishing, 2002) 139.

By contrast, the driver has infringed a right of the claimant by his driving. If he had not driven the car, the claimant would not have been injured ... The driver would not have infringed the claimant's right to bodily safety 'but for' his careless driving.[100]

Nobody has a right not to be exposed to an increased risk. The right is not to be harmed as a result of that increased risk.[101] The NBA takes the same basic view: to say that X created a risk of Y happening is in no way synonymous with saying that X caused Y.

As an example to illustrate the practical importance of this approach, take a modified version of *Saunders*. The driver failed to brake and so hit the claimant. Prior to that, a mechanic working for the car hire company had negligently failed to check the brakes on the vehicle. That same mechanic, however, had not been trained properly to fix defective brakes and so, even had he made the necessary checks, he would have been unable safely to rectify the problem, meaning the brakes would still be faulty following his (hypothetical) non-negligent action. The mechanic's deficient training was a result of his tutor's negligent instruction 18 months before. This latter fact is not before the court because it has not yet come to light; not even the mechanic himself knows that his technical ability falls short of a reasonable standard. On the basis of the NBA, this does not matter because the effect of his tutor's breach was never allowed to become operative. In fact, it was two steps removed from being operative; the mechanic's own negligence could have pre-empted it, and that of the driver of the car did in fact pre-empt it. On the basis of the orthodox But For test, however, there is nothing in substantive causal terms to distinguish the tutor's breach from the mechanic's; it is merely heuristic opportunity which dictates that the court gets to consider the latter but not the former (and, non-operative causes are, by definition, far less likely to be forensically traced). Nevertheless, the orthodox But For test would, were the tutor's breach forensically available, deem that to be a causative factor as well. This casts significant doubt on the ability of the But For test to be an authentic means of identifying legally relevant causes. Potential causal factors are pre-empted all the time, often without recognition or acknowledgement. It is essential, therefore, for the law not arbitrarily to recognise this in relation to some factors and not others. Potential causal factors which have been pre-empted are those that never reach the stage of being operative in bringing about the claimant's damage. They can, therefore, only ever be *hypothetical* causes and should not, merely by virtue of capricious submission to the But For test, be thereby converted into legally relevant causes. The But For test, applied in the orthodox manner, actually has the effect of converting legally irrelevant factors into legally relevant causes if it is used to analyse pre-empted cause situations. It is not, therefore, fit for purpose in this context.

This is the area in which the NBA differs most significantly from NESS. The difference between the two can be accounted for by the relative weight each

---

[100] Stevens, *Torts and Rights*, above n 11 at 136.
[101] ibid, 43–52.

attaches to the concept of sufficiency. Using sufficiency to identify relevant causal factors, as the NESS test does, means that, despite its most famous proponent's avowed intention,[102] it inevitably strays from the actual to the hypothetical in certain multiple factor situations; notably, those involving pre-emptive causes. The requirement that a factor be a necessary element of *a* sufficient set of antecedent conditions leading to a claimant's injury means that the NESS test can be too inclusive. In order to be relevant in NBA terms, a breach needs to be *necessary to the injury which in fact occurred* or, in Hart and Honoré's terms, 'necessary on the particular occasion'.[103] By contrast, giving causal recognition to factors *necessary to any sufficient set* (which is what occurs in both Professor Wright's formulation of NESS and Hart and Honoré's earlier version of it[104]) can lead to the inclusion of hypothetical, as opposed to actual, factors. According to Beever: 'The problem with the NESS test is the same as the problem with the but for ... test: the reliance on necessity'.[105] On the contrary, the problem with the NESS test is its concession to *sufficiency* as an instrumental means of establishing liability.[106] This is the crux of the division between (all versions of) NESS and NBA; NESS insists on the primacy of sufficiency and NBA on the primacy of necessity. It is in situations of duplicative and potentially duplicative cause that the distinction is both apparent and definitive.

A foundational principle of NESS is the concept of weak necessity:

> Under this weak sense of necessity, which is also referred to as strong sufficiency, necessity is subordinated to sufficiency: a causally relevant factor need merely be necessary for the sufficiency of a set of conditions sufficient for the occurrence of the consequence, rather than being necessary for the consequence itself as in the *sine qua non* account.[107]

The problem with this aspect of NESS is exacerbated by the fact that Wright's own interpretation of it does not align with others' understanding of it, and this is a particularly acute problem where pre-emptive cause situations are concerned. Wright himself asserts that NESS, applied to the facts of *Saunders*, would give the result that the driver's failure to brake pre-empted the faulty brakes factor, meaning that *only* the driver should be deemed a cause.[108] Beever, on the other hand, understands NESS as leading in that case to the exoneration of the driver. Whilst he has reached the wrong conclusion on the facts, Beever is right in detecting an inconsistency in NESS's treatment of pre-emptive cause situations. Beever's mistake lies in his[109] seeing the 'sufficient set' element of NESS as being singular and

---

[102] See Wright's own reference to his claims in R Wright, 'The NESS Account of Natural Causation: A Response to Criticisms' in R Goldberg (ed), *Perspectives on Causation* (Oxford, Hart Publishing, 2011) n 14.
[103] Hart and Honoré, *Causation in the Law*, above n 11 at 113.
[104] ibid, 122–28, 235–53.
[105] A Beever, *Rediscovering Negligence* (Oxford, Hart Publishing, 2009) 425.
[106] See Ch 2, under heading 'Wright'.
[107] Wright, 'The NESS Account', above n 102 at 286.
[108] The NBA would reach the same result.
[109] And not only his: Fumerton and Kress make a similar error in 'Causation and the Law', above n 98 at 92–95.

concrete, when it is in fact plural and abstract: for NESS, a factor has only to be a necessary element of *a* sufficient set,[110] rather than *the* sufficient set which was instantiated when the injury occurred:

> The essence of the concept of causation under this philosophic account is that a particular condition was a cause of (condition contributing to) a specific consequence if and only if it was a necessary element of a set of antecedent actual conditions that was sufficient for the occurrence of the consequence. (Note that the phrase 'a set' permits a plurality of sufficient sets.)[111]

This is fundamental to NESS, and is the feature which led to its being described in Chapter 2 as 'longitudinal and algorithmic' as opposed to 'latitudinal and explanatory'. The problem with this aspect of NESS, however, is that it allows digression into the realms of hypothesis. Worse is that, according to Wright's use of it, it does not even do so consistently. The merit of NESS is that, in being algorithmic, it is able to provide an account of factual causation which is broader than the bespoke legal inquiry with which the NBA is concerned. Applied consistently, however, any algorithm which inculpates negligent driving on the facts of *Saunders* should equally inculpate the faulty brakes. Both are necessary elements of *a* sufficient set antecedent to the harm caused; the driving to a set with functional brakes and the brakes to a set with non-negligent driving, and yet Wright maintains that only the driving factor should be deemed a relevant cause. Beever's criticism is, therefore, correct in form but not in substance: the NESS test, properly applied, fails to differentiate between pre-empting and pre-empted causes, but it over-inculpates, rather than over-exculpates, as Beever suggests.[112] Wright's answer to this is to appeal to causal priority, so that, whilst occurring later in temporal terms, the negligent driving was *causally* prior to the instantiation of the defective brakes. Consequently, according to Wright: 'The failure to attempt to use the brakes pre-empted the potential negative causal effect of the other non-instantiated conditions in the braking-stops-car causal process'.[113] He then adds: 'This conclusion is based on our knowledge of the sequence of events that must take place for the occurrence of the braking-stops-car causal process . . .'.[114]

If NESS really does 'capture the concept of causation',[115] as Wright claims, it is difficult to see why its operation should require an appeal to extrinsic knowledge

---

[110] Or, as Wright has put it in response to Fumerton and Kress, 'necessary for the sufficiency of a sufficient set' in R Wright, 'Once More into the Bramble Bush: Duty, Causal Contribution and the Extent of Legal Responsibility' (2001) 54 *Vanderbilt Law Review* 1071, 1102.

[111] Wright, 'Causation', above n 90 at 1790.

[112] See also the examples given in Fumerton and Kress, 'Causation and the Law', above n 98 at 100.

[113] R Wright, 'Acts and Omissions as Positive and Negative Causes' in Neyers, Chamberlain and Pitel (eds), *Emerging Issues in Tort Law*, above n 15 at 304. See also Wright, 'Once More into the Bramble Bush' above n 110 at 1130. For a reiteration and defence of the same position, see also Wright, 'The NESS Account', above n 102 at 316–21.

[114] Wright, 'Acts and Omissions' above n 113 at 304. This has something in common with Bagshaw's point about legal causation having to be consistent with our understanding of natural mechanisms. See Bagshaw, 'Causing the Behaviour of Others', above n 15 at 366.

[115] Wright, 'Once More into the Bramble Bush' above n 110 at 1107.

in this of all situations; one of the particularly thorny problems in which it is supposed to come into its own.[116] As Fischer points out:

> If Professor Wright's claim is that causal priority is a causal law, then his reasoning does appear circular because he merely asserts the existence of the causal law. He does not explain the source of the causal law or how others might verify its existence. In essence, he would be claiming the second omission is the NESS cause because the second omission caused the accident.[117]

Fumerton and Kress, in their 'fundamental and devastating'[118] criticism of Wright's formulation of the NESS test, make a similar point:

> Professor Wright is caught between a rock and a hard place. To avoid the charge that he cannot handle certain cases of Pre-emption, he must come up with a way to distinguish *lawful* (or law-like) sufficiency from *causal* sufficiency without relying on the concept of causation, a task that has eluded all philosophers to date.[119]

This, however, is only 'fundamental and devastating' in terms of NESS's ability to provide a *legally relevant* causal conclusion in pre-emptive cause situations. The tenor of Fumerton and Kress's argument, with which Stapleton agrees,[120] is not that the approach has nothing to contribute at all, but that Wright overpromises on behalf of a theory that underdelivers.[121] Wright variously describes NESS as 'a comprehensive, factual test of causation'[122] and 'the essence of the concept of causation',[123] and argues that 'the concept of causation is embodied in the NESS test'.[124] It is true that the longitudinal nature of NESS means that, properly interpreted, it is an effective means of identifying physical causes (referred to by Wright as instantiations of causal laws, or, somewhat misleadingly, 'lawful causes'). Applied mechanically, and not how Wright intended, it would, as we have seen, identify both the faulty brakes and the negligent driving as NESS causes of the claimant's injury in *Saunders*.[125] Since, in philosophical terms, this has never been conclusively refuted, Wright's claims for NESS with regard to its potential for establishing causation in a broader sense are not obviously wrong. Unfortunately,

---

[116] See Wright, 'Causation in Tort Law', above n 90 at 1792. See also Wright, 'The NESS Account', above n 102 at 289.

[117] Fischer, 'Insufficient Causes', above n 68 at 312. Fischer goes further in his criticism, pointing out that, although appeal to extrinsic causal intuition can be forensically acceptable, there is a lack of consensus with regard to causal intuitions, which makes such reliance problematic. His article contains some empirical support for this assertion. In 'Choosing', Stapleton also casts doubt on the efficacy of the causal priority argument, making the point that the assignment of such priority is not self-evident – see Stapleton, 'Choosing What We Mean by "Causation" in the Law', above n 15 at 478.

[118] Stapleton, 'Choosing What We Mean by "Causation" in the Law', above n 15 at 472.

[119] Fumerton and Kress, 'Causation and the Law', above n 98 at 102. Wright uses 'lawful sufficiency' to refer to those factors which are sufficient according to empirical generalisations, and 'causal sufficiency' to cover those factors which are legally relevant on the basis of a given state of affairs.

[120] Stapleton, 'Choosing What We Mean by "Causation" in the Law', above n 15 at n 145.

[121] See ibid, 473 and Fumerton and Kress, 'Causation and the Law', above n 98 at 89.

[122] Wright, 'Causation in Tort Law', above n 90 at 1774.

[123] ibid, 1805.

[124] Wright, 'Liability for Possible Wrongs', above n 82.

[125] See also Fischer, 'Insufficient Causes', above n 68.

however, his claims for NESS go further than this, and he has also said that NESS is 'a test for causal contribution that is applicable to the entire spectrum of causation cases'.[126] As the pre-emptive cause situation shows, this latter claim is not true or, rather, it is not true if what we are seeking is an answer with legal significance.

What Constitutes Pre-emption?

In order for a situation to involve at least one factor which has been pre-empted by another, the relevant multiple factors must:

- have a non-duplicative effect on the claimant (ie the risks generated by at least one factor must never eventuate in damage to the claimant);
- be independent of one another.

Non-duplicative Effect on the Claimant

This is the key characteristic that separates pre-empted cause situations from overdetermined events and, as such, has been substantially dealt with in the previous section. Essentially, only the pre-empting factor[127] must actually affect the claimant, in that his injury must eventuate from the risk created by it. By contrast, any pre-empted factors will only ever have exposed the claimant to a *risk* of injury which does not ultimately materialise.[128] Whilst this is of no relevance to the first stage of the NBA,[129] it is crucial at the second stage. In *Saunders*, for example, the claimant was injured because the driver's negligence meant that the car did not stop before it reached him, so the risk of injury created by negligent driving was the risk that was operative on the claimant. The risk created by letting out a car with faulty brakes did not lead to any injury because the driver's failure to use the brakes prevented that risk from having any effect on the claimant in the actual course of events. Where an event is overdetermined, on the other hand, the risks created by all overdetermining factors must remain operative, and therefore *capable* of being the source of the injury which actually occurred.[130] We know in *Saunders*[131] that the faulty brakes were not, in the non-hypothetical world, *capable* of causing the injury since they were not used, and were not therefore brought into contact with the claimant's course of events:

---

[126] Wright, 'Causation in Tort Law', above n 90 at 1788.
[127] Or factors.
[128] But see M Moore, *Causation and Responsibility* (Oxford, OUP, 2010) 248, where he refers to pre-empted factors having started to do their work before the pre-empting factor takes effect. Once a claimant has actually been affected by any factor, however, this makes the situation rather different, and usually more straightforward to analyse. Such situations are be analysed in Chs 3 and 5.
[129] Because *at least one* breach made a difference to the claimant's normal course of events.
[130] We can say no more than this, however, since this is precisely the conceptual problem with overdetermined events; that the forensic process is unable definitively to identify whether one risk materialised and another did not. This point also explains why both 'single-hit' and 'double-hit' hunters cases are both appropriately described as examples of overdetermined harm.
[131] 117 So 72 (Ala 1928).

There was evidence tending to show that plaintiff's hurt was caused, not by reason of any defect in the brakes, but because the driver of the automobile, going at a high rate of speed across a crowded crossing, failed to make any use, or attempted use, of the brakes with which the car was equipped.[132]

By contrast, in an overdetermined case, such as *Summers*,[133] we know that the risks created by both negligent shots were still operative when the harm occurred because, in the non-hypothetical world, both were capable of being the risks to affect the claimant's course of events. This explains why, on the basis of the NBA, all overdetermining causes will be deemed relevant, but only pre-empting (as opposed to pre-empted) factors will have any legal bearing for the purposes of negligence.

Had this analysis been followed, it would have led to more constructive reasoning in the Australian decision of *Elayoubi v Zipser*.[134] In that case, the claimant suffered from spastic quadriplegia and intellectual disability as a result of being deprived of oxygen during his birth. Since his mother's previous experience of childbirth had involved a particular type of surgical intervention, she should not have been allowed to proceed with any subsequent vaginal deliveries. That she attempted to do so was a result of several factors: first, that the hospital at which she had had her previous surgery failed to warn her that no future vaginal delivery should be attempted; second, that the same hospital failed to record this procedure correctly in her notes; and third, that the (different) hospital at which the claimant was delivered failed to contact the first hospital in order to get an accurate medical history. The New South Wales Court of Appeal found for the claimant, and apportioned damages one-third to the first hospital and two-thirds to the second, on the basis that each had materially contributed to the ultimate injury. On those facts, the NBA would similarly conclude that both defendants should be liable to some extent, but the reasoning would differ on one significant point: whilst both the first hospital's failure to warn the claimant's mother and the second hospital's failure to inquire about her medical history did indeed both materially contribute to his injury, the first hospital's omission in not recording accurate information contributed nothing to the outcome (a point that was not recognised in the *Elayoubi* judgment). Much like the faulty brakes in *Saunders*, the risk created by the absence of patient information was not one which materialised because its effects were pre-empted by the failure to inquire. The specific risk created by this breach – that correct information would not be provided in response to any subsequent inquiry – remained dormant and at no point operated on the claimant. It *would* have done so, hypothetically, had such an inquiry been made, but such hypothetical possibilities are of no relevance to the NBA.[135] *Elayoubi*, therefore, is a case in which both defendants materially contributed to the outcome, but the first defendant did

---

[132] 117 So 72 (Ala 1928) at [6] 74.
[133] 199 P 2d 1 (Cal 1948).
[134] *Elayoubi v Zipser* [2008] NSWCA 335.
[135] In contrast to NESS, for instance, which would attribute causal relevance to such a factor because it is necessary to *a* set sufficient to cause the outcome.

not contribute wholly in the manner suggested by the court,[136] since one of its breaches was pre-empted by that of the second hospital.

Factors Independent of One Another

Unless two factors are independent of one another, the effects of one cannot be said to pre-empt those of the other. In conceptual terms, factors which are somehow connected cannot easily accommodate the idea of causal substitution which pre-emption necessarily entails. It is simply not open to a defendant to argue that an injury did not materialise from the risk created by his agency if one of the risks of his breach was the commission of another, and where that other breach created a risk more immediately associated with a claimant's damage. Although, as we have seen, a pre-empted factor must do no more than expose the claimant to a risk of harm, where that risk can be *connected to* the claimant's harm, it amounts to more than risk exposure. Consider as an example a slightly (but significantly) modified version of the facts of *Saunders*.[137] In this example, the driver of the car has realised, prior to seeing the claimant cross the road, that the car's brakes are defective, and, when she encounters the claimant, she is trying to get the car safely back to the rental company. In this hypothetical, she fails to brake perhaps because she knows there is no point in doing so, or perhaps because she is so distressed by her knowledge of the faulty brakes that she is not thinking straight. On such facts, the faulty brakes factor is not pre-empted by the driver's conduct (even though the latter may well be judged to have been negligent on its own terms) because it cannot, in risk terms, be disassociated from that conduct. As the case law shows, however, the simplicity of this as a general proposition belies the difficulty of its application in marginal cases.

The tort of negligence, requiring as it does particular consequences to result in order for it to avail a claimant, inevitably involves some measure of moral luck.[138] Negligence which causes no damage is not actionable, so two people can behave in precisely the same way, but only one turn out to be liable if only that one causes damage by so behaving. This is trite negligence law,[139] but since it is crucial to an understanding of the difference between overdetermination and pre-emption, it would seem apt to draw together some of the most relevant views on the concept here.

---

[136] In *Elayoubi*, however, the first hospital's failure to warn the claimant's mother did have an actual effect on the claimant because, had she been warned, she would have been able to inform the second hospital of her position. Since it is likely that this would have prompted the second hospital to contact the first for full details, the two factors cannot be said to be independent from one another, which explains why this case fits within the category of material contribution to injury and, as such, will be dealt with in Ch 5.
[137] 117 So 72 (Ala 1928).
[138] See below, at n 140.
[139] See, eg, *Browning v War Office* [1963] 1 QB 750 (CA) at 765 (per Diplock LJ): 'A defendant in an action in negligence is not a wrongdoer at large; he is a wrongdoer only in respect of the damage which he actually causes to the plaintiff'.

Moral Luck[140]

Where a significant aspect of what someone does depends on factors beyond his control, yet we continue to treat him in that respect as an object of moral judgment, it can be called moral luck.[141]

Since causally determined liability of the type employed by the common law does not exculpate defendants on the basis that the outcomes of their actions might (and in fact probably will) have been determined by factors extrinsic to their agency, it is fair to say that the concept of moral luck is engaged, to a greater or lesser extent, across its jurisdiction.[142] Within the tort of negligence particularly, with its explicit actionable damage requirement, moral luck could be said to play a major role in determining who is liable and who is not, and this particular means of exposure to moral luck has been famously analysed by Honoré as 'outcome responsibility'.[143] Professor Gardner has identified the moral implications of this:

> Depending as it does on the moral defensibility of outcome responsibility, the tort of negligence is in the same boat, morally speaking, as a strict liability tort. The tort of negligence is essentially a variation on the theme of a strict liability tort, in which the basic wrong lies in D's actually injuring P.[144]

Nagel provides a vivid illustration of this in his example of a truck driver who negligently causes an accident in which he injures a child:

> [W]hat makes this an example of moral luck is that he would have to blame himself only slightly for the negligence itself if no situation arose which required him to brake suddenly and violently to avoid hitting a child. Yet the *negligence* is the same in both cases, and the driver has no control over whether a child will run into his path.[145]

In turn, Moore's analysis of such a situation is an experiential one:

> [W]e *are* more blameworthy when we cause some evil, than if we merely try to cause it, or unreasonably risk it. The reason we *feel* so guilty in such cases is because we are so guilty. Some no doubt are troubled by the shortness of the inference chain here – although the longer one stays in philosophy the shorter seems to be the argument chains one finds acceptable.[146]

---

[140] See B Williams, 'Moral Luck' (1976) 75 *Proceedings of the Aristotelian Society* (Supp) 115.
[141] Nagel, 'Moral Luck', above n 99 at 26. For a criticism of the use of the term, see Moore, *Causation and Responsibility*, above n 128 at 23.
[142] What follows, it should be made clear at the outset, is a discussion of exposure to moral luck as a result of the machinations of the tort of negligence, rather than the law in a more general sense. The arguments have, for example, a different emphasis in criminal law in particular. See, for example, A Ashworth, 'Taking the Consequences' in S Shute, J Gardner and J Horder (eds), *Action and Value in the Criminal Law* (Oxford, Clarendon Press, 1993).
[143] T Honoré, *Responsibility and Fault* (Oxford, Hart Publishing, 2002).
[144] J Gardner, 'Obligations and Outcomes in the Law of Torts' in Cane and Gardner (eds), *Relating to Responsibility: Essays in Honour of Tony Honoré on his 80th Birthday*, above n 99 at 126.
[145] Nagel, 'Moral Luck', above n 99 at 25 and 28–29.
[146] Moore, *Causation and Responsibility*, above n 128 at 30.

In one sense, it is not unreasonable to expect some justification, or at least explanation, for the law's willingness to allow such resource allocation (as legal liability has been characterised)[147] to depend on the relative fortunes of individuals, rather than purely on the manner in which they conduct their lives.[148] In another, undoubtedly less obvious sense, any such justification could be regarded as redundant, as we shall see below.

Broadly, three strands of analysis can be identified in the literature concerned with responsibility[149] for outcomes:

i. moral;
ii. definitional;
iii. functional.

Moral

It has been no small part of Honoré's intellectual objective to account for the way in which the law, through the medium of outcome responsibility, holds people liable for their bad luck. Two principal forms have been identified within Honoré's scholarship.[150] The first of these has a moral basis and, characterised by Perry as the 'personhood' understanding,[151] is founded on the following basic premise:

> outcome-allocation is crucial to our identity as persons; and, unless we were persons who possessed an identity, the question of whether it was fair to subject us to responsibility could not arise. If actions and outcomes were not ascribed us on the basis of our bodily movements and their mental accompaniments, we could have no continuing history or character. There would indeed be bodies, and associated with them, minds. Each would possess a certain continuity. They could be labelled A, B, C. But having decided nothing, and done nothing, these entities would hardly be people.[152]

He has also pointed out that to 'bear the risk of bad luck is inseparable from being a choosing person'.[153] To this moral argument, Cane has suggested a further dimension:

---

[147] For the argument that legal liability is a resource to be allocated, see Honoré, *Responsibility and Fault*, above n 143 at 78.

[148] See J Waldron, 'Moments of Carelessness and Massive Loss' in D Owen (ed), *Philosophical Foundations of Tort Law* (Oxford, OUP, 1997) 387.

[149] Used here in its most generic sense. The exact use of this term varies between authors.

[150] Perry, 'Honoré on Responsibility for Outcomes', above n 99. It should be made clear at this point that the distinction between the personhood and the social understanding is *not* Honoré's own, and so not organic to his theory, but is the result of Perry's analysis of it. I have adopted that analysis here because it illustrates well the two elements of Honoré's theory, as they map onto the three headings used herein.

[151] ibid, 63. The other is the 'social' understanding, for which, see below, under the sub-heading 'Functional'.

[152] T Honoré, 'Responsibility and Luck' (1988) 104 *LQR* 530, 543 (footnotes omitted).

[153] ibid, 553.

If people could disown responsibility for harm done to others whenever bad luck played a part in producing it, this would not only destroy our sense of personal identity as agents, but also our sense of humanity as victims. On the other hand, responsibility to others for all the contributions of fate would not only debilitate us as agents, but also as victims. It is an important part of our identity as individuals to feel that we have some control not only over what others suffer at our hands, but also over what we suffer at the hands of others. It would damage our sense of personal identity to feel that all the harm we suffered was fated, as much as to feel that all the harm we caused was fated. It is not only agents who need to own outcomes. Victims do too.[154]

Such arguments draw attention to a moral phenomenon which transcends any system of legal rules or political distributions:

[O]ne is morally at the mercy of fate, and it may seem irrational upon reflection, but our ordinary moral attitudes would be unrecognisable without it. We judge people by what they actually do or fail to do, not just for what they would have done if circumstances had been different.[155]

The tenor of these arguments is that, irrespective of the decisions we might wish to make about responsibility in law, we are bound, as autonomous persons, to hold ourselves and others morally responsible for the outcomes of our behaviour, as well as that behaviour itself.[156] If the events of our lives, bad and good, were no more than predetermined phenomena, we would be no more autonomous as individuals than Gloucester famously feared:

As flies to wanton boys are we to th' gods
They kill us for their sport.[157]

And we all know what happened to him.

Such moral arguments as these are less justifications of outcome responsibility than they are a phenomenological acknowledgement of a basic truth. It is perhaps in this sense that Honoré regards outcome responsibility as a 'pre-moral' notion.[158] Whilst this in itself may well not be sufficient grounds for defending a legal mechanism which uses those same conclusions to implement sanctions, there also exist functional and definitional analyses to supplement the argument.

---

[154] P Cane, 'Responsibility and Fault: A Relational and Functional Approach to Responsibility' in P Cane and J Gardner (eds), *Relating to Responsibility: Essays in Honour of Tony Honoré on his 80th Birthday* (Oxford, Hart Publishing, 2011) 93.

[155] Nagel, 'Moral Luck', above n 99 at 34.

[156] There is then a question of whether we are more blameworthy when our proscribed behaviour has adverse results, or just blameworthy for more. This question is beyond the scope of the current discussion, but see, eg, Moore, *Causation and Responsibility*, above n 128; J Gardner, 'Wrongdoing by Results: Moore's Experiential Argument' (2012) 18 *Legal Theory* 459; C Sartorio, 'Two Wrongs Do Not Make a Right: Responsibility and Overdetermination' (2012) 18 *Legal Theory* 473; and M Moore, 'Four Friendly Critics: A Response' (2012) 18 *Legal Theory* 491.

[157] W Shakespeare, *King Lear*, Act IV, Scene 1, 32–37.

[158] Honoré, 'Responsibility and Luck', above n 152 at 544.

Definitional

The definitional form of analysis reveals the futility inherent in trying to defend something to which there is no substantive alternative; only different arrangements of the same elements, and in each case configured around a bedrock of person and their ineradicable connection to their imprints in the world. Whilst Fischer and Ennis have described liability for outcomes as being not 'ideally fair',[159] the concept of fairness in such a context is in truth both elusive and protean. Rather like energy, it does not exist or cease to exist, but merely changes shape and form depending on the conditions to which it is exposed. As Gardner explains:

> There can be no such thing as a coherent moral objection to our being exposed to moral luck. Attempts to explicate such an objection are an object lesson in the hazards of argumentative overkill. For what counts as luck is always, Nagel shows, luck *only relative to some baseline or other*. Whenever something is held to be luck, there is necessarily something else that is held *not* to be luck, it is only relative to this second thing that the first counts as lucky or unlucky. The problem with a general objection to our exposure to moral luck is that *everything* we do is entirely a matter of luck relative to *some* baseline or other ... It follows that to object to moral luck *tout court* is to object to morality *tout court* ... Indeed, it is to object to judging people's actions by any standards at all. As Nagel himself puts it: 'The area of genuine agency, and therefore of legitimate moral judgment, seems to shrink under this scrutiny to an extensionless point'.[160]

Cane makes a similar argument:

> Once the role of dispositional luck in people's lives is understood, the role of circumstantial luck in relation to the outcomes of conduct ceases to provide a convincing rationale for ignoring outcomes when determining responsibility and culpability. If circumstantial luck negatived culpability in relation to outcomes, then equally dispositional luck would negative culpability in relation to conduct and mental states. Because circumstantial luck, like dispositional luck, is ubiquitous, it is necessary, both in morality and law, to adopt principles to determine when people can fairly be held responsible for outcomes that are, in some respect(s) or other, outside their control.[161]

Objections to reliance on moral luck raise the question of what alternative discriminatory measures are available and appropriate. The alternative to liability based on outcomes is, presumably, liability based on conduct,[162] but, as Cane explains in the extract above, the latter is just as much the product of fortune as the former: our 'dispositional luck'[163] determines what we, as individuals, are able

---

[159] J Fischer and R Ennis, 'Causation and Liability' (1986) 15 *Philosophy and Public Affairs* 33.
[160] Gardner, 'Obligations and Outcomes in the Law of Torts', above n 144 at 127, quoting Nagel, 'Moral Luck', above n 99 at 35. See also B Williams, 'Moral Luck' in *Moral Luck* (Cambridge, CUP, 1981) 20–21.
[161] Cane, *Responsibility in Law and Morality*, above n 99 at 138.
[162] Inherent in this is the corresponding thorny question of 'control'. See, eg, Moore, *Causation and Responsibility*, above n 128 at 24–29.
[163] Or 'constitutive luck', as Nagel might call it (Nagel, 'Moral Luck', above n 99 at 28).

or prone to achieve. It explains why, for example, some people will never attain the reasonable person standard in negligence, and others will exceed it with ease. What makes us act, and determines how we do so, is, in its most primal form, constitutive and not rational. I don't open the batting for England because I am not capable of doing so, regardless of how long I might spend in coaching sessions or nets. For the same reason, I will never be a concert pianist, run the Marathon des Sables, defraud an elderly woman of her savings, or calmly accept that some people like to drive continuously and slowly in the middle lane of motorways. I do not, however, generally find it difficult to commit things to memory, make friends or lead a healthy lifestyle. None of these innate abilities or inabilities is the result of my reason or choosing, and yet they all determine much about my life. As characteristics, they are results of my dispositional luck and therefore play a significant role in how I conduct myself, and how I am consequently judged. To insist that individuals are judged on the basis of their behaviour, therefore, rather than on how it turns out, is merely to swap one measure of fate for another.

Functional

Under this heading come arguments which take their force from the practical and logistical demands of administering a system of liability rules. It is this strand of analysis to which Honoré's 'social' understanding of outcome responsibility[164] belongs:

> Any principle which can justify responsibility for bad luck must be fair. If it is to be fair, it must entail that when we bear the risk of bad luck we also benefit if our luck is good. Allocation according to luck must cut both ways. So the system will be fair only if there are situations in which we implicitly bet on the outcomes of our actions.[165]

For Honoré, legal liability is an enhanced form of outcome responsibility, in that such responsibility should attract sanctions only if the relevant behaviour was unjustified. By the time an inquiry in negligence has reached the causal stage, however, only 'unjustified' behaviour (in the form of breaches of duties of care) will remain under consideration, meaning that, at the causation stage and beyond, outcome responsibility amounts to the same thing as legal liability: 'Causation merely provides the link between proscribed conduct and proscribed outcomes'.[166] There is, it would seem, in Honoré's conception, a functional layer to add to the moral framework in order to reach a legal conclusion:[167] not only must the defendant be outcome-responsible in the personhood sense, but there must also exist a

---

[164] Interpreted and labelled as such by Perry. See Perry, 'Honoré on Responsibility for Outcomes', above n 99 at 63.
[165] Honoré, 'Responsibility and Luck', above n 152 at 541.
[166] Cane, *Responsibility in Law and Morality*, above n 99 at 116.
[167] In this, according to Perry, it differs from the Libertarian approach, in which the duty to compensate is morally fundamental and so internally sufficient: Perry, 'Honoré on Responsibility for Outcomes', above n 99 at 69.

'principle of corrective justice [which] applies so as to place the defendant under an agent-relative obligation to compensate the plaintiff for her loss'.[168]

There are, therefore, two distinct dimensions to Honoré's account of the law's willingness to allow moral luck to play a role in founding liability. The first is a recognition of this role as a moral inevitability, and the second is a *conditional* defence of adopting the phenomenological conclusion as a legal one, as long as its inherent moral reciprocity is allowed to stand. This is the critical point: we are all potential claimants, but we are also all potential defendants;[169] what makes us winners also makes us losers. So, every time we act negligently but cause no harm, we chalk up a credit to the account we must draw upon when our sub-standard behaviour leads to loss.

Ripstein's idea of risk ownership also belongs under this heading in terms of its justification of moral luck as a discriminator, although it provides a particularly acute example of the difficulties of categorising concepts such as these. In essence, the theory is that

> the person who exposes another to a risk 'owns' the risk, and if the risk ripens into an injury, that person owns the injury. The basic idea is simple: In assigning liability, the fault system determines whose problem a certain loss is When a risk ripens into an injury, the injury belongs to whomsoever the risk in question belongs. Reasonable risks – those risks the imposition of which is compatible with appropriate regard for the interests of others – lie where they fall. Unreasonable risks belong to those who create them; as a result, the injuries that result from unreasonable risk imposition belong to the injurers.[170]

This idea of risk ownership, which has much in common with Honoré's approach,[171] is arguably a functional concept with a moral basis, of which there is some recognition in the following:

> One thing is clear. There is no way of drawing boundaries around risks in terms of natural causation, because the risks of causing things are even more indeterminate than the facts of causing things. There are too many chances involved in any outcome to simply assign those chances on the basis of some natural feature. Of course we might *pick* some natural feature to mark which chances belong to whom, just as we use natural features such as rivers and mountains to draw boundaries between states. The point is that natural features will not select themselves for us. We need some rationale for picking the particular features we do.[172]

Ripstein explains his preferred rationale thus:

> [T]he idea of risk ownership acknowledges that the effects of chance cannot be eliminated. It supposes instead that those who have a fair chance to avoid liability are not

---

[168] ibid, 70.
[169] This is a recurring theme of justification: see Ch 8.
[170] A Ripstein, *Equality, Responsibility and the Law* (Cambridge, CUP, 2001) 54.
[171] See, for instance, T Honoré, 'The Morality of Tort Law: Questions and Answers' in T Honoré, *Responsibility and Fault*, above n 143 at 79.
[172] Ripstein, *Equality, Responsibility and the Law*, above n 170 at 46.

treated arbitrarily if they are held liable even if they might have had better luck it is luck all right, but it is *their* bad luck, and as responsible agents, they must accept its consequences, just as responsible agents must accept the consequences of foolish choices they make for themselves . . . The duty to compensate is not only the result of an injured person's need for compensation, but of the fact that the loss properly belongs with the person who imposed the risk. *On the risk ownership conception of liability, chance is allowed a role, but that role is not arbitrary.*[173] (Emphasis supplied)

As the previous section suggests, luck of some type or other is a constant and universal discriminator. The arguments put forward here by both Honoré and Ripstein justify the law's choice of one type of luck – 'resultant' or 'causal' in Nagel's terms – over another type – 'constitutive' (Nagel) or 'dispositional' (Cane) – for the purposes of allocating legal liability. In functional terms, this is, they say, the fairest and most coherent way to proceed, given the potential reciprocity inherent in the distributive mechanism.[174]

---

[173] ibid, 84.
[174] Honoré, *Responsibility and Fault*, above n 143 at 78.

5

Material Contribution to Injury

Illustrative Cases: *Bonnington, Bailey, O2, Rahman, Fitzgerald*

Factual Basis

This analysis is applied where:

- there are multiple factors (not all of which need to be tortious);
- the case is not one of overdetermination or pre-emption;
- it has been established that the tortious factor/s have had an actual effect[1] on the claimant's position at trial; and either
- the injury is divisible in principle, but it is not possible to attribute constituent parts to particular factors on the facts of a given case (*Bonnington*); or
- the injury is indivisible (*Bailey, O2, Rahman*).

A 'material contribution to injury' analysis is appropriate where it is more likely than not that at least one defendant's breach has made a difference to the claimant's outcome, but it is not possible to isolate the physical effects of individual breaches from one another. This impossibility precludes the application of basic causal principles. For instance, in a factual scenario like the one in *Performance Cars Ltd v Abraham*,[2] there were two physically distinct instances of damage to the claimant's car, each of which could be linked discretely to a particular defendant.[3] By contrast, those situations in which a court should resort to a 'material contribution to injury' analysis are those where the fact[4] of a defendant's contribution to

---

[1] The question of what this amounts to is a contentious one and will be examined in detail below. In summary, under the NBA, a factor must have been a necessary one in the claimant's injury occurring as and when it did in order to amount to a material contribution. This does not accord with the idea, expressed in *Sienkiewicz v Greif* [2011] UKSC 10, [2011] 2 AC 229 at [90] 265 (per Lord Phillips) and in *Bonnington Castings v Wardlaw* [1956] AC 613 (HL) at 621 (per Lord Reid) that such a factor need only have made a contribution in excess of a *de minimis* level in order to be deemed a material contribution.

[2] *Performance Cars Ltd v Abraham* [1956] 1 QB 33 (CA).

[3] In this case, such an exercise was a straightforward one because the damage to the claimant's car occurred in two consecutive chronological stages, but this is not necessary for an injury to be divisible. Were a car to have been hit by two other vehicles simultaneously, it might of course still be possible to link separable parts of the total damage to each particular collision.

[4] In the sense of the claimant having proven, on the balance of probabilities, that but for the defendant's breach, her natural course of events would have been unaffected.

damage has been established, but there are other factors involved, and there is simply no means of discretely assigning constituent parts of that damage to particular factors. In the case of indivisible injuries, this is because such damage is binary in nature and so cannot be broken down into constituent parts. In the case of injuries divisible in theory, the problem arises because it is not possible as a matter of evidence to identify any clean correspondence between constituent parts and multiple potential factors.

Injury Is Divisible in Principle but It Is Not Possible to Attribute Constituent Parts to Particular Factors

The classic example of this category of case is *Bonnington Castings v Wardlaw*,[5] in which the claimant contracted pneumoconiosis during the course of his employment by the defendants. There were two factors identified as contributing to this disease: first, those particles of silica dust in the workplace atmosphere which had emanated from swing grinders and second, those particles of silica dust which had come from pneumatic hammers. Whilst both types of workplace machinery were the legal responsibility of the defendants,[6] they had only breached their duty in relation to the first, since there was no known or practicable means of reducing the dust escaping from the latter. This meant that the particular question for the court in this case was whether the dust resulting from the defendant's breach could be causally linked to the claimant's injury. The essence of this problem was identified by Lord Keith:

> The disease is a disease of gradual incidence. Small though the contribution of pollution may be for which the defenders are to blame, it was continuous over a long period. In cumulo, it must have been substantial, though it might remain small in proportion. It was the atmosphere inhaled by the pursuer that caused his illness and it is impossible, in my opinion, to resolve the components of that atmosphere into particles caused by the fault of the defenders and particles not caused by the fault of the defenders, as if they were separate and independent factors in his illness. Prima facie the particles inhaled are acting cumulatively, and I think the natural inference is that had it not been for the cumulative effect the pursuer would not have developed pneumoconiosis when he did and might not have developed it at all.[7]

The specific problem posed for the causal inquiry by *Bonnington* stems from the fact that each potential causal factor (ie the 'innocent' dust and the 'guilty' dust)

---

[5] *Bonnington Castings v Wardlaw* [1956] AC 613 (HL). Although, as Lord Rodger states in *Fairchild v Glenhaven Funeral Services* [2002] UKHL 22, [2003] 1 AC 32 at [129] 100: 'The idea of liability based on wrongful conduct that had materially contributed to an injury was ... established long before *Wardlaw*. But *Wardlaw* became a convenient point of reference, especially in cases of industrial disease'.
[6] Covered specifically by regulation 1 of the Grinding of Metals (Miscellaneous Industries) Regulations 1925.
[7] [1956] AC 613 (HL) at 626 (per Lord Keith).

was operating on the claimant concurrently.[8] As a direct result of such concurrence, the causal inquiry could not rely on the incremental nature of the disease's development to attribute causal valence to particular factors in the way that it could if the factors had been operating consecutively.

> Bonnington may represent a departure from the ... orthodox approach in the context of a particular evidentiary gap: namely, where it is known that the victim's total condition is a divisible one but there is no acceptable evidentiary basis on which the disability due to the separate insults to the body could be apportioned to the individual sources, the claimant is allowed to recover for the total condition ... the pursuer could prove an orthodox causal connection between breach and a part of the divisible injury, he just could not quantify it.[9]

An 'acceptable evidentiary basis' of the type lacking in Bonnington would exist where the relevant potential causal factors affected the claimant during separate consecutive periods, meaning that each could be linked to a particular stage of development, and therefore divisible part, of the injury. Such is the factual basis of cases in which orthodox apportionment can be carried out. Take, for example, *Thompson v Smiths Shiprepairers (North Shields) Ltd*,[10] in which Mustill J (as he then was) reasoned as follows about the situation which would represent *Bonnington* were the innocent and guilty dust factors to have operated consecutively, as opposed to concurrently:

> Next, one must consider how this approach can be applied to a case where either (a) there are two successive employers, of whom only the second is at fault, or (b) there is a single employer, who has been guilty of an actionable fault only from a date after the employment began ... Employer B has ... 'inherited' a workman whose hearing is already damaged by events with which that employer has had no connection, or at least no connection which makes him liable in law. The fact that, so far as the worker is concerned, the prior events unfortunately give him no cause of action against anyone should not affect the principles on which he recovers from employer B. Justice looks to the interests of both parties, not to those of the plaintiff alone.[11]

This distinction between the concurrent nature of the factors in *Bonnington* and the consecutive operation of the factors in *Thompson* explains why, although both injuries are divisible in principle, only the damage in the latter case was divisible in practice. Therefore, *Bonnington* requires a material contribution to injury analysis, whereas *Thompson* does not. Looked at in this way, the material contribution analysis appears to be relatively simple. There are, however, two complications that have arisen in relation to it. The first is a question mark over what effect the imposition of liability for material contribution to injury should have on a conse-

---

[8] '[C]oncurrent in effect, if not necessarily in time' – *Thompson v Smiths Shiprepairers (North Shields) Ltd* [1984] QB 405 (QB) at 442 (Mustill J).
[9] J Stapleton, 'Unnecessary Causes' (2013) 129 *LQR* 39, 52–53. See also J Stapleton, 'Lords a'Leaping Evidentiary Gaps' (2002) 10 *Torts Law Journal* 276, 283 onwards.
[10] [1984] QB 405 (QB). See Ch 3, text to n 24.
[11] [1984] QB 405 (QB) at 438.

quent award of damages, and the second is whether a material contribution to injury analysis is an application of, or an exception to, the But For test.

The first question no doubt arises because, as we see from *Bonnington*, injuries which are divisible in principle will sometimes call for liability to be assessed on a material contribution to injury basis where that divisibility is not possible in practice, but where there have been multiple potential causal factors. Facts such as those arising in *Bonnington* therefore occupy something of a halfway house. It is trite negligence law that, where possible, defendants should only be held liable for that part of the claimant's ultimate damage to which they can be causally linked, as is clear from cases such as *Thompson* and *Performance Cars*. It is equally trite that, where a defendant has been found to have caused or contributed to an indivisible injury, she will be held fully liable for it, even though there may well have been other contributing causes:[12]

> [I]t is . . . hard – and settled law – that a defendant is held liable in solidum even though all that can be shown is that he made a material, say 5%, contribution to the claimant's indivisible injury. That is a form of rough justice which the law has not hitherto sought to smooth, preferring instead, as a matter of policy, to place the risk of the insolvency of a wrongdoer or his insurer on the other wrongdoers and their insurers.[13]

Since the injury in *Bonnington* is theoretically divisible, it seems not to fit into this second category, but it can no more fit into the first because there is, as we have seen, no practical basis on which any sensible division could be made.[14] This was not a question which troubled the Court in *Bonnington*, since the defendants' case was that they were not liable for the damage at all; they made no plea for any apportionment to be made on the basis that their breach of duty was not the only causal factor involved in triggering pneumoconiosis. The very fact, however, that no apportionment was made, has led to questions being asked subsequently as to whether that conclusion was the purely the result of the conduct of that particular case, or whether the same outcome would have occurred for substantive reasons, had the defendants requested that it be considered.[15]

The answer is that there should be no apportionment in cases which require, on the basis outlined in this chapter, a material contribution to injury analysis. The argument that apportionment of damages is not appropriate in cases in which a defendant has materially contributed to an *indivisible* injury is both well established and easy to justify.[16] There is, however, no obvious or defensible reason why a different approach should be taken where a *practical* segmentation is no more feasible, despite the fact that the damage in question might be *theoretically* divisible

---

[12] See, eg, *Dingle v Associated Newspapers* [1961] 2 QB 162 (CA) at 188 (per Devlin LJ), *Hotson v East Berkshire Health Authority* [1987] AC 750 (HL) at 783 (per Lord Harwich) and J Smith, 'Causation – the Search for Principle' [2009] *Journal of Personal Injury Law* 101, 103. See also *Baldwin & Sons Pty Ltd v Plane* (1998) 17 NSWCCR 434 (NSWCA), *Bendix Mintex Pty Ltd v Barnes* (1997) 42 NSWLR 307 (NSWCA) and *Gates v Howard Rotavator Pty Ltd* (2000) 20 NSWCCR 7 (NSWCA).
[13] *Barker v Corus* [2006] UKHL 20, [2006] 2 AC 572 at [90] 607–08 (per Lord Rodger).
[14] Stapleton, 'Lords a' Leaping', above n 9 at 283.
[15] See, for example, Stapleton, 'Unnecessary Causes', above n 9 at 52.
[16] See nn 12 and 13 above.

in nature.[17] As illustrated above, had the factors in *Bonnington* operated consecutively, the damage therein would have been both theoretically and practically divisible. This is because the cumulative nature of pneumoconiosis (which is what makes it divisible in theory) would have lent itself to being divided up chronologically according to the extent of its development during the period of exposure to each successive factor. A material contribution to injury approach would not therefore have been necessary because such facts would have been amenable to conventional causal analysis along the same lines as *Thompson*.[18] The very fact that such division was not possible is what makes it an appropriate case for analysis on the basis of material contribution to injury. Once this practical impossibility exists, there is no effective means of distinguishing between divisible and indivisible injuries, since a court is no more able sensibly to divide up the one than the other. A defendant who has been found to have materially contributed to such an injury, therefore, should be held liable for 100 per cent of the claimant's damages.[19]

The answer to the second question, of whether the material contribution to injury analysis is an application of, or an exception to, the But For test, can be made equally emphatically: it adheres to, and does not depart from, the basis of But For causation. As will become clear, the first stage of the NBA, which is based on aggregate But For causation, has still to be satisfied where a material contribution to injury analysis is applied, just as it does in other types of case.

1 – Is it more likely than not that *a* defendant's breach of duty changed the claimant's normal course of events so that damage (including constituent parts of larger damage) occurred which would not otherwise have done when it did?

Under this analysis, a defendant's breach has either to have *part-caused* an indivisible injury, or *caused part of* a (theoretically) divisible injury. Unless, however, a defendant has made a difference to the claimant's course of events in this way, there will be no liability.[20]

Injury Is Indivisible

An indivisible injury is one which cannot be broken down into separable constituent parts. It is obvious, given this characteristic, why a quantification of respective

---

[17] But see *Sienkiewicz v Greif* [2011] UKSC 10, [2011] 2 AC 229 at [90] 265 (per Lord Phillips). And, in order for a material contribution to injury analysis to be appropriate, the injury must by definition be one which cannot practically be divided up amongst discrete causal factors.

[18] [1984] QB 405 (QB) at 438.

[19] According to Jane Stapleton, this is the approach adopted by the US courts in asbestosis claims – see J Stapleton, 'The Two Explosive Proof-of-Causation Doctrines Central to Asbestos Claims' (2009) 74 *Brooklyn Law Review* 1011. This is also subject, as outlined below (see text to n 38) to a possible reduction in certain heads of damage if the court decides that the injury might have occurred at some point in the future owing to factors unrelated to the defendant's breach of duty.

[20] This argument will be addressed fully below, where the case of *Bailey v Ministry of Defence* [2008] EWCA Civ 883, [2009] 1 WLR 1052 is discussed.

contributions cannot be carried out amongst multiple factors.[21] Some of the more obvious examples of indivisible injury include a limb broken in one place at a time, psychiatric injury[22] and death. Psychiatric injury was the damage for which the claimant sought recovery in a case which has come to be regarded as one of the most difficult in this area: *Rahman v Arearose Ltd*.[23] An NBA, however, clarifies the main issues.

In *Rahman*, the claimant was working for the first defendants as the manager of a fast food restaurant when he was attacked by two black youths. The injuries he sustained during that attack necessitated his having a bone graft in his eye at the second defendant's hospital. The first defendants breached their duty by not providing sufficient protection for their employee, whilst the second defendant's breach lay in the negligent performance of the operation on his eye. Ultimately, there were several dimensions to the claimant's damage. First, he was blind in the eye on which the operation had been performed. In addition, he developed post-traumatic stress disorder, a severe depressive disorder of psychotic intensity, a specific phobia of Afro-Caribbean people with paranoid elaboration, and an enduring personality change; all of which left him unable to work, or even to function normally in society. It was this psychiatric aspect of his damage which was forensically contentious. As a result of the Court of Appeal's consideration of the case,

> the claimant got judgment against each defendant for part of his loss only. For his pain and suffering/loss of amenity he obtained judgment for £7.5k against the employer and £55k against the hospital, and for his economic loss of about £500k, one quarter against the employer and three-quarters against the hospital . . .[24]

Laws LJ's judgment, with respect, makes the causation issues in the case appear to be far more complex than in fact they are. An application of the NBA leads to the simple conclusion that both defendants should have been jointly and severally liable for the whole of the damage following the negligent operation.[25] The first

---

[21] This is the correct way to analyse the well-known US case of *Ybarra v Spangard* 154 P 2d 687 (Cal 1945), in which the defendant suffered a traumatic arm injury whilst under sedation for an unrelated surgical procedure. The alleged difficulty in his case was that he could not identify exactly how or by which member of the team in theatre his injury was caused. Applying *res ipsa loquitur*, however, the Supreme Court of California decided in favour of the claimant, and held, inter alia, that he did not have to prove which party was responsible for the physical trauma because, in any event, each practitioner had breached a duty to ensure his safety whilst in theatre. This is correct, and precisely the result that would be reached on an NBA: since all members of the surgical team had breached a duty, there was *at least* aggregate But For causation and, since his safety was compromised in theatre, every breach of that duty was operative on him at the time his injury occurred. On this view, it is a straightforward case of material contribution to an indivisible injury.
[22] As a result of *Rahman v Arearose Ltd* [2001] QB 351 (CA), some doubt has been cast on this classification. As the analysis below will make clear, however, psychiatric injury is indeed indivisible.
[23] [2001] QB 351 (CA).
[24] T Weir, 'The Maddening Effect of Consecutive Torts' (2001) 60 *CLJ* 237, 238.
[25] As conceded by counsel for the first defendants, and acknowledged at [2001] QB 351 (CA) at [36] (per Laws LJ), a distinction must be made between the pre-operation and post-operation losses: since the effect of the second defendant's breach could clearly not be operative on the claimant until it occurred, the question of multiple factors simply does not arise prior to the second breach occurring.

stage of the NBA is clearly satisfied, since it is more likely than not that at least one defendant's breach of duty changed the claimant's normal course of events for the worse. The second stage, asked of the two defendants in turn, would also be satisfied in each case, since the effects of both were still operative on the claimant when he suffered the damage for which he claimed. Contrary to the view of Laws LJ, the claimant's psychiatric injury should properly have been classed as indivisible harm, since his ongoing suffering could not be disaggregated and assigned to separate causes. Whilst Laws LJ gave detailed consideration to the question of whether the psychiatric injury resulting from the two defendants' breaches could be said to be 'the same damage' for the purposes of the Civil Liability (Contribution) Act 1978, he ultimately held that it was not. His Lordship's decision appeared to give much weight to

> an absurd report confected jointly by the experts for the three parties, who tentatively divided up the victim's present condition in terms of the two causes. They should not have been asked to do this, and their answer should have been ignored, for there is no scientific basis for any such attribution of causality: the claimant is not half-mad because of what the first defendant did and half-mad because of what the second defendant did, he is as mad as he is because of what both of them did. His mania is aetiologically indiscerptible ... [26]

The conclusion then reached was that

> on the evidence the respective torts committed by the defendants were the causes of distinct aspects of the claimant's overall psychiatric condition, and it is positively established that neither caused the whole of it ... one cannot ... draw a rough-and-ready conclusion to the effect that this is really an indivisible injury and therefore 'same damage' within section 1(1) of the 1978 Act.[27]

With respect, it is difficult to see how the claimant's ultimate psychiatric damage could be divided into 'distinct aspects', and Weir's analysis is the more coherent. Whilst Rahman did indeed suffer from more than one manifestation of mental trauma, in that he had PTSD, depression and phobia, it would be neither authentic nor feasible to regard these as having been the separate and discrete results of individual breaches of duty. Not only is it far more likely that the two defendants' actions worked synergistically to bring about the claimant's ongoing injury, but the tenor of more recent case law appears to support the view that such damage is indivisible.[28] For instance, Hale LJ in *Hatton v Sutherland*[29] refers to *Rahman* as a

---

On basic causation principles (see Ch 3), the first defendant is liable for all of the claimant's damage up until the time the negligent operation was performed. Applying the NBA confirms this since, under its second stage of analysing the pre-operation damage (loss of earnings and removal expenses), the second defendant's breach would be found (obviously) to be not yet operative on the claimant. See also *Wright v Cambridge Medical Group* [2011] EWCA Civ 669, [2013] QB 312 at [52] 328 and [129] 347.

[26] Weir, 'The Maddening Effect', above n 24 at 238.

[27] [2001] QB 351 (CA) at [23]–[24] 364–365.

[28] Although some heads of damage, such as the first three years' loss of earnings and reasonable removal expenses, were attributed solely to the defendant employer since they pre-dated the effects of the negligent medical treatment. This is standard practice and, as such, is unremarkable.

[29] *Hatton v Sutherland* [2002] EWCA Civ 76, [2002] 2 All ER 1 at [37]–[40] 17–18.

case involving indivisible injury,[30] as does Smith LJ in *Dickins v O2 Plc*.[31] In an extra-judicial context, this latter view is reiterated in no uncertain terms, when psychiatric damage is described as '*par excellence* an indivisible injury'.[32] In one sense, Laws LJ concedes this in an indirect way when he says of the contentious report of the experts:

> It is true that this agreed evidence does not purport to distribute causative responsibility for the various aspects of the claimant's psychopathology between the defendants with any such degree of precision as would allow for an exact quantification by the trial court; no doubt any attempt to do so would be highly artificial. But the lack of it cannot drive the case into the regime of the 1978 Act to which, in principle, it does not belong... The fact-finding court's duty is to arrive at a just conclusion on the evidence as to the respective damage caused by each defendant, even if it can only do it on a broad-brush basis which then has to be translated into percentages.[33]

In essence, what his Lordship prepares to do here is to apportion liability between defendants on a basis apparently unconnected to the factual matrix of the case. Such apportionment (one-quarter to the employers and three-quarters to the health authority) was instead carried out according to an apparently impressionistic account of the relative culpability of the defendants. Despite the strikingly heterodox nature of this approach, it was validated by *obiter* remarks made by Hale LJ in the Court of Appeal in *Hatton*[34] and, although the House of Lords expressly declined to offer a view on this point when they considered the same case,[35] the suggestion was implemented in *Dickins v O2 Plc*.[36]

The claimant in *Dickins* was suing her former employer for psychiatric injury caused by excessive stress at work. The evidence suggested that she had been promoted to a position beyond her natural capabilities and that this had, over time, led to her suffering from mental health problems, variously characterised as anxiety and depression. Once the Court of Appeal accepted that her work problems stemmed from a breach of duty, it was faced with a potential problem of causation in that, besides the situation at work, the claimant had a vulnerable personality, and had suffered from mental health issues in the past. In addition to this, she suffered from IBS, was at the material time experiencing difficulties in her relationship with her partner (although the evidence was inconclusive as to whether these phenomena were causes or effects of her stress), and had also, during the relevant period, had to move out of her home for nine months as a result of flooding. There existed, therefore, several non-breach factors which could potentially have

---

[30] [2002] EWCA Civ 76, [2002] 2 All ER 1 at [37] 17 and again at [40] 18.
[31] *Dickins v O2 Plc* [2008] EWCA Civ 1144, [2009] IRLR 58 at [45] 64. See also Weir, 'The Maddening Effect', above n 24 at 239, where he says of *Rahman* that 'the harm was not incremental, but the indivisible result of a synergistic or catalytic concatenation of events'.
[32] Smith, 'Causation', above n 12 at 103.
[33] [2001] QB 351 (CA) at [23] 364.
[34] [2002] EWCA Civ 76, [2002] 2 All ER 1.
[35] [2004] UKHL 13, [2004] 2 All ER 385 at [63] 405 (per Lord Walker).
[36] [2008] EWCA Civ 1144, [2009] IRLR 58.

contributed to the illness to which she eventually succumbed. In the Court of Appeal, Smith LJ summarised the trial judge's means of dealing with this point:

> Following the guidance given in Hatton, the judge took those other matters into account when apportioning the damages as to 50% being due to the tort and 50% due to the non-tortious factors. Before the judge, both parties had accepted that it was right to apportion the damages. The dispute between them was only as to how they should be apportioned.[37]

As a result of the nature of the dispute between the parties, and therefore of the grounds of appeal, the Court of Appeal in *Dickins* was unable conclusively to rectify the trial judge's mistake, which lay in his apportionment of liability in respect of an indivisible injury to which the defendant made a material contribution. As in *Hatton* and *Rahman*, this was apparently done on a 'broad brush' basis, according to intuitive estimations of relative culpability, and was as inappropriate in this case as it was in the judgments from which it took its lead. Fortunately, those who heard the *Dickins* appeal seemed minded to do what they could to arrest the development of this novel and ill-advised practice. For instance, although their comments could only be *obiter*, there was commendable force in the remarks of two of their Lordships on the apportionment point. Smith LJ stated:

> I respectfully wish (obiter) to express my doubts as to the correctness of Hale LJ's approach to apportionment. My provisional view (given without the benefit of argument) is that, in a case which has to be decided on the basis that the tort has made a material contribution but it is not scientifically possible to say how much that contribution is (apart from the assessment that it was more than de minimis) and where the injury to which that has led is indivisible, it will be inappropriate simply to apportion damages across the board. It may well be appropriate to bear in mind that the claimant was psychiatrically vulnerable and might have suffered a breakdown at some point in the future even without the tort. There may then be a reduction in some heads of damage for future risks of non-tortious loss. But my provisional view is that there should not be any rule that the judge should apportion the damages across the board merely because one non-tortious cause has been in play.[38]

In concurring with that judgment, Sedley LJ added:

> I am troubled by the shared assumption about the appropriateness of apportionment on which the case has proceeded. While the law does not expect tortfeasors to pay for damage that they have not caused, it regards them as having caused damage to which they have materially contributed. Such damage may be limited in its arithmetical purchase where one can quantify the possibility that it would have occurred sooner or later in any event; but that is quite different from apportioning the damage itself between tortious and non-tortious causes. The latter may become admissible where the aetiology of the injury makes it truly divisible, but that is not the case.[39]

---

[37] [2008] EWCA Civ 1144, [2009] IRLR 58 at [39] 63.
[38] [2008] EWCA Civ 1144, [2009] IRLR 58 at [46] 64.
[39] [2008] EWCA Civ 1144, [2009] IRLR 58 at [53] 65.

Furthermore, Tony Weir points out that this is not 'just a matter of aesthetics. Consequences ensue. If, in the present case, either defendant had been insolvent, the claimant would not have been fully indemnified'.[40] As both Smith and Sedley LJJ make clear,[41] full indemnity in such a case as this may well take the form of reduced damages, to account for the possibility of the same injury occurring in the future as a result of non-tortious causes. This is the orthodox legal approach, and is far superior to one which employs apportionment carried out on 'the basis of speculation or guesswork':[42]

> It is important conceptually to differentiate apportionment or divisibility from another perfectly common process in the assessment of loss, which is to take account of the vicissitudes of life or contingencies as applicable to the individual claimant. If, for example, a claimant suffered from a natural disease from which he was likely to die in five years, the court would take that fact into account when limiting damages to a period of five years ... In truth, ... [this] is not apportionment at all – what the court is doing is taking a snapshot of the claimant, at a point immediately prior to the accident, which incorporates at that time all the particular negative or positive factors in the claimant's own past or future, as well as the future factors which might afflict persons generally. The court, when taking into account the contingencies and vicissitudes of life, makes its award in an attempt to reproduce the snapshot, the object being to restore the claimant, warts and all, to the position he was in before the tort was committed.[43]

To understand exactly why this is so, it is necessary to consider the temporal dimension of the causal inquiry. When we ask whether the claimant's damage would have occurred but for a defendant's breach, this can only meaningfully be understood as meaning 'But for a defendant's breach, would the claimant's damage have occurred *when it did*?'[44] Unless it is understood in this way, a But For inquiry is either impossible to answer, or legally meaningless, or both. Consider, as an example, a claim in which death forms the gist of the damage. Clearly, here, but for the defendant's breach, the claimant would definitely have died. At some stage.[45] The point of legal relevance is of course whether the claimant would have died when she did but for the defendant's breach.[46] Where death is concerned,

---

[40] Weir, 'The Maddening Effect', above n 24 at 238.
[41] And Smith LJ reiterates extra-judicially in Smith, 'Causation', above n 12 at 103.
[42] The correct approach was taken in *Fitzgerald v Lane* [1989] AC 328 (HL), in which two defendants, acting independently, were found to have been equally responsible for the ultimate indivisible injury (partial tetraplegia), alongside a substantial contribution from the claimant himself. After reducing the total damages by 50% for contributory negligence, the defendants were held jointly and severally liable for the remaining 50%.
[43] L Caun, 'Multiple Causes of Injury' [2003] *Journal of Personal Injury Law* 96, 107–08.
[44] As already apparently recognised in academic literature, but rarely translated into express practical applications – see R Wright, 'The NESS Account: Response to Criticisms' in R Goldberg (ed), *Perspectives on Causation* (Oxford, Hart Publishing, 2011) n 48 and J Stapleton, 'Choosing What We Mean by "Causation" in the Law' (2003) 73 *Missouri Law Review* 433, 452–53.
[45] See D Lewis, 'Causation as Influence' in J Collins, N Hall and LA Paul (eds), *Causation and Counterfactuals* (Cambridge, MA, The MIT Press, 2004) 86. The significance of this point becomes even more acute in cases involving epidemiological evidence; see below, text to n 50.
[46] Although it is sometimes obvious that this is really what the But For inquiry is asking, this is by no means universally the case (nor is it universally accepted as being appropriate).

therefore, a non-temporally specific But For inquiry will always be easy to answer, but it will tell us nothing of value to the causal inquiry.[47] Where, on the other hand, the gist of the claim is damage of some other type, such as a broken leg, it will be impossible to predict whether a claimant would ever have suffered such an injury at any point in their life, were it not for the defendant's breach. An inquiry such as this one is therefore both impossible to conduct *and* lacks any legal purchase. Facile examples these may be, but they illustrate clearly how important it is that any But For inquiry is imbued with temporal specificity. Without this, as we have seen, no such test will be able to distinguish between overdetermined and pre-empted causal inquiries.[48] It is difficult to discern a reason, therefore, for omitting this crucial qualification from the express formulation of the test.[49]

In addition to this conceptual argument, there are practical justifications for why the causal inquiry should have as its focus *the stage* at which a claimant incurred the damage of which she ultimately complains. First, the specific question of whether an individual would have incurred damage *when she did* is an established part of epidemiological causal theory:

> One definition of the cause of a specific disease occurrence is an antecedent event, condition or characteristic that was necessary for the occurrence of the disease at the moment it occurred, given that other conditions are fixed. In other words, a cause of a disease occurrence is an event, condition or characteristic that preceded the disease onset and that, had the event, condition or characteristic been different in a specified way, the disease either would not have occurred at all or would not have occurred until some later time.[50]

It is easy to identify several significant reasons why claimants should want to remain undamaged for as long as possible. Clearly, most of us would want to die later rather than sooner, and this applies not only to our emotional and physiological perspectives, but also to the material effect that a longer life will often have on our estate. Where illness is concerned, similar arguments apply, in that people generally will want to have as much of their life as possible unaffected by pain, suffering and infirmity. In the context of property damage, particularly where that property is fungible and replaceable, the significance of timing is not necessarily so striking from the subjective viewpoint of the claimant. Nevertheless, it remains legally pertinent because, as with personal injury, determining the point at which a defendant's breach affected the claimant allows a court either to divide damage up amongst several factors (where this is possible) or, in any event, to determine

---

[47] See text to n 124 below.

[48] Wright, 'The NESS Account', above n 44 at n 125.

[49] Which is what Richard Wright has long argued for in relation to the *US Restatement*: American Law Institute, *82nd Annual Meeting, Proceedings 2005* (Philadelphia, PA, American Law Institute, 2006) 81–84 (thus far in vain – see American Law Institute, *Restatement of the Law, Third, Torts: Liability for Physical and Emotional Harm* (St Paul, MN, American Law Institute, 2010) § 26).

[50] K Rothman, S Greenland, T Lash, *Modern Epidemiology* 3rd edn (Philadelphia, PA, Wolters Kluwer, 2008) 6. The statistical implications of this for the legal test are explored further below.

exactly what a claimant has lost as a result of her injury.[51] The first stage of the NBA explicitly includes this temporal condition by asking 'but for at least one breach of duty, would the claimant's damage have occurred *when it did*?'

This timing issue is the principal difference between material contribution to injury cases and pre-empted cause situations. As we have seen,[52] the latter type of situation is characterised by the non-duplicative effects of the factors concerned, since pre-empted factors are those which *never* have an effect on the claimant. A defining feature of material contribution to injury cases, on the other hand, is the multiple effect of different factors on the claimant, and there is no requirement for all of the factors concerned to be breaches of duty. So, even if a claimant is actually affected by both breach factors and non-breach factors, and even if non-breach factors would eventually have led to the same damage occurring at any point in the future,[53] a defendant will remain liable if her breach hastened the occurrence. The quantification of *what* such a factor has caused a claimant to lose is a *subsequent*, and non-causal, question of damages.[54] In the well-known US case of *Dillon v Twin State Gas & Electric Co*,[55] for example, the defendant was held liable for negligently allowing a 14-year-old boy to be electrocuted by its electric cables. The fact that, had the deceased not grabbed a cable, he would have fallen to his death or at least to serious injury, made no difference to the causal question. It was relevant only to the conceptually distinct issue of how much the defendant had to pay in order to redress its wrong.[56]

Consequently, in *Dickins*,[57] where the claimant had been affected by stress at work, her own vulnerable personality, IBS, relationship issues and domestic flooding, and where her depressive illness was ongoing, as it was at trial, the better analysis of the case would have been that suggested in the *obiter* remarks of Smith and Sedley LJJ.[58] The defendant employer's breach led to Ms Dickins' breakdown happening when it did, but the other factors affecting her concurrently made it likely that she would have suffered the same damage at some point in the future anyway, meaning that the defendant employer's liability could be described as hastening damage to which she was anyway vulnerable.[59] It would, in such

---

[51] If it were possible to determine that a breach of duty accelerated the occurrence of damage that would have occurred at some time in the future in any event, the quantification of the claimant's damages should reflect this – see M Green, 'The Intersection of Factual Causation and Damages' (2006) 55 *DePaul Law Review* 671, 677–80.
[52] See Ch 4.
[53] Even a moment later.
[54] See Green, 'The Intersection', above n 51.
[55] *Dillon v Twin State Gas & Electric Co* 163 A 111 (NH 1932).
[56] See also R Stevens, *Torts and Rights* (Oxford, OUP, 2007) 134.
[57] [2008] EWCA Civ 1144, [2009] IRLR 58.
[58] See above, text to n 39.
[59] [2008] EWCA Civ 1144, [2009] IRLR 58 at [46] 64. See also J King Jr, 'Causation, Valuation and Chance in Personal Injury Torts Involving Pre-Existing Conditions and Future Consequences' (1981) 90 *Yale Law Journal* 1353.

circumstances, be open to a court to reduce the damages payable accordingly.[60] This has been referred to as the 'crumbling skull doctrine':

> The so-called 'crumbling skull rule' simply recognizes that the pre-existing condition was inherent in the plaintiff's 'original position'. The defendant need not put the plaintiff in a position *better* than his or her original position. The defendant is liable for the injuries caused, even if they are extreme, but need not compensate the plaintiff for any debilitating effects of the pre-existing condition which the plaintiff would have experienced anyway. The defendant is liable for the additional damage but not the pre-existing damage ... if there is a measurable risk that the pre-existing condition would have detrimentally affected the plaintiff in the future, regardless of the defendant's negligence, then this can be taken into account in reducing the overall award ... This is consistent with the general rule that a plaintiff must be returned to the position he would have been in, with all of its attendant risks and shortcomings, and not a better position.[61]

In the case from which this excerpt is taken, *Athey v Leonati*,[62] the Supreme Court of Canada dealt meticulously with the particularities of the material contribution to injury means of analysis. In that case, the claimant, who had a history of back problems, had been injured in two successive car accidents, each resulting from the defendants' breach of duty.[63] Subsequently, whilst performing a routine stretch, he suffered a disc herniation for which he required surgery. As a result, he was forced to take lower-paid employment so that he could avoid heavy manual work. The trial judge awarded the claimant 25 per cent of the global damages figure on the basis that the two accidents were not the sole cause of the herniation, (because of the pre-existing back problems) but that they played a causative role, estimated to be in the region of 25 per cent. The Court of Appeal agreed with this assessment, but the Supreme Court rectified this mistake conclusively and constructively in awarding full damages against the defendants. It did so primarily on the basis that the trial judge had concluded that 'the plaintiff has proven, on a balance of probabilities, that the injuries suffered in the two earlier accidents contributed to some degree to the subsequent disc herniation'.[64] Major J, in giving the judgment of the Court, went on to say:

> Had the trial judge concluded (which she did not) that there was some realistic chance that the disc herniation would have occurred at some point in the future without the accident, then a reduction of the overall damage award may have been considered. This is because the plaintiff is to be returned to his 'original position', which might have included a risk of spontaneous disc herniation in the future. However, in the absence of such a finding, it remains 'speculative' and need not be taken into consideration.[65]

---

[60] [2008] EWCA Civ 1144, [2009] IRLR 58 at [47] 64. See also A Dugdale and M Jones, *Clerk & Lindsell on Torts*, 20th edn (London, Sweet & Maxwell, 2010) 2-161.
[61] *Athey v Leonati* [1996] 3 SCR 458 (SCC) at [35].
[62] *Athey v Leonati* [1996] 3 SCR 458 (SCC).
[63] Although there were two defendants in fact, both were represented as one at the trial.
[64] [1996] 3 SCR 458 (SCC) at [44].
[65] [[1996] 3 SCR 458 (SCC) at [48]. See also *Graham v Rourke* (1990) 75 OR (2d) 622 (Ont CA), *Malec v JC Hutton Proprietary Ltd* [1990] HCA 20, (1990) 169 CLR 638 and *Schrump v Koot* (1977) 18 OR (2d) 337 (Ont CA).

Even more helpfully, the Supreme Court explicitly summarised the principles on which it reached its decision:

> If the injuries sustained in the motor vehicle accidents caused or contributed to the disc herniation, then the defendants are fully liable for the damages flowing from the herniation. The plaintiff must prove causation by meeting *the 'but for' or material contribution test*.[66] Future or hypothetical events can be factored into the calculation of damages according to degrees of probability, but causation of the injury must be determined to be proven or not proven. This has the following ramifications:
>
> 1. If the disc herniation would likely have occurred at the same time, without the injuries sustained in the accident, then causation is not proven.
> 2. If it was necessary to have *both* the accidents *and* the pre-existing back condition for the herniation to occur, then causation is proven, since the herniation would not have occurred but for the accidents. Even if the accidents played a minor role, the defendant would be fully liable because the accidents were still a *necessary* contributing cause.
> 3. If the accidents alone could have been a sufficient cause, and the pre-existing back condition alone could have been a sufficient cause, then it is unclear which was the cause-in-fact of the disc herniation. The trial judge must determine, on the balance of probabilities, whether the defendant's negligence materially contributed to the injury.[67]

This last paragraph is slightly ambiguous. If it means that, because the judge in such a situation is unable to make such a determination on the balance of probabilities there can be no liability, it is correct. Otherwise, it is difficult to follow. Nonetheless, the *Athey* judgment as a whole is to be welcomed for its generally concise (at only 53 short paragraphs), straightforward and accurate exposition of how to analyse a situation in which there have been material contributions to an injury.

At the other end of the spectrum, a judgment apt to cause much confusion is that of the English Court of Appeal in *Bailey v Ministry of Defence*.[68] The claimant in this case suffered from severe brain damage, resulting from a cardiac arrest brought on by her aspirating her own vomit. She had attended a hospital managed by the defendants in order that she might undergo a procedure, known as an ERCP, to examine and treat a suspected gall stone in her bile duct. It was in relation to the claimant's post-operative care that the defendant breached its duty of care, since there was a failure to resuscitate the claimant during the night following the procedure, leading to her being very unwell by the following morning. At the same time, and unrelated to the defendant's breach of duty, the claimant also developed pancreatitis, an illness which is known to occur in some patients following an ERCP. More than a fortnight after her initial operation, the claimant

---

[66] Emphasis added – these italics highlight an unfortunate flaw in the Court's otherwise coherent reasoning: as should now be apparent, the But For and material contribution to injury analyses are not mutually exclusive. Rather, the latter is a specialised application of the former.
[67] [1996] 3 SCR 458 (SCC) at [41].
[68] *Bailey v Ministry of Defence* [2008] EWCA Civ 883, [2009] 1 WLR 1052.

aspirated her vomit, leading to her cardiac arrest and, ultimately, to hypoxic brain damage. For the purposes of the appeal, the pertinent issue was simply that of whether or not the defendant's breach of duty in failing to give proper postoperative care, was causative of the claimant's brain damage.

There is no doubt, according to the criteria outlined above, that the facts of this case require a material contribution to injury analysis. For a start, there were multiple factors in the form of the defendant's negligent aftercare and the naturally occurring pancreatitis. Furthermore, this was not a case of overdetermination or pre-emption.[69] The potential causal factors in *Bailey* could only be described as *inter*dependent because it is not possible to establish what effects negligent aftercare and pancreatitis would have if suffered separately. Their effects are best described as synergistic, or at least potentially so, meaning that *Bailey* belongs outside of the duplicative cause category. Since brain damage is indivisible in principle, it was therefore necessary to establish whether the defendant's breach of duty had been a partial cause of the ultimate injury. Unfortunately, although the facts of *Bailey* were correctly deemed to require a material contribution to injury analysis, this is not what followed. Whilst, with respect, there is much to lament in the judgment as a whole,[70] the essence of the mistake is encapsulated in the following statement by Waller LJ:

> In a case where medical science cannot establish the probability that 'but for' an act of negligence the injury would not have happened but can establish that the contribution of the negligent cause was more than negligible, the 'but for' test is modified, and the claimant will succeed.[71]

This proposition, which formed the basis of the Court of Appeal's dismissal of the defendant's appeal against liability, is dangerously misleading. As Stapleton points out,

> until that flawed proposition is disapproved it threatens to have an explosive impact in the field of medical negligence. This is because, for example, it may often be the case that a breach by a medical provider increases the weakness of a patient, by some non-negligible but un-assessable degree, before the patient suffers an indivisible injury that would have been avoided had the patient been of adequate strength ... Clearly, if such claimants are entitled to succeed under that proposition it would expose medical providers to a radically expanded realm of liability.[72]

This explains why, in order to be deemed a legally relevant cause, a factor must either be a necessary part-cause of an indivisible injury, or the necessary cause

---

[69] Since it does not fit the criteria outlined in Ch 4.

[70] Such as equating material contribution to injury with the exceptional material contribution to risk analysis developed in *Fairchild v Glenhaven Funeral Services* [2002] UKHL 22, [2003] 1 AC 32 and a description of that case as one in which the evidence established that one fibre caused the claimant's injury. For a more detailed criticism of these arguments, see Ch 6.

[71] [2008] EWCA Civ 883, [2009] 1 WLR 1052 at [46] 1069.

[72] Stapleton, 'Unnecessary Causes', above n 9 at 58. See also M Stauch, '"Material Contribution" as a Response to Causal Uncertainty: Time For a Rethink' (2009) 68 *CLJ* 27, 28–29.

of part of a (theoretically) divisible injury. The first stage of the NBA *must* be satisfied in order for there to be liability on the grounds of a material contribution to injury.

1 – Is it more likely than not that *a* defendant's breach of duty changed the claimant's normal course of events so that damage (including constituent parts of larger damage) occurred which would not otherwise have happened when it did?

In his proposition, Waller LJ appears to conflate But For with *de minimis non curat lex*, by seeing them as mutually exclusive, or at least as alternatives. Since both are necessary, this may well be the source of his confusion. Where indivisible injury, such as that suffered in *Bailey*, is concerned, a claimant must establish, on the balance of probabilities, that the defendant's breach was a part-cause of that damage. In other words, she must prove that, but for the defendant's breach, her injury would not have occurred when it did. It may well be the case that, even where a claimant has established that a defendant's breach was, in But For terms, a part-cause of her injury, a court may decide that that contribution was so small as to absolve the defendant from liability on the basis of the *de minimis* principle, but the two principles are complementary and *not* alternative. The facts in *Bailey* should have led to a conclusion of no liability because the experts therein were unable to say that, on the balance of probabilities, but for the negligent care, Ms Bailey would have avoided having the cardiac arrest (which led to the brain damage) when she did.[73] The medical evidence did, however, suggest that the defendant's breach had a more than negligible *chance of being a cause* of the claimant's ultimate damage and this is what, erroneously, Waller LJ regarded as being sufficient to find liability on a material contribution to injury basis. As a result, *Bailey* is a confused, confusing, and ultimately unhelpful decision.

Medical Negligence

The performance of a material contribution to injury analysis requires particular care in situations involving medical negligence. The medical context provides, in any event, a special kind of challenge for the causal inquiry because, almost by definition, medical practitioners deal with individuals who are already injured or damaged in some way. Often, it is the task of extricating the breach from the non-breach factors which makes this area of the law so difficult. An added dimension to this problem is the fact that human physiology is unpredictable, imperfectly understood and often makes it very hard to determine where the effect of one factor ends and another begins.

---

[73] For an Australian perspective, see *Tubemakers of Australia Ltd v Fernandez* (1976) 50 AJLR 720 (HCA) at 724.

110 *Material Contribution to Injury*

Wright v Cambridge Medical Group provides an example of these difficulties.[74] In this case, the claimant's GP, who worked for the defendant partnership, was consulted by the claimant's mother by telephone, and subsequently failed to see the claimant or to refer her to hospital as (the defendants conceded) he should have done. The claimant, who was 11 months old, was actually suffering from a bacterial super-infection contracted during an earlier visit to hospital for chicken pox treatment. She was finally referred to hospital two days later. Once there, however, she received inadequate medical care, which would doubtless have been deemed to be in breach of duty had the hospital been joined as a defendant to the action which, inexplicably, it was not. The causal question for the court, therefore, was whether the GP's breach of duty was causative of the claimant's ultimate injury (a permanently unstable hip, restricted movement range, leg length discrepancy and restricted mobility). It had to consider whether the hospital's inadequate treatment broke the chain of causation between the defendant's breach and the claimant's injury, and whether it would be reasonable to hold, as the trial judge had done, that even a timely referral would have made no difference to the claimant's position, since the hospital would have treated her inadequately, leading to her injuries occurring in any event.

Ultimately, the Court of Appeal (Elias LJ dissenting) allowed the claimant's appeal and found the defendants liable in full for the claimant's permanent damage.[75] The essential basis of this decision can be found in the judgment of Lord Neuberger of Abbotsbury MR:

> In the present case, I consider that the defendants' negligence was a causative factor of the claimant's permanent injury. In other words, as in *Rahman's* case [2001] QB 351, para 34, I have concluded that the negligence of the defendants and the failings of the hospital had a 'synergistic interaction, in that each tends to make the other worse', and accordingly it seems appropriate to proceed on the basis that both were causative of the damage suffered by the claimant.
>
> I do not consider the hospital's failure to treat the claimant properly once she was admitted ... was of such significance that it justifies a finding that the defendant's negligence was not causative of the claimant's injury – or indeed a finding that it broke the chain of causation between the defendant's negligence and the claimant's injury. It was not such an egregious event, in terms of the degree or unusualness of the negligence, or the period of time for which it lasted, to defeat or destroy the causative link between the defendant's negligence and the claimant's injury.[76]

Although there was no mention in any of the majority judgments[77] of a material contribution to injury analysis, the facts of the case meant that this would have

---

[74] *Wright v Cambridge Medical Group* [2011] EWCA Civ 669, [2013] QB 312.
[75] Although it also made the point that the pain and suffering endured by the claimant *after* the time of the negligence but before her eventual admission to hospital did not form part of the ultimate permanent damage, and the defendants should not therefore be liable for it. See [2011] EWCA Civ 669, [2013] QB 312 at [52]–[53] 328 and [92] 336–37.
[76] [2011] EWCA Civ 669, [2013] QB 312 at [36]–[37] 325 (per Lord Neuberger).
[77] Elias LJ refers to it implicitly, see [2011] EWCA Civ 669, [2013] QB 312 at [92] 336–37 and explicitly at [96] 337–38.

been appropriate: the claimant's permanent damage was indivisible, there were multiple potential factors, and this was not a case of duplicative causation because of the lack of simultaneity and the interdependence of the effects of those factors (meaning that there was no several sufficiency).[78] It should have fallen, therefore, to establish, as the final part of the jigsaw, whether the defendant's breach of duty had, on the balance of probabilities, been a part-cause of that damage. The Court decided it had, but it was a conclusion not couched in material contribution to injury terms. So, whilst the Court of Appeal's analysis and one based on material contribution to injury would have led to the same outcome, use of the latter approach would probably have led to the case being more easily aligned with the line of authority to which it belongs. In any event, on an NBA, the same answer is easily reached.

1 – Is it more likely than not that *a* defendant's breach of duty changed the claimant's normal course of events so that damage (including constituent parts of larger damage) occurred which would not otherwise have happened when it did?

This question is easy to answer on the facts of *Wright* because it was clear on the evidence that, in the absence of both breaches, the claimant's damage would not have occurred when it did.

2 – Was the effect of this defendant's breach operative when the damage occurred?

This was less easy to answer than the first because of the interdependence of the factors in this case. The defendant's breach of duty and the consequent failure of the hospital to diagnose and treat Wright accordingly were interdependent because, first, it was not possible to say that, had the defendant made the referral at the correct time, the hospital's treatment would have been as bad as it was when the referral was eventually made. There might, for instance, have been more competent staff available on the earlier day. In fact, given that the claimant should have been referred on a Wednesday, but was eventually referred on a Friday, and that the Court found there were more consultants available during the week than there were on weekends, the effects of the defendant's failure to refer appear likely to have continued, and operated in combination with those created by the hospital's poor treatment.[79] Secondly, it was generally accepted on the evidence that the claimant's damage became permanent after six days of insufficient treatment. Had the defendant made an earlier referral, therefore, the hospital would have had a larger 'window' during which to reach a proper diagnosis and begin treatment accordingly. Moreover, Lord Neuberger also made the point that

> the judge's conclusion did not take into account the agreed expert evidence . . . which, in fairness to him, seems to have been overlooked in the argument before him (and, indeed, the argument before us). In my judgment, the effect of that evidence is to

---

[78] See Ch 4, under sub-heading 'Several Sufficiency'.
[79] See [2011] EWCA Civ 669, [2013] QB 312 at [68] 331 and [72] 332.

establish, at the lowest, that it is more likely than not that any permanent damage the claimant would have suffered due to the inept treatment, which the judge found that she would have received if she had been referred on 15 April [the Wednesday], would have been significantly less than that which she did suffer. Indeed, I think that this evidence established that there is a reasonable chance that she would have suffered relatively little long-term damage if she had been referred in the late afternoon or early evening of 15 April.[80]

When the claimant's damage became permanent, the risk created by the defendant's breach was still operative.[81] It is clear, therefore, why Lord Neuberger considered the effects of the two factors to have been operating synergistically. As a factor which partially contributed thereby to an indivisible injury, the breach of duty was rightly held to be a legally relevant cause and the defendant liable for full damages. Had the hospital been joined as a defendant, it would have been appropriate for the Court to have held both parties jointly and severally liable, since both would have materially contributed to the claimant's injury.

It is worth re-emphasising a point already made in Chapter 4 on Duplicative Causation. Like the case of *Elayoubi*[82] mentioned there,[83] *Wright* is a case properly categorised as one involving a material contribution to injury because the multiple factors therein were not independent from one another. If, however, the GP's failure to refer in *Wright* had in no way affected the treatment provided by the hospital, and if the delay had made no difference to the ultimate outcome, the factors would have been independent, which would have made the case one of pre-empted cause. Had this been the case, the defendants would not have been liable because the second stage of the NBA would not have been satisfied in relation to it: the effects of its breach would not ever have affected the claimant, since they would have been pre-empted by the effects of the hospital's sub-standard treatment. It is easy to see, therefore, how significant interaction between factors (or lack of it) can affect the outcome of the causal inquiry.

A material contribution to injury analysis is also appropriate to the harrowing facts of *Paroline v United States et al*.[84] Although a case principally about criminal restitution in the US,[85] the judgment makes significant reference to causation in tort law. The claimant in that case had as a young girl been the victim of sexual abuse, which had been filmed and distributed extensively online. Her hurt and humiliation were therefore set to continue into the future, as more and more individuals witnessed the material on the Internet. The defendant, whilst not the individual who produced the material or first put it into circulation, was charged with possessing child pornography, including images of the claimant, and the US Supreme Court was asked to consider

---

[80] [2011] EWCA Civ 669, [2013] QB 312 at [73] 332 (per Lord Neuberger).
[81] [2011] EWCA Civ 669, [2013] QB 312 at [65]–[79] 330–34 (per Lord Neuberger).
[82] *Elayoubi v Zipser* [2008] NSWCA 335.
[83] See Ch 4, text to n 134.
[84] *Paroline v United States et al* No 12-8561, April 23, 2014 (USA).
[85] Under the Violence against Women Act 1994.

the theory of 'aggregate causation,' one formulation of which finds factual causation satisfied where a wrongdoer's conduct, though alone 'insufficient . . . to cause the plaintiff's harm,' is, 'when combined with conduct by other persons,' 'more than sufficient to cause the harm.' 1 Restatement (Third) of Torts: Liability for Physical and Emotional Harm §27, Comment *f*.[86]

Since the claimant's mental distress, as well as her need for ongoing counselling and lost income, amounted to indivisible damage, and since the act of each individual viewing the material contributed to that damage in an interdependent way, this was a case in which the defendant had materially contributed to the claimant's injury.[87]

The 'Doubling of the Risk' Test[88]

Cases classified as those in which a defendant has materially contributed to a claimant's injury are just one of the instances in which an application of the 'doubling of the risk' (DTR) test has been mooted. This test, conceived as an epidemiological device, performs a very specific function, and has unfortunately been misapplied in a forensic context several times. The following excerpt from *Novartis Grimsby Ltd v Cookson*[89] exemplifies a factual phenomenon which is particularly common in cases in which the claimant's damage takes the form of a disease.

> Besides the occupational exposure to aromatic amines in the dyestuffs industry, which has been recognised as a cause of bladder cancer for many years, it is also known that cigarette smoking can be a cause, as can the ingestion of certain drugs. More than one potential cause was present in this case. As well as working for the Appellant for many years, Mr Cookson had been a moderate cigarette smoker (10 to 20 a day) for about 20 years. He had given up the habit in about 1980. He had also taken potentially harmful drugs for a time but it was not suggested by either side that those drugs had had any significant effect on the causation of his bladder cancer. Cigarette smoke contains amines and the amines from both sources act on the body in the same way. It was accepted by both sides that the two forms of exposure would have had at least an additive, if not multiplicative, effect. The argument between the parties was about the relative potency of the effects of smoking and occupational exposure. In essence, the argument was whether the occupational exposure was sufficient to have caused or materially contributed to the development of the cancer.[90]

---

[86] No 12-8561, April 23, 2014 (USA) at B.
[87] The defendant in that case was not held liable for all of the restitution claimed by the defendant. Instead, the Court attempted to establish what was the defendant's relative causal contribution – see No 12-8561, April 23, 2014 (USA) at B.
[88] With thanks to Mark Ingham for checking the statistical accuracy of what follows.
[89] *Novartis Grimsby Ltd v Cookson* [2007] EWCA Civ 1261, [2007] All ER (D) 465 (Nov).
[90] [2007] EWCA Civ 1261, [2007] All ER (D) 465 (Nov) at [44] (per Smith LJ).

The essence of the issue from a forensic point of view is that there are not only multiple potential causes, but that those causes might operate synergistically, rather than independently. 'Interaction is common, and exposures that simultaneously protect against the effect of other exposures when they cause a disease are rare'.[91]

The significance of such 'synergistic' effects has unfortunately been exaggerated in some recent decisions, and attempts to apply the DTR test to deal with the perceived problem are misguided. The practice appears to have started in Smith LJ's judgment in *Novartis*. As far as the extract above goes, the issue was correctly identified but it is not clear that the DTR test was in any way determinative of, or necessary for, the result:

> The evidence ... was that occupational exposure accounted for 70% to 75% of the total. Put in terms of risk, the occupational exposure had more than doubled the risk due to smoking. In my view, if ... the correct test for causation in a case such as this is the 'but for' test and nothing less will do, that test is plainly satisfied on the facts as found. The natural inference to draw from the finding of fact that the occupational exposure was 70% of the total is that, if it had not been for the occupational exposure, the Respondent would not have developed bladder cancer. In terms of risk, if occupational exposure more than doubles the risk due to smoking, it must, as a matter of logic, be probable that the disease was caused by the former.[92]

Even without any consideration of whether a risk had been doubled, it seems likely that the defendant in *Novartis* would have been held liable, on the basis that the occupational exposure accounted for 70–75 per cent of the *total* risk involved.[93] This is strong evidence on which a tribunal could base a belief probability[94] that the occupational exposure was more likely than not to have part-caused the cancer and so formed a material contribution to the claimant's injury. Given that the DTR test was irrelevant to that result, therefore, it was unfortunate that it was mentioned at all because the 'test' is simply not independently determinative of But For causation; it merely provides one indication of a potential statistical association between a given factor and a given result.[95] Sadly, however, the *Novartis* decision has led to the DTR test being either applied or requested in other, equally unsuitable, contexts since. It was, for instance, applied in *Shortell v BICAL Construction*,[96] and received judicial acceptance at some point (albeit in both cases

---

[91] A Broadbent, 'Epidemiological Evidence in Proof of Specific Causation' (2011) 17 *Legal Theory* 237, 259.

[92] [2007] EWCA Civ 1261, [2007] All ER (D) 465 (Nov) at [74] (per Smith LJ).

[93] Although the Court of Appeal decision in *AB v Ministry of Defence* [2012] UKSC 9, [2013] 1 AC 78 at [153] 135 (per Smith LJ) suggests at [153] that the DTR test was in fact determinative of this case, such an approach is, with respect, not the most helpful analysis of the evidence therein.

[94] See Ch 2, under heading 'The Balance of Probabilities'.

[95] C McIvor, 'The "Doubles the Risk" Test for Causation' in S Pitel, J Neyers and E Chamberlain (eds), *Tort Law: Challenging Orthodoxy* (Oxford, Hart Publishing, 2013).

[96] *Shortell v BICAL Construction*, Liverpool District Registry (unreported) 16 May 2008. The defendant's employee had died from lung cancer, having been exposed at work to asbestos and having been a smoker for a number of years of his life. On the basis that the exposure to asbestos *more than doubled* the risk of contracting lung cancer, Mackay J found for the claimant.

by the same person) in both the Court of Appeal in *Sienkiewicz v Greif*[97] and *AB v Ministry of Defence*.[98]

As the name suggests, the DTR test purports to equate a finding that a factor had the effect of exposing a claimant to twice the risk of suffering damage, relative to a claimant not exposed to that factor, with a finding of probable causation. It is traceable to Mackay J's judgment in *XYZ v Schering*.[99] That case, however, posed a highly specific question, *expressly concerned with relative instances of injury*. In *Sienkiewicz v Greif*, a decision which suggests an appropriate limitation on the use of the DTR test, Lord Philips recognises this fact:

> *XYZ v Schering Health Care Ltd* 70 BMLR 88 is a lengthy and complex judgment devoted exclusively to a preliminary issue on the effect of epidemiological evidence. The issue was whether a second generation of oral contraceptives more than doubled the risk of causing deep vein thrombosis (DVT) that was created by the first generation of oral contraceptives. It was common ground that, if the claimants in this group litigation could not establish this, their claims under the Consumer Protection Act 1987 were doomed to failure. I do not believe that Smith LJ has correctly identified the relevance of this issue. It was *not whether the DVT suffered by the claimants had been caused* by the second generation of oral contraceptives which they had taken. It was whether the second generation of contraceptives created a significantly greater risk than the first. The experts appear to have been in agreement that the 'doubles the risk' test was the proper one to apply in order to resolve this issue. Thus I do not believe that that decision affords any direct assistance to the question of whether the 'doubles the risk' test is an appropriate test for determining causation in a case of multiple potential causes. (Emphasis added)[100]

In other words, the explicit question for the Court in *XYZ*, a product liability case, was whether the product which had been prescribed for the claimants was more than twice as likely as its earlier incarnation (the risks of which the claimants were taken to have accepted) to cause its users harm. Here, asking whether the risk has been doubled is clearly and directly appropriate because it answers this specific question. As Lord Philips identified in *Sienkiewicz*, however, this is not the same as asking whether it is more likely than not that one of several potential factors caused a particular injury, and his Lordship was correct in ruling that the DTR 'test' is not, therefore, the correct approach to take on those facts:[101]

> For reasons that I have already explained, I see no scope for the application of the 'doubles the risk' test in cases where two agents have operated cumulatively and simultaneously in causing the onset of a disease. In such a case the rule in *Bonnington* applies. Where the disease is indivisible, such as lung cancer, a defendant who has tortiously contributed to the cause of the disease will be liable in full. Where the disease is divisible,

---

[97] *Sienkiewicz v Greif* [2009] EWCA Civ 1159, [2010] QB 370 at [23] 379 (per Smith LJ).
[98] *AB v Ministry of Defence* [2012] UKSC 9, [2012] 1 AC 78 at [132] 130, [140] 132, [146] 134, [151] 135, and [153] 135.
[99] *XYZ v Schering* [2002] EWHC 1420, 70 BMLR 88 (QB).
[100] [2011] UKSC 10, [2011] 2 AC 229 at [74] 261–62.
[101] See also J Stapleton, 'Factual Causation, Mesothelioma and Statistical Validity' (2012) 128 *LQR* 221, 223.

such as asbestosis, the tortfeasor will be liable in respect of the share of the disease for which he is responsible.[102]

With respect, whilst his Lordship's conclusion is undoubtedly correct on this point, the reasons he gives are neither complete nor fully accurate. In order to understand why, the function of this epidemiological tool must be fully understood.[103] There are two principal reasons why the DTR test is not, without more, an appropriate device for calculating probability of causation. The first is that the relative 'risk' on which it is based is not in any technical sense a risk at all, but merely a measure of incidence over a defined population:

> [I]t is crucial to realize that 'risk' in this context means nothing more than incidence over a specified time interval. It does not denote the product of the probability that the harm will occur and the gravity of that harm; in fact it does not even denote a probability, strictly speaking . . . It is a purely statistical measure of the relative frequency with which a disease occurs in exposed and unexposed populations.[104]

To treat it as substitutive for probability, therefore, is to confuse two distinct concepts, with potentially adverse consequences for the causal inquiry.

The second reason is that the result reached by the relative risk calculation involved gives a 'net result'. That is, it tells us how many *extra* cases of a disease are caused by the factor being tested. It does not, therefore, include those cases in which the adverse outcome would have happened anyway, just not when it did. In other words, it only identifies cases which would *never* otherwise have happened, and does not account for those which the factor merely accelerated. This is significant since, as we have seen above,[105] suffering injury sooner rather than later is undoubtedly damage for which the law should provide compensation.

To illustrate this point, Alex Broadbent has provided a simple and effective example:

> Consider, for example, a fictitious group of Himalayan porters. One might imagine that carrying heavy loads up and down mountains increases their risk of back injury. Among those porters suffering back injury will be some who for anatomical reasons would have developed a back injury anyway. But having a weak back does not protect these porters from the effects of carrying heavy loads. On the contrary, it is quite possible that carrying a heavy load will be a cause of back injury among many or even all those porters whose backs were such that they would have developed a back injury even in a less physically demanding profession. [Equating relative risk incidence with probability of causation] amounts to endorsing the astonishing view that a person who would develop a disease without the exposure in question is thereby protected from the effects of that exposure: that having a weak back will protect a Himalayan porter from the harm that carrying a

---

[102] [2011] UKSC 10, [2011] 2 AC 229 at [90] 265.
[103] For a highly useful account, specific to its use in negligence, see McIvor, 'The "Doubles the Risk" Test', above n 95. There also exists the question of how far such epidemiological evidence is relevant to the causal inquiry in negligence, and what its limits are. This will be addressed in Ch 7.
[104] Broadbent, 'Epidemiological evidence', above n 91 at 240.
[105] See also *Smith v Leech Brain* [1962] 2 QB 405 (QB) at 413 (per Lord Parker CJ).

heavy load would otherwise do to his back. It does not take any great conceptual sophistication to see that this is an error.[106]

What this does tell us is that the DTR test and the measures on which it relies can *underestimate* a factor's effect on a population. What it cannot do, however, is overestimate that effect:

> The ... problem is that the exposure dose at which the probability of causation exceeds 50% (the point at which exposure causation is more likely than not) may fall well below the 'doubling dose' (the dose at which the incidence of the disease is doubled) ... When an effect of exposure is to accelerate the time at which disease occurs, the rate fraction[107] ... will tend to underestimate the probability of causation because it does not fully account for the acceleration of disease occurrence. In particular, and contrary to common perceptions, a rate fraction of 50% [which equates to a finding that the 'risk' has been doubled by the factor in question][108] does not correspond to a 50% probability of causation.[109]

The conclusion, therefore, that the incidence of injury has been doubled by the factor being tested amounts to a single indication that there might exist a statistical association between that factor and the injury.[110] It should also be clear, however, that this is not the same thing as establishing on the balance of probabilities that the factor caused the injury: first, such a conclusion does not amount to a 'fact probability' of 50 per cent or more and second, for reasons examined elsewhere in this book, fact probabilities are themselves not independently sufficient to establish causation in negligence.[111] Rather, such statistics are just one component of the range of evidence on which a court reaches its decision, and, *eo ipso*, tell us nothing about what happened *in this particular case*. Wright is insistent, for example, that non-particularised statistics are 'mere ad hoc distributions not related to any

---

[106] Broadbent, 'Epidemiological Evidence', above n 91 at 256.
[107] That proportion of the totality of disease within a population which is attributable to the factor being tested ([incidence rate among exposed population minus incidence rate among non-exposed population] divided by incidence rate among exposed population).
[108] For example, suppose that 40% of those exposed to brick dust get dermatitis and 20% of those not so exposed suffer the same fate. On these figures, it is easy to see how the incidence or 'risk' is doubled by exposure and how 50% of cases are attributable to exposure, which is what Greenland's 'rate fraction' refers to. For a more detailed account of the relationship between the two measures, see A Broadbent, *Philosophy of Epidemiology* (Basingstoke, Palgrave Macmillan, 2013) ch 3.
[109] S Greenland, 'Relation of Probability of Causation to Relative Risk and Doubling Dose: A Methodology Error That Has Become a Social Problem' (1999) 89 *American Journal of Public Health* 1166, 1166 and 1168–69. See, for instance, *Jones v Secretary of State for Energy and Climate Change* [2012] EWHC 2936 (QB), [2012] All ER (D) 271 (Oct).
[110] Conditional language is employed here deliberately to account for all of the potential errors to which such a conclusion is in any event vulnerable, such as poorly designed experiments and sampling errors (both of which are independent dangers – see D Barnes, 'Too Many Probabilities: Statistical Evidence of Tort Causation' (2001) 64 *Law and Contemporary Problems* 191). This is a topic on which there has been extensive academic discussion in the US in particular: for both a contribution and a brief review of some of the arguments, see MD Green, 'The Future of Proportional Responsibility' in S Madden (ed), *Exploring Tort Law* (New York, NY, CUP, 2005).
[111] See, for instance, discussion of belief probabilities in Ch 2, under heading 'The Balance of Probabilities'.

causal generalisation'.[112] Indeed, the fallacy of equating these to conclusions about individualised instances of causation is now well known, thanks to the famous examples of the taxi cab problem[113] and the Gatecrasher paradox.[114]

Broadbent, however, has different concerns:

> There are two serious mistakes in judicial and academic literature on this topic. The first is that RR > 2 is necessary for proof of specific causation.[115] The second is that epidemiological evidence is never sufficient for proof of causation: that no matter how strong the evidence for a general causal link between a wrong and a harm, we are never thereby warranted in deciding that a particular claimant's harm was caused by the wrong. Both errors have led to injustices . . . if epidemiological evidence is not capable of bearing on individual cases then it would lead to the radical misuse of evidence that is clearly admissible, such as that generated by medical tests. It follows that, where epidemiological evidence to the effect that RR > 2 is the only evidence, it is capable of proving causation as more likely than not . . .[116] What epidemiological evidence cannot do, however, is disprove a causal claim where RR < 2.[117]

In referring to the situation in which epidemiological evidence is the 'only' evidence available, Broadbent draws attention to an area in which Wright's hostility to the independent influence of such evidence requires the most scrutiny.[118] If all a court has at its disposal is that with which epidemiology provides it, should it use this as a basis for its decision? The following extract from one of Wright's earliest and most comprehensive works on causation suggests that there is less disagreement on this point than might first appear:

---

[112] R Wright, 'Proving Causation: Probability v Belief' in Goldberg (ed) *Perspectives on Causation*, above n 44 at 210.

[113] In which an individual is knocked down by a taxi whose colour is not observed. The incident occurs in a town where there are only two taxi firms: one which has three blue cabs, and another which has one yellow cab. The example is supposed to show that it would be inappropriate to infer from those facts alone that it was more probable than not that the accident involved a blue cab, despite the existence of a 75% statistical probability of this. The example can be found in the dissenting judgment of Brachtenbach J in *Herskovits v Group Health Cooperative of Puget Sound* 664 P 2d 474 (Wash 1983), a decision of the Supreme Court of Washington. A more complex example, to which particularistic evidence and reliability estimates have been added, can be found in A Tversky and D Kahneman, 'Evidential Impact of Base Rates' in D Kahneman, P Slovic and A Tversky (eds), *Judgement Under Uncertainty: Heuristics and Biases* (Cambridge, CUP, 1982) 153. See also Ch 7.

[114] In which there is a rodeo with 1,000 attendees, 499 of whom have paid for admission. There are no issued tickets, nor testimony as to whether A climbed over a fence to gain entry. There is, however, a mathematical probability of .501 that A did not pay, which would mean, on a mathematical interpretation of the forensic standard, that the organisers would be entitled to judgment against him. The problem here is that they would also be entitled, on those grounds, to judgment against every attendee on the same basis. LJ Cohen, *The Probable and the Provable* (Oxford, OUP, 1977) 75.

[115] This refers to the distinction sometimes made between the question of whether an agent is capable of causing an injury of the type in question (general causation) and the question of whether a particular claimant's injury was caused by a particular defendant's breach (specific causation). See *Seltsam Pty Ltd v McGuiness* (2000) 49 NSWLR 262 (NSWCA) at [22] (per Spigelman CJ).

[116] Since courts have to make a decision one way or another, any relevant evidence is *capable* of helping to reach a belief probability.

[117] Broadbent, *Philosophy of Epidemiology*, above n 108 at 206. Chapter 7 will elaborate further on the general applicability of epidemiological evidence to the forensic exercise.

[118] See MD Green, 'The Future of Proportional Liability: The Lessons of Toxic Substances Causation' in Madden (ed), *Exploring Tort Law*, above n 110 at 352.

A judgment on what actually happened on a particular occasion is a judgment on which causal generalization and its underlying causal law was fully instantiated on the particular occasion. Particularistic evidence connects a possibly applicable causal generalization to the particular occasion by instantiating the abstract elements in the causal generalization, thereby converting the abstract generalization into an instantiated generalization. Without such particularistic evidence, there is no basis for applying the causal generalization to the particular occasion.[119]

In other words, on Wright's analysis, evidence of a claimant's exposure to a given agent *is* particularised evidence sufficient to link epidemiological information about a condition to that claimant.[120] Where this is all that is available,

> then the only possibly applicable causal generalisation with at least some particularistic instantiation in the particular situation is the toxic-agent causal generalisation, which fact could support the formation of a belief that it was the causal process actually at work in the particular situation.[121]

Wright's point is not, therefore, that epidemiology cannot be used or useful. Rather, his entire conceptualisation of causation demands that there be a distinction between 'mere ad hoc distributions' and particular instances of causation.[122] Logically, this must be correct. One means of illustrating this is to show how a RR of < 2 is equally compatible with a 0 per cent and 100 per cent probability of causation. Greenland does this using the following illustration:

> As an extreme example, suppose the damage done by exposure was that of accelerating the development of disease in all individuals destined to contract disease. Then, when considering the lifetime experience of the exposed cohort, all of the exposed occurrences of disease would be accelerated cases . . . In other words . . . the excess fraction would be 0 [and the RR would be 1], incorrectly suggesting that there was no exposure effect, and yet the probability of causation would be 100%.[123]

In conclusion, it would seem that the epidemiological device of estimating the effect of a given factor can be forensically useful where that effect is shown *at least* to have doubled the rates of incidence as against background factors. It is equally clear, however, that this method should in no way constitute a routine or stand-alone test which is treated as conclusive of the causal inquiry.[124] The decision of the High Court of Australia in *Amaca Pty Ltd v Ellis*[125] is an object lesson in how to deal

---

[119] R Wright, 'Causation, Responsibility, Risk, Probability, Naked Statistics, and Proof: Pruning the Bramble Bush by Clarifying the Concepts' (1988) 73 *Iowa Law Review* 1001, 1051.
[120] Although see *Sienkiewicz v Greif* [2011] UKSC 10, [2011] 2 All ER 857 at [158] 906–07 (per Lord Rodger) and at [170] 910 (per Baroness Hale).
[121] Wright, 'Proving Causation', above n 112 at n 67.
[122] See also *Sienkiewicz v Greif* [2011] UKSC 10, [2011] 2 All ER 857 at [96] 888.
[123] Greenland, 'Relation of Probability', above n 109 at 1168. Greenland further points out that, whilst these facts may seem farfetched, it is precisely what will be seen if the outcome in question is death rates in a population followed for its entire lifetime, such as the atomic bomb survivors in Japan.
[124] See *Seltsam Pty Ltd v McGuiness* (2000) 49 NSWLR 262 (NSWCA) at [78]–[89] and [102] (per Spigelman CJ). Although this was a material contribution to risk case, the point remains relevant whatever the causal question.
[125] *Amaca Pty Ltd v Ellis* [2010] HCA 5, (2010) 240 CLR 111.

appropriately with such epidemiological evidence.[126] The deceased in that case had died from lung cancer, having been a smoker for over 25 years and also having been exposed by the defendant employers to asbestos in breach of their duty of care. Appropriately, the High Court made no mention of the DTR test, but instead considered the epidemiological evidence presented to it as an evidential component in its inference from the facts[127] as to what, on the balance of probabilities, caused the deceased's cancer:

> If the relative risks and probabilities derived from epidemiological studies were to be treated as revealing what was a probable explanation of what caused Mr Cotton's cancer, those analyses support two conclusions. First, it is more probable than not that smoking was a cause of (in the sense that it was a necessary condition for) Mr Cotton's cancer. Second, the risks and probabilities associated with asbestos, whether alone or in conjunction with smoking, are low and not sufficient to found the inference which the plaintiff sought to have made: that it is more probable than not that exposure to respirable asbestos fibres was a cause of Mr Cotton's cancer.[128]

Since the findings of liability in both *Novartis*[129] and *Shortell*[130] were based on evidence that the respective breaches of duty *more than doubled* the risk of the harm occurring, the formation of a belief probability that those breaches more likely than not caused the injuries seems unobjectionable.[131] Nevertheless, the presentation of these results as being predicated on satisfaction of the DTR test is regrettable.

> The relative risks permits such an inference only if it is credible, which means that the belief probability is greater than fifty percent. The preponderance of the evidence standard is not met without a belief probability of greater than fifty percent *and* a risk ratio greater than 2.0.[132]

For the reasons just outlined, therefore, courts should refrain from asking experts to present their evidence in the form of an answer to the DTR test, but should instead ensure that, where possible, such evidence takes account of what epidemiologists refer to as the *etiologic* fraction. In essence, this covers both the excess

---

[126] See also D Hamer, 'Mind the "Evidential Gap": Causation and Proof in *Amaca Pty Ltd v Ellis*' (2009) 31 *Sydney Law Review* 465 (criticising the contrasting approach of the Court of Appeal).

[127] See *Clements v Clements* [2012] SCC 32, [2012] 2 SCR 181 at [38] (per McLachlin CJ).

[128] [2010] HCA 5, (2010) 240 CLR 111 at [64] 134. These facts would not, therefore, satisfy the first stage of the NBA: see above, text to n 20. See also *Evans v Queanbeyan City Council* (2011) 9 DDCR 541, [2011] NSWCA 230 which follows *Ellis* in all aspects material to the current discussion. In *Evans*, the Court of Appeal made a misleading and unfortunate reference to the *Fairchild* principle, in saying that it could not be applied by them, but would be a matter for the High Court. The facts of *Evans* would not, in any event, be suitable for a correct application of that principle – see Ch 6 generally, and Ch 7, n 31.

[129] [2007] EWCA Civ 1261, [2007] All ER (D) 465 (Nov).

[130] Liverpool District Registry, (unreported) 16 May 2008.

[131] See also J Stapleton, 'Factual Causation, Mesothelioma and Statistical Validity', above n 101 at 223 and 227 for an explanation of why the DTR test is only applicable where conclusions are sought as to which of several *mutually exclusive* or 'competing alternative' mechanisms caused the injury in question.

[132] Barnes, 'Too Many Probabilities', above n 110 at 207.

fraction already referred to (those cases which would never have happened but for the exposure) *as well as* that fraction of cases which would still have happened but for the exposure, but not until later (those accelerated by the exposure). This paints a comprehensive picture of all of those cases affected by the exposure, and does not underestimate it in the way that DTR does.[133]

> To explain the problem in algebraic terms, suppose that A_T exposed persons contracted the disease during the time period in question and that, of these individuals, A_0 are unaffected, A_1 were accelerated by exposure, and A_2 represented all-or-none occurrences of disease [those that would not have occurred at all but for the exposure]. By definition, exposure ... harmed persons whose disease was either accelerated or all or none. Hence, the fraction of exposed persons with the disease who were harmed by the exposure is $(A_1 + A_2)/A_T$. This quantity is the etiologic fraction. Furthermore, if we randomly select an exposed person with disease from the total A_T, the chance that exposure harmed that person (i.e., the chance that the person had an accelerated or all-or-none occurrence) is also $(A_1 + A_2)/A_T$. The latter quantity is thus also the probability of causation.[134]

In *Merrell Dow Pharmaceuticals v Havner*,[135] the Supreme Court of Texas, in a judgment referred to with apparent approval by Lord Philips in *Sienkiewicz*,[136] made the methodologic error to which Greenland refers:

> Assume that a condition naturally occurs in six out of 1,000 people even when they are not exposed to a certain drug. If studies of people who *did* take the drug show that nine out of 1,000 contracted the disease, it is still more likely than not that causes other than the drug were responsible for any given occurrence of the disease since it occurs in six out of 1,000 individuals anyway. Six of the nine incidences would be statistically attributable to causes other than the drug, and therefore, it is not more probable that the drug caused any one incidence of disease. This would only amount to evidence that the drug *could* have caused the disease. However, if more than twelve out of 1,000 who take the drug contract the disease, then it may be *statistically* more likely than not that a given individual's disease was caused by the drug.[137]

This statement assumes, of course, that anyone who was ever likely to have suffered from the condition in question would have been immune to the effects of the drug (and thereby fails to recognise those cases where the drug would accelerate the development of the disease in those who would have contracted it at some point). Whilst this biological pattern can probably not be ruled out as ever occurring, it should certainly not be treated as if it were a normal occurrence. The excerpt above, therefore, demonstrates a forensic underestimation of the effects of the drug in question. This error is unfortunate, not only for its intrinsic logical flaw, but also because it detracts from a decision which is otherwise constructive in its treatment of epidemiological evidence (and which, anyway, did not rely on

---

[133] It also emphasises the importance of the temporal dimension to the But For question, as discussed above.
[134] Greenland, 'Relation of Probability', above n 109 at 1167–68.
[135] *Merrell Dow Pharmaceuticals v Havner* 953 SW 2d 706 (Tex 1997).
[136] [2011] UKSC 10, [2011] 2 AC 229 at [85] 264.
[137] 953 SW 2d 706 (Tex 1997) at 717.

this mistaken interpretation of the evidence for its ultimate result of no liability). Elsewhere, for instance, Owen J states:

> We recognize, as does the federal *Reference Manual on Scientific Evidence*, that a disease or condition either is or is not caused by exposure to a suspected agent and that frequency data, such as the incidence of adverse effects in the general population when exposed, cannot indicate the actual cause of a given individual's disease or condition ... But the law must balance the need to compensate those who have been injured by the wrongful actions of another with the concept deeply imbedded in our jurisprudence that a defendant cannot be found liable for an injury unless the preponderance of the evidence supports cause in fact. The use of scientifically reliable epidemiological studies and the requirement of more than a doubling of the risk strikes a balance between the needs of our legal system and the limits of science ... We do not hold, however, that a relative risk of more than 2.0 is a litmus test or that a single epidemiological test is legally sufficient evidence of causation. Other factors must be considered. As already noted, epidemiological studies only show an association. There may in fact be no causal relationship even if the relative risk is high. .... Likewise, even if a particular study reports a low relative risk, there may in fact be a causal relationship. The strong consensus among epidemiologists is that conclusions about causation should not be drawn, if at all, until a number of criteria have been considered.[138]

This excerpt forms only part of the judgment's considerable evaluation of the use of epidemiological evidence, and the various ways in which it can be interpreted and used. Unlike the English cases which rely on epidemiological data, this decision cites much academic commentary on the issue, and highlights points of disagreement and uncertainty.[139] This explains, at least in part, the measured and cautionary approach taken by the court to the use of such information, and it is unfortunate that the English courts have instead chosen to implement the specialised DTR test without any comparable reference to expert evaluation.

---

[138] 953 SW 2d 706 (Tex 1997) at 718. Although the *Reference Manual* referred to also makes the interpretative mistake highlighted here, as Alex Broadbent points out in Broadbent, 'Epidemiological Evidence', above n 91 at 254. See, however, *Restatement (Third) of Torts: Liability for Physical Harm* § 28 comment c.

[139] 953 SW 2d 706 (Tex 1997) at 715–21 in particular.

6

Material Increase in Risk

Illustrative Cases: *McGhee, Fairchild, Barker, Sienkiewicz*

Factual Basis

This analysis is applied where:

- there are multiple potential *sources* of the risk which has eventuated in the claimant's damage;
- there is a single agent responsible for that risk;
- there exists a 'rock of uncertainty' such that causation is impossible to prove *in principle*.

This analysis does *not* apply where causation is theoretically possible to prove, but where evidence is not available in a particular case.

Exceptionally, a link between a defendant's breach and a claimant's damage can be established by finding that the actions of the defendant materially increased the risk of the claimant's damage occurring.[1] This method started life in *McGhee v National Coal Board*[2] and mutated into the approach taken in *Fairchild v Glenhaven Funeral Services*.[3] In the latter case, but not the former, this was done to good effect.

---

[1] An approach which has been expressly rejected in Australia: see *Evans v Queanbeyan City Council* [2011] NSWCA 230, (2011) 9 DDCR 541, particularly at [53], *Amaca Pty Ltd (formerly James Hardie and Co Pty Ltd) v Hannell* [2007] WASCA 158, (2007) 34 WAR 109 at [31]. The approach favoured there, which retains the balance of probabilities standard of proof, is outlined in *Amaca* at [395]: 'As we understand the law in Australia, once a plaintiff demonstrates that a breach of duty has occurred followed by injury within the area of foreseeable risk, a prima facie causal connection will be established and the defendant has an evidential burden to adduce evidence that the breach had no effect or that the injury would have occurred even if the duty had been performed. If there is evidence sufficient to displace the prima facie case, it remains for the plaintiff upon the whole of the evidence to satisfy the tribunal of fact that the injury was caused by the defendant's negligence: *Purkess v Crittenden* (1965) 114 CLR 164 at 168'. This derives from the leading Australian authorities on the point: *Seltsam Pty Ltd v McGuiness* (2000) 49 NSWLR 262 (NSWCA) and *Bendix Mintex Pty Ltd v Barnes* (1997) 42 NSWLR 307 (NSWCA).

[2] *McGhee v National Coal Board* [1972] 3 All ER 1008 (HL).

[3] *Fairchild v Glenhaven Funeral Services* [2002] UKHL 22, [2003] 1 AC 32. Or, as Lord Hoffmann has put it, 'The intellectual prop . . . in *McGhee v National Coal Board* underwent a ghastly exhumation at

The claimant in *McGhee*, whose complaint was dermatitis, was employed by the defendants to work in brick kilns. As a result of this occupation, his skin was inevitably in contact with brick dust, which was the agent identified as the cause of his illness. His employers, however, were not in breach of their duty of care for exposing him to brick dust whilst he was physically in the kilns, but for failing to provide washing facilities to enable him to wash the brick dust off his skin before cycling home. Medical evidence was unable to say whether the dermatitis would have occurred but for the defendant's breach of duty, because it was unable to determine whether, and if so to what extent, the disease was triggered by brick dust attributable to the breach of duty, as opposed to brick dust exposure which would have occurred even if showers had been provided for use after work:

> The medical witnesses are in substantial agreement. Dermatitis can be caused, and this dermatitis was caused, by repeated minute abrasion of the outer horny layer of the skin followed by some injury to or change in the underlying cells, the precise nature of which has not yet been discovered by medical science. If a man sweats profusely for a considerable time the outer layer of his skin is softened and easily injured. If he is then working in a cloud of abrasive brick dust, as this man was, the particles of dust will adhere to his skin in considerable quantity and exertion will cause them to injure the horny layer and expose to injury or infection the tender cells below. Then in some way not yet understood dermatitis may result.[4]

In order to deal with this 'evidentiary gap',[5] the House of Lords held the defendants liable for having materially *increased the risk* of the claimant's developing dermatitis, since risk was the only medium through which such a conclusion could be reached. Lord Wilberforce summarised the issue in the following way:

> [T]he question remains whether a pursuer must necessarily fail if, after he has shown a breach of duty, involving an increase of risk of disease, he cannot positively prove that this increase of risk caused or materially contributed to the disease while his employers cannot positively prove the contrary. In this intermediate case there is an appearance of logic in the view that the pursuer, on whom the onus lies, should fail – a logic which dictated the judgments below. The question is whether we should be satisfied in factual situations like the present, with this logical approach. In my opinion, there are further considerations of importance. First, it is a sound principle that where a person has, by breach of duty of care, created a risk, and injury occurs within the area of that risk, the loss should be borne by him unless he shows that it had some other cause. Secondly, from the evidential point of view, one may ask, why should a man who is able to show that his employer should have taken certain precautions, because without them there is a risk, or an added risk, of injury or disease, and who in fact sustains exactly that injury or disease, have to assume the burden of proving more: namely, that it was the addition to the risk, caused by the breach of duty, which caused or materially contributed to the

the hands of all five Law Lords in *Fairchild*, after which it was declared to be the *fons et origio* of the principle we adopted'. L Hoffmann, '*Fairchild* and After' in A Burrows, D Johnston and R Zimmermann (eds), *Judge and Jurist: Essays in Memory of Lord Rodger of Earlsferry* (Oxford, OUP, 2013) 64.

[4] [1972] 3 All ER 1008 at 1010 (per Lord Reid).
[5] So characterised by J Stapleton in 'Lords a'Leaping Evidentiary Gaps' (2002) 10 *Torts Law Journal* 276.

injury? In many cases of which the present is typical, this is impossible to prove, just because honest medical opinion cannot segregate the causes of an illness between compound causes. And if one asks which of the parties, the workman or the employers should suffer from this inherent evidential difficulty, the answer as a matter in policy or justice should be that it is the creator of the risk who, ex hypothesi, must be taken to have foreseen the possibility of damage, who should bear its consequences.[6]

The idea of a defendant being liable in negligence for materially increasing the risk of a claimant incurring damage was a novel development. It was also of fairly minor importance for nearly three decades until the House of Lords resurrected it in the landmark decision of *Fairchild v Glenhaven Funeral Services*.[7] The defendants in *Fairchild* were former employers of mesothelioma victims, who had exposed their employees to asbestos in breach of their duties of care. It was accepted that, for the purposes of the case, any exposure to asbestos other than that for which the defendants were responsible (such as general environmental exposure) could be discounted. The major complications, however, were first, that each individual had been exposed to asbestos by more than one employer, and second, that medical knowledge about mesothelioma was incomplete, and therefore unable to associate the development of the disease with any particular source of asbestos (in other words, with any particular employer):

> The mechanism by which a normal mesothelial cell is transformed into a mesothelioma cell is not known. It is believed by the best medical opinion to involve a multi-stage process, in which six or seven genetic changes occur in a normal cell to render it malignant. Asbestos acts in at least one of those stages and may (but this is uncertain) act in more than one. It is not known what level of exposure to asbestos dust and fibre can be tolerated without significant risk of developing a mesothelioma, but it is known that those living in urban environments (although without occupational exposure) inhale large numbers of asbestos fibres without developing a mesothelioma. It is accepted that the risk of developing a mesothelioma increases in proportion to the quantity of asbestos dust and fibres inhaled: the greater the quantity of dust and fibre inhaled, the greater the risk. But the condition may be caused by a single fibre,[8] or a few fibres, or many fibres: medical opinion holds none of these possibilities to be more probable than any other, and the condition once caused is not aggravated by further exposure. So if C is employed successively by A and B and is exposed to asbestos dust and fibres during each employment and develops a mesothelioma, the very strong probability is that this will have been caused by inhalation of asbestos dust containing fibres. But C could have inhaled a single fibre giving rise to his condition during employment by A, in which case his exposure by B will have had no effect on his condition; or he could have inhaled a single fibre giving rise to his condition during his employment by B, in which case his exposure by A will have had no effect on his condition; or he could have inhaled fibres during his employment by A and B which together gave rise to his condition; but medical science cannot support the suggestion that any of these possibilities is to be regarded

---

[6] [1972] 3 All ER 1008 at 1012 (per Lord Wilberforce).
[7] [2002] UKHL 22, [2003] 1 AC 32.
[8] Although medical science has now effectively rejected this 'single fibre theory'; see *Amaca Pty Ltd v Booth* [2011] HCA 53, (2011) 246 CLR 36 and below, text to n 38.

as more probable than any other. There is no way of identifying, even on a balance of probabilities, the source of the fibre or fibres which initiated the genetic process which culminated in the malignant tumour. It is on this rock of uncertainty, reflecting the point to which medical science has so far advanced, that the three claims were rejected by the Court of Appeal and by two of the three trial judges.[9]

In this sense, there was an evidentiary gap comparable to that in *McGhee*. It is hardly surprising, therefore, that, after detailed consideration of the earlier case, the House of Lords applied the material increase in risk approach in *Fairchild* in order to decide in favour of the claimants. In the process, the Court emphasised the exceptional nature of what it was doing,[10] and limited it by reference to, inter alia, the intractable epistemic problems associated with mesothelioma:

> The essential question underlying the appeals may be accurately expressed in this way. If (1) C was employed at different times and for differing periods by both A and B, and (2) A and B were both subject to a duty to take reasonable care or to take all practicable measures to prevent C inhaling asbestos dust because of the known risk that asbestos dust (if inhaled) might cause a mesothelioma, and (3) both A and B were in breach of that duty in relation to C during the periods of C's employment by each of them with the result that during both periods C inhaled excessive quantities of asbestos dust, and (4) C is found to be suffering from a mesothelioma, and (5) any cause of C's mesothelioma other than the inhalation of asbestos dust at work can be effectively discounted, but (6) C cannot (because of the current limits of human science) prove, on the balance of probabilities, that his mesothelioma was the result of his inhaling asbestos dust during his employment by A or during his employment by B or during his employment by A and B taken together, is C entitled to recover damages against either A or B or against both A and B? To this question (not formulated in these terms) the Court of Appeal (Brooke, Latham and Kay LJJ), in a reserved judgment of the court reported at [2002] 1 WLR 1052, gave a negative answer. It did so because, applying the conventional 'but for' test of tortious liability, it could not be held that C had proved against A that his mesothelioma would probably not have occurred but for the breach of duty by A, nor against B that his mesothelioma would probably not have occurred but for the breach of duty by B, nor against A and B that his mesothelioma would probably not have occurred but for the breach of duty by both A and B together. So C failed against both A and B. The crucial issue on appeal is whether, in the special circumstances of such a case, principle, authority or policy requires or justifies a modified approach to proof of causation.[11]

This is the crux of the issue in both *McGhee* and *Fairchild*, characterised by Lord Bingham as the 'rock of uncertainty'.[12] Specifically, the crucial uncertainty here is that the nature of mesothelioma makes it (currently) impossible to discern *in principle* which exposures contributed to the disease and which did not. In these 'rock

---

[9] [2002] UKHL 22, [2003] 1 AC 32 at [7] 43 (per Lord Bingham).
[10] Extra-judicially, Lord Hoffmann has said of this development: 'We were very conscious in *Fairchild* that we were taking a nibble out of the principle that a claimant who sues for negligence has to prove his case and, in particular, has to prove that the defendant's negligent act had the necessary causal connection with his injury'. Hoffmann, '*Fairchild* and After', above n 3 at 64.
[11] [2002] UKHL 22, [2003] 1 AC 32 at [2] 40 (per Lord Bingham).
[12] [2002] UKHL 22, [2003] 1 AC 32 at [7] 43 (per Lord Bingham).

of uncertainty' cases, therefore, the issue is not that factual evidence is *unavailable* in any given instance, but that, on the basis of current knowledge, factual evidence is *not capable* of answering the causal question.[13] It is essential that this specific type of impossibility is distinguished from the impossibility which can arise on the facts of *any* given case, when the available evidence is unable to provide the court with a scientifically conclusive answer, despite the fact that it is in principle possible to provide such an answer on such facts. In the absence of this critical distinction, the material contribution to risk analysis loses one of its principle justificatory bases, and becomes potentially applicable to any case in which proof of causation is physically (as opposed to theoretically) impossible:[14]

> [T]he option of finding that a material contribution to risk approach is available whenever proof of 'but for' causation cannot be made on the facts is equally problematic. First, how does one distinguish between a case of true impossibility of factual proof and a situation where the plaintiff simply fails to meet her burden of establishing 'but for' causation on the evidence? Unless one can make a clear distinction, one effectively undermines the requirement that the plaintiff bears the burden of showing that, 'but for' the defendant's negligence, she would not have been injured. In any difficult case, the plaintiff would be able to claim impossibility of proof of causation. Such a result would fundamentally change the law of negligence and sever it from its anchor in corrective justice that makes the defendant liable for the consequences, but only the consequences, of his negligent act.[15]

This, the foundational difficulty in evidentiary gap cases, was outlined in its updated[16] form by Lord Rodger in *Sienkiewicz v Greif*, the most recent Supreme Court case to engage the *Fairchild* principle:

> In the case of a disease like mesothelioma the claimant will be able to prove on the balance of probability that he is suffering from mesothelioma and that he has suffered loss as a result. He may also be able to prove, on the balance of probability, that a defendant or a number of defendants negligently exposed him to asbestos in the course of his employment with them . . . What, however, the claimant will be quite unable to prove, on the balance of probability, in the present state of medical knowledge, is that

---

[13] It is entirely possible, according to our current state of knowledge, that all but one defendant on the facts of *Fairchild* (or, indeed, every defendant, if Stapleton's point that environmental exposure cannot defensibly be discounted, despite what their Lordships decided in *Fairchild* – see Stapleton, 'Lords a' Leaping', above n 5 at 17–20) could have had no effect whatsoever on the claimant. This is what makes the issue so difficult; a point which is missed by the example of the multiple stabbing given in C Miller, 'Causation in Personal Injury After (and Before) Sienkiewicz' (2012) 32 *Legal Studies* 396, 400. His example is one of material contribution to injury, since all of those stabbing Caesar on the facts there given caused him some injury, regardless of whether they triggered his death by dealing the fatal blow. This is an orthodox causal situation, and is not subject to the evidentiary gap which is so intractable in the mesothelioma cases.

[14] This also explains why the exceptional material contribution to risk analysis is not applicable to cases in which claimants claim to have 'lost a chance' of avoiding an adverse physical outcome; but see A Burrows, 'Uncertainty About Uncertainty: Damages for Loss of a Chance' [2008] *Journal of Personal Injury Law* 31 and E Peel, 'Lost Chances and Proportionate Recovery' [2006] *LMCLQ* 289. See Ch 7.

[15] *Clements v Clements* [2012] SCC 32, [2012] 2 SCR 181 at [37] (per McLachlan CJ).

[16] Necessary because, in the decade separating *Fairchild* and *Sienkiewicz*, the 'single-fibre theory' had been effectively discredited. See above, n 8.

he developed mesothelioma as a result of inhaling any particular fibre or fibres and that, therefore, a particular defendant was responsible for exposing him to the fibre or fibres that caused his illness. Moreover, medical experts are no more able to tell whether the fibre or fibres which triggered the claimant's mesothelioma came from the general atmosphere than they can tell whether they came from exposure during the claimant's work with one or other of a number of employers.[17]

In *Sienkiewicz*, the victim of mesothelioma had been exposed to asbestos by the defendants during her employment by them. The context differed, however, from *Fairchild* in a significant way, in that the court accepted that her occupational exposure had been very light, and that her environmental exposure (which was not associated with any breach of duty) should not be discounted. In fact, the trial judge accepted evidence to the effect that her asbestos exposure by the defendants increased her overall risk of developing mesothelioma by only 18 per cent, and that the rest was due to background environmental factors.[18] The major question facing the Supreme Court was, therefore, whether the exceptional *Fairchild* principle should be applied in situations in which there had been only one occupational exposure to risk. As Lord Brown pointed out, this question had not avoided scrutiny in *Fairchild* itself:

> Lord Rodger of Earlsferry, for example, expressly recognised (at para 170 of his speech) that 'it can also apply where, as in *McGhee*, the other possible source of the injury is a similar, but lawful, act or omission of the same defendant'. But he immediately then 'reserve[d] [his] opinion as to whether the principle applies where the other possible source of injury is a similar but lawful act or omission of someone else or a natural occurrence'.[19]

The same point called for consideration in *Barker v Corus*,[20] but this time it was of central importance: *Barker* raised two issues that were bound to require clarification in the wake of *Fairchild*. The late Mr Barker had been exposed to asbestos during three periods of his working life; one whilst working for an employer other than the defendant;[21] one whilst working for the defendant; and another whilst he was working for himself. The issue, therefore, was whether the *Fairchild* principle should apply to situations where an individual has been exposed to asbestos dust during periods of self-employment as well as during periods of employment by others. Significantly, the House of Lords was also asked to address the question of whether or not a defendant in a *Fairchild*-type situation should be entitled to apportionment of his liability so that the award made against him would reflect the extent of exposure for which he was responsible.[22] Ultimately, the *Fairchild* prin-

---

[17] [2011] UKSC 10, [2011] 2 AC 229 at [139] 280.
[18] In Australia (a jurisdiction which has rejected the *Fairchild* principle) explicit consideration is given to the *additional* risk created by the breach of duty, *relative* to background risks. See *Amaca Pty Ltd (formerly James Hardie and Co Pty Ltd) v Hannell* [2007] WASCA 158, (2007) 34 WAR 109 at [44].
[19] See *Sienkiewicz v Greif* [2011] UKSC 10, [2011] 2 AC 229 at [179] 292–293 (per Lord Brown).
[20] *Barker v Corus* [2006] UKHL 20, [2006] 2 AC 572.
[21] Who had since become insolvent.
[22] See R Merkin and J Steele, *Insurance and the Law of Obligations* (Oxford, OUP, 2013) 366.

ciple was held to apply in such circumstances, but on the basis, the House suggested, of liability aliquoted according to relative risk contributions.[23]

In some ways, the mesothelioma cases look like instances of overdetermined harm. To recall, situations in which an injury has been overdetermined are those in which all factors:[24]

- are severally sufficient;
- are independent of one another;
- lead to indivisible injury;
- lead to the same type of damage;
- affect the claimant simultaneously.

In *McGhee* and *Fairchild*, the factors under scrutiny (the defendant's breach and the non-negligent exposure in the former, and all of the defendants' breaches in the latter) were independent of one another,[25] mesothelioma and dermatitis are indivisible, and there was only one type of damage being claimed for in each case. The other criteria, several sufficiency and simultaneous effects, cannot definitively be said to be *in*applicable. It is simply not possible, given the evidentiary gap in *McGhee* or in the asbestos/mesothelioma cases, to establish whether each breach was severally sufficient, or whether the effects of each breach affected the claimants simultaneously.[26] In this sense, these cases look different to the classic overdetermined factual situations involving hunters and votes, of which we can and do know that both several sufficiency and simultaneous effects are features.[27] What is certainly true of the evidentiary gap cases is that they are subject to even more uncertainty than are situations of overdetermined harm: where overdetermination is concerned, we know what we can't know, but in evidentiary gap situations, we are not even sure of that.[28] It might be the case, for example, that the facts of *Fairchild* amount to a situation in which the defendant employers all materially contributed to the mesothelioma because the disease actually requires a cumulative build-up of asbestos fibres to trigger it. On the other hand, it might be that it is really a situation of overdetermination, in that the exposure from each defendant would have been sufficient to trigger the disease on its own, and all of the inhaled fibres began their destructive work at the same time. Similarly, if each exposure were severally sufficient, but the earlier exposure triggered the disease before the others could take effect,[29] *Fairchild* would be a case of pre-emption.

---

[23] The effect of the latter part of this ruling was reversed in relation to asbestos and mesothelioma by s 3 Compensation Act 2006 – see discussion below.
[24] See Ch 4, text to n 21.
[25] See, eg, *Barker v Corus* [2006] UKHL 20, [2006] 2 AC 572 at [61] 598–99 (per Lord Scott).
[26] See, eg, *Barker v Corus* [2006] UKHL 20, [2006] 2 AC 572 at [29] 588 (per Lord Hoffmann).
[27] See Ch 4 and Stapleton, 'Lords a'Leaping', above n 5 at 53.
[28] Which is why the *Fairchild* principle is not generally applicable to any case in which there is a causal difficulty, but in which there is no evidentiary gap or 'rock of uncertainty'.
[29] See, for instance, the statement of Lord Phillips in [2011] UKSC 10, [2011] 2 AC 229 at [103] 268 to the effect that exposure after a cell has become malignant is not regarded as causative. If, therefore, it were possible to identify that point, it would also be possible in principle to distinguish pre-empted from pre-emptive factors.

As it happens, one of the experts who assisted the court in *Fairchild* seemed to suggest in the much more recent case of *Jones v Secretary of State for Energy & Climate Change* that medical understanding of mesothelioma might now point more towards a possible analysis of the facts on the basis of a material contribution to injury:

> Dr Rudd was asked why, if his thesis was correct, there had been any need for the *Fairchild* exception. His response was that, in *Fairchild*, he and the other experts had been instructed to consider from what source the asbestos fibre(s) that had caused the final step in the production of the malignant cell had come. They were unable to do so; hence the impossibility of establishing causation and the necessity for the creation of the *Fairchild* exception. He said that the expert evidence in *Fairchild* was given in the light of the knowledge of carcinogenesis at that time. Judgment at first instance was given in *Fairchild* in February 2001 and, since then understanding of the molecular basis of carcinogenesis has progressed considerably. Dr Rudd said that, if he were asked the same questions now as he had been asked in *Fairchild*, he would say that it was probable that the asbestos fibres from each source had contributed to the carcinogenic process.[30]

Nevertheless, this is of only incidental *analytical* significance. As Lord Phillips recognised in *Sienkiewicz*:

> Of course, the Fairchild exception was created only because of the present state of medical knowledge. If the day ever dawns when medical science can identify which fibre or fibres led to the malignant mutation and the source from which that fibre or those fibres came, then the problem which gave rise to the exception will have ceased to exist. At that point, by leading the appropriate medical evidence, claimants will be able to prove, on the balance of probability, that a particular defendant or particular defendants were responsible. So the Fairchild exception will no longer be needed. But, unless and until that time comes, the rock of uncertainty which prompted the creation of the Fairchild exception will remain.[31]

It is difficult to imagine, however, that there will ever exist medical knowledge so comprehensive that the *Fairchild* principle will come to be redundant. There are signs that it might not be necessary for much longer in cases involving asbestos exposure and mesothelioma but, just as there was an evidentiary gap in relation to dermatitis 40 years ago, there is bound always to be a similar issue with some disease or other. An appropriate analysis and understanding of the *Fairchild* principle is, therefore, of more than transitory importance because evidentiary gaps are unlikely ever to stop posing difficulties for the forensic process.

---

[30] [2012] EWHC 2936 (QB) [2012] All ER (D) 271 (Oct) at [8.21].
[31] [2011] UKSC 10, [2011] 2 AC 229 at [142] 281.

The Necessary Breach Analysis and Evidentiary Gaps

The effect of applying the NBA to the *McGhee/Fairchild* line of authority would be to reach different results in some, but not all, of those cases. Essentially, this is because the starting point for an NBA is that at least one breach of duty must be necessary, on the balance of probabilities, for the claimant's damage to have occurred. This analysis would not give consistent results across *McGhee*, *Fairchild*, *Barker* and *Sienkiewicz*. Only on the facts of *Fairchild* and *Barker* would the NBA come to a conclusion in favour of liability. *Fairchild* is the most straightforward example of this because of the fact that it was accepted therein that only exposures to asbestos in breach of the defendants' duties of care were potential causal factors of the employees' mesothelioma. It is clear, therefore, that, on the balance of probabilities, but for at least one breach of duty, those employees would not have contracted mesothelioma. *Barker* is hardly more complicated. Vernon Barker was exposed to asbestos by two employers in addition to the period of his exposure whilst self-employed. Two-thirds of the potential factors in his illness were, therefore, breaches of others' duties to him. In the absence of any evidence suggesting that any of the three periods of exposure had any greater causal valence than any other, the principle of indifference[32] would suggest that equal significance is ascribed to each.

> This principle eliminates informational open-endedness that frustrates any attempt at determining probability. This principle of indifference replaces this open-endedness with informational closure, easily governed by mathematical logic... Artificial as it may sound, this assumption may be justified in the long run. The unknown possibilities, some favourable and some unfavourable to the examined hypothesis, can be expected to cancel each other out... We can, therefore, exercise control over the ratio of right versus wrong decisions by ignoring these possibilities.[33]

On this basis, where two-thirds of his exposure resulted from others' breaches of duty and a third from his own actions, the formation of a belief probability that it is more likely than not that at least one breach of duty caused Mr Barker's mesothelioma is entirely reasonable. It could of course be argued that such an approach is unnecessary, given that liability apportionment has been possible since 1945 on the basis of contributory negligence.[34] That scheme, however, is incomplete in its *causal* reach because it allows for apportionment only where a claimant has herself been negligent. It does not, therefore, cover situations in which a claimant has

---

[32] See Ch 4, text to n 43.
[33] A Porat and A Stein, *Tort Liability Under Uncertainty* (Oxford, OUP, 1999) 46–47.
[34] Law Reform (Contributory Negligence) Act 1945 (particularly ss 1 and 4). Whilst s 4 makes it clear that 'fault' for these purposes is broader than negligence in its technical sense, it is by no means a definition which covers all manner of behaviour: '"fault" means negligence, breach of statutory duty, or other act or omission which gives rise to a liability in tort or would, apart from this Act, give rise to the defence of contributory negligence'. See R Stevens, *Torts and Rights* (Oxford, OUP, 2007) 125–26.

brought about her own adverse outcome, but has done so non-negligently. The manner of her actions says nothing about their causal valence, and harming oneself, negligently or otherwise, is, in negligence terms, part of a claimant's 'normal course of events' because it does not amount to being made worse off as a result of another's breach of duty.

A theoretical objection to this simple analysis of *Barker* (but one which would make no difference to the outcome on the facts) could be made on the grounds that it might be possible to ascribe relative causal valence to each exposure on the basis of its duration, intensity or the type of asbestos concerned (some types being more harmful to health than others).[35] Indeed, these were the criteria on which the House of Lords suggested that liability in that case should be aliquoted.[36] The problem with such an approach, however, is that if division along such lines were to be possible, the *Fairchild* principle would in itself be redundant. It was devised, as we have seen, to deal with a situation in which neither the existence, nor extent, of contributions can be known. Situations to which it applies, therefore, must by definition be those in which any discriminatory apportionment is impossible. This point was recognised by Lord Phillips in *Sienkiewicz*:

> Why, if it was possible to equate increasing exposure to increasing risk, could one not postulate that, on balance of probabilities, where one employer had caused over 50% of a victim's exposure, that employer had caused the victim's mesothelioma? Why could one not, by the same token, postulate that where over 50% of the victim's exposure was not attributable to fault at all, on balance of probability, the victim's mesothelioma had not been caused tortiously? In short, why was there any need to apply the Fairchild/Barker rule where epidemiological evidence enabled one to use statistics to determine causation on balance of probability?[37]

and

> It used to be thought that mesothelioma was probably triggered by a single asbestos fibre and that the cause of the disease could be attributed exclusively to that one fibre. Were that the case it would be reasonable to postulate that the risk of contracting the disease was proportional to the exposure... The single fibre theory has, however, been discredited. The amount of exposure does not necessarily tell the whole story as to the likely cause of the disease. There may well be a temporal element. The Peto Report also raised the possibility (but no more) of synergistic interaction between early and later exposures. Causation may involve a cumulative effect with later exposure contributing to

---

[35] See *Bolton Metropolitan Borough Council v Municipal Mutual Insurance Ltd* [2006] EWCA Civ 50, [2006] 1 WLR 1492 at [7]–[12] (per Longmore LJ).

[36] See [2006] UKHL 20, [2006] 2 AC 572 at [35]–[37] 589–90 and [48] 593–94 (per Lord Hoffmann), at [61]–[62] 599–600 (per Lord Scott) and at [109] 612 (per Lord Walker). In *Sienkiewicz*, Lord Phillips disapproved of such an approach, see [2011] UKSC 10, [2011] 2 AC 229 at [101]–[106]. See also J Morgan, 'Lost Causes in the House of Lords: *Fairchild v Glenhaven Funeral Services*' (2003) 66 *MLR* 277, 281 and *International Energy Group Ltd v Zurich Insurance Plc UK* [2013] EWCA Civ 39, [2013] 3 All ER 395. See also Merkin and Steele, *Insurance and the Law of Obligations*, above n 22 at 377–78 on how such criteria are anyway those used in claims amongst insurers for contribution.

[37] [2011] UKSC 10, [2011] 2 AC 229 at [94] 266. See also S Steel and D Ibbetson, 'More Grief on Uncertain Causation in Tort' (2011) 70 *CLJ* 451, 461.

causation initiated by an earlier exposure. Applying the conventional test of causation, the relevant question is, on balance of probability, which exposures in an individual case may have contributed to causing the disease? Epidemiology does not enable one to answer that question by considering simply the relative extent of the relevant exposures.[38]

In an article written four years before *Barker* was decided, Stapleton identified this specific impossibility as being the essence of the *Fairchild* problem:

> [T]here is as yet no means of telling whether the mesothelioma of a person subjected to a sequence of asbestos exposures was due to all exposures, only some or one, let alone which one. Importantly, there is no direct basis for saying that longer, more intense exposures are more likely to have been the cause of a case of mesothelioma than much shorter exposures, nor is there any basis for saying that earlier exposures are more likely to have been the cause than later exposures ... [S]ince the aetiology of mesothelioma is unknown, the extent of a defendant's contribution to the total risk is, as a matter of science, unknown. This then means that the relative culpability, which will include a consideration of the defendant's contribution to the total risk, is unknown. As a matter of scientific logic, we cannot, for example, equate quantity of dust inhaled with the extent of contribution to risk.[39]

It is, therefore, clearly not possible to make any *qualitative* evaluation of the respective contributions to risk made by different periods of exposure. Indeed, the very nature of this particular evidentiary gap suggests that any one period of exposure could have the same potential causal valence as any other.[40] So significant, however, is this epistemic limitation, at least in terms of its practical effects, that its extent can actually be regarded as an analytical strength rather than a weakness. In some branches of science, for instance, as the well-known theories of Heisenberg and Gödel demonstrate,[41] such complete and inevitable uncertainty is not only explicitly recognised, but used as a positive basis of analysis. Firestein, who devoted a whole book to the subject of Ignorance,[42] makes the following point:

> Some things can never be known and, get this, it doesn't matter. We cannot know the exact value of pi. That has little practical effect on doing geometry ... the problem of the unknowable, even the really unknowable, may not be a serious obstacle. The unknowable may itself become a fact. It can serve as a portal to deeper understanding.[43]

---

[38] *Sienkiewicz v Greif* [2011] UKSC 10, [2011] 2 AC 229 at [101]–[102] 267–68 (per Lord Phillips).
[39] Stapleton, 'Lords a'Leaping', above n 5 at 26 and 83. As someone who has a Phd in chemistry, Professor Stapleton is better placed than most lawyers to make such statements with conviction. Unsurprisingly, she states in the same paper (at 90) that any purported apportionment in combination with the *Fairchild* principle would be 'artificial' (at 26).
[40] ibid, 63: 'the principle will only be triggered where there is a gap in medical knowledge of aetiology such that it cannot be said that the cause-in-fact of a disease was more likely to be one source of risk of the outcome than another source of risk, howsoever trivial'.
[41] Heisenberg's Uncertainty Principle tells us that, for instance, we can never know both the position and momentum of a subatomic particle simultaneously. Gödel's Incompleteness Theorems establish that some mathematical problems are quite impossible to solve.
[42] S Firestein, *Ignorance: How It Drives Science* (Oxford, OUP, 2012). I have wondered, but been unable to establish, why lawyers use 'uncertainty' to describe the concept scientists know as 'ignorance'.
[43] ibid, 43–44.

In other words, the giants on whose shoulders we stand need not be omniscient. What matters is that we know that they are not. There is, however, an important limitation to this idea, because when 'ignorance is managed incorrectly or thoughtlessly, it can be limiting rather than liberating'.[44] It is crucial, therefore, for the law to deal correctly and reflectively with the things it cannot know. In this context, the NBA does this by *utilising* the completeness of the evidentiary gap in mesothelioma cases, rather than attempting to find an appropriate work-around. As we have already seen,[45] it does this simply by assigning equal potential causal valence to each source of asbestos, regardless of its duration, intensity, length, type or author. To combine any other method with the *Fairchild* principle would be oxymoronic, since any relative means of apportionment would be conceptually incompatible with the reason for applying the exception in the first place. It is of course true that the 'mechanism of mesothelioma is unknown so it cannot be said that the sources of risk are "rivals" or in "competition" with each other'.[46] Since, however, it equally cannot be said that they are not, an approach which assumes indifference between the two possibilities would seem to be the least objectionable one to adopt unless and until we know more.

Essentially, this means eschewing any attempt to *evaluate* the contributions made by different sources, in favour of merely counting each as a single and discrete factor, with a uniform causal valence. Then, as it would with overdetermined events,[47] the first stage of the NBA asks whether the number of sources that emanate from a breach of duty exceeds the number that emanate from non-breaches. Only where there are more breach than non-breach sources will this first stage be satisfied. On facts such as those in *Fairchild*, any breach which is then subjected to the second stage of the NBA will be deemed to remain operative in terms of a claimant's mesothelioma, because every exposure to asbestos creates a risk of that disease, and this was precisely the risk which eventuated.[48] This is why, as outlined above, an NBA would concur with the result reached by the House of Lords in both *Fairchild* and *Barker*, but would come to a different conclusion in both *McGhee* and *Sienkiewicz*. The reason is arithmetically simple: in *McGhee* one source of brick dust resulted from a breach of duty (that which remained on James McGhee's skin whilst he cycled home, having been denied a shower), and one from the defendants' non-negligent action in exposing him to dust whilst he worked in the kilns. Since the number of non-breach sources equals the number of breach sources, McGhee could not establish, on the balance of probabilities, that it was more likely than not that his dermatitis was caused by at least one breach of duty. The same argument applies to *Sienkiewicz*, in that there was one source emanating from a breach of duty (Enid Costello's exposure to asbestos at

[44] ibid, 44.
[45] Text to n 25.
[46] J Stapleton, 'Factual Causation, Mesothelioma and Statistical Validity' (2012) 128 *LQR* 221, 230.
[47] See Ch 4, under heading 'Combination of Tortious and Non-Tortious Factors'.
[48] See Ch 3, under heading 'Operative: the second stage of the NBA', for a detailed discussion of what amounts to 'operative'. This is also the point at which the 'single agent' limitation on the *Fairchild* principle becomes pertinent – see below, under heading 'Single Agent'.

work) and one source emanating from a non-breach (her environmental or background exposure to the same substance).[49]

The adoption of an arithmetical means of dealing with uncertainty is not a novel concept. A similar method was employed by Morland J in *The Creutzfeldt-Jakob Disease Litigation, Groups A and C Plaintiffs*,[50] a case involving the contraction of CJD by individuals who had been injected with Human Growth Hormone (HGH) as a treatment for stunted growth in childhood. The uncertainty arose because, although the accepted scientific view was that the CJD was caused by a single injection or dose containing a sufficient titre of the CJD agent, each individual had received large numbers of injections. It was also accepted that there was no issue of cumulative cause, and that no individual was any more susceptible to the illness than any other. Those claiming were the patients whose treatment straddled the date (1 July 1977) after which continuation of treatment was deemed to amount to a breach of a duty of care.[51] Any patients whose treatment had ceased by that date were obviously not harmed *by any breach of duty*, whilst those who received all of their injections after that date clearly were. The 'straddlers' however, posed a problem for the causal inquiry in negligence. This is a question for the first stage of the NBA: is it more likely than not that the claimants' CJD would not have occurred when it did but for at least one breach of duty?

Morland J decided that any straddler victim would succeed on causation if it was proved that he received the majority of doses after the cut-off date, thereby using a straightforward mathematical approach to the balance of probabilities. In reaching this conclusion, he approved of the analogy made by Professor Grimmett, the statistical expert called by the claimants:

> A pack of cards is dealt into two piles containing 25 and 27 cards respectively. Under reasonable assumptions about the random order of the pack it is a matter of mathematical calculation and common sense that there is a probability, 27 out of 52, that the larger pack contains the ace of spades. In two equal piles there are equal probabilities, one out of two, that the ace of spades is in the left pile or the right pile. Now if one card is moved from the left pile to the right pile we do not know what that card is, but there is a probability of one in 52 that it is the ace of spades. This grain of probability tips the balance

---

[49] The same also applies to *Willmore v Knowsley Borough Council*, the case heard with *Sienkiewicz*, in that the claimant therein had been exposed to asbestos at school, as a result of the local council's breach of duty. Whilst it was accepted in that case that there were multiple 'sources' of asbestos within the school (such as in a corridor and in the girls' lavatories), these do not form separate sources for the purposes of current analysis, since they were both simultaneous and resulted from the same, single breach of duty (which occurred because the council failed to protect those who attended its schools from asbestos exposure). Of *Sienkiewicz*, however, Stapleton points out that '[t]hough not alluded to by the court,...it was highly probable that some of the environmental asbestos pollution...had been due to some tortious, albeit unidentified, cause', Stapleton, above n 46 at 222. The problem with allowing this supposition to alter the legal analysis of the facts is that it is too open-ended, since it could be said of many toxic-agent cases, and the incorporation of unidentified hypothetical defendants into the causal inquiry has nothing to commend it. (This was not, in any event, the approach which Stapleton was advocating therein.).
[50] *The Creutzfeldt-Jakob Disease Litigation, Groups A and C Plaintiffs* 54 BMLR 100 (QB).
[51] Because studies adverting to the risks of transmitting CJD through such treatment had by then been made known to the defendants.

of probabilities from one out of two to 25 out of 52 in one pile, and 27 out of 52 in the other pile.[52]

This is not, of course, wholly analogous to the material increase in risk cases, because the very reason for the exceptional approach in those situations is that we simply do not know whether the aetiology of the illnesses therein was characterised by a single dose or by a cumulative process. It serves, however, as an example of a method which, although criticised by the defendants in that case as being 'mechanistic or simple',[53] is exactly that, and effective for it. In an area in which the intrinsic nature of evidence is potentially highly complex and almost inevitably incomplete, there is much to be said for a legal response which is clear, consistent and predictable. The need for such qualities exists a fortiori, therefore, in situations in which the evidential uncertainty is greater.

This purely arithmetical means of satisfying the balance of probabilities standard is less heterodox than it might first appear, once the nature of the forensic exercise is considered in practical terms. In an observation about civil law in general, Peter Cane has written:

> In practice ... reliable evidence is typically unavailable to support assigning a numerical value to causal factors ... in which case, what matters is not whether the conduct more probably than not caused the outcome, but whether the court is satisfied that it did.[54]

This fits with the 'belief probability' approach first outlined in Chapter 2,[55] and summed up concisely in a well-known article by Steve Gold, which was cited with approval in *Sienkiewicz*:[56]

> The 'more likely than not' test should be applied only to the requisite strength of the fact-finder's belief, and not to the definition of the factual elements which the parties

---

[52] 54 BMLR 100 (QB) at 102: an analogy which the defendants attempted to refute on the basis that there were multiple infected (ie potential) single fatal doses and so there was more than one ace of spades in the metaphorical pack. This makes no difference, however, as long as there was only one single dose which *actually* triggered the disease, and this was what the accepted medical evidence said. There might, therefore, have been many aces, but only one ace of spades.

[53] 54 BMLR 100 (QB) at 103.

[54] P Cane, *Responsibility in Law and Morality* (Oxford, Hart Publishing, 2002) 124.

[55] See Ch 2, under heading 'The Balance of Probabilities'..

[56] [2011] UKSC 10, [2011] 2 AC 229 at [156] 285 (per Lord Rodger), at [194] 296 (per Lord Brown) and, most notably, at [217]–[222] 302–03 (per Lord Dyson). At [143]–[159] 281–86, Lord Rodger is at pains to point out that the *McGhee/Fairchild* principle, applied in cases where epidemiological evidence holds sway, allows a claimant to succeed where a court is satisfied on the balance of probabilities not that the defendant's breach caused the damage, but that it *probably* caused the damage. With respect, His Lordship is here doing exactly what the article he cites with approval warns against: collapsing the belief and fact probabilities. The belief probability standard which the court has to reach in any context is synonymous with its being able to say 'This is what probably happened', which is then treated as if that is what actually happened. There is, therefore, no distinction to be made *in terms of belief* between 'probably what happened' and 'what probably happened'. Lord Rodger makes this point in the context of his concern about the use of epidemiological evidence, however, and, in terms of how persuasive such evidence is, and therefore how likely it is to engender a reasonable belief, there may well be a difference between saying what probably happened in the relationship between A and B and what happened in the probable relationship between A and B. See also Ch 7 and Steel and Ibbetson, 'More Grief on Uncertain Causation', above n 37 at 464.

must prove. Refusing to collapse burden and standard into a single test is essential to restoring rationality to doctrine on causation in toxic torts. Juries should not be instructed that the preponderance standard means that plaintiffs must prove a fact probability greater than 50%. Rather, they should be instructed that plaintiffs must establish certain facts ... by evidence which convinces juries that the fact is more likely than not.[57]

In adopting this approach, Lord Dyson has made the following point:

> Our civil law does not deal in scientific or logical certainties. The statistical evidence may be so compelling that, to use the terminology of Steve Gold, the court may be able to infer belief probability from fact probability. To permit the drawing of such an inference is not to collapse the distinction between fact probability and belief probability. It merely recognises that, in a particular case, the fact probability may be so strong that the court is satisfied as to belief probability. Whether an inference of belief probability should be drawn in any given case is not a matter of logic. The law does not demand absolute certainty in this context or indeed in any context.[58]

If the fact probability can be so strong that the court is satisfied as to belief probability, then the converse must also true. So, even in a factual situation in which the evidentiary gap is as insurmountable as it currently is in mesothelioma cases, if the number of breach sources is smaller than the number of non-breach sources, it would make sense for a court to conclude that the requisite belief probability has not been established. On this, Wright makes the following point:

> When the coherence of the particularistic evidence with one of the possibly applicable causal generalisations and the overall causal story in which it is embedded is sufficiently greater than its coherence with competing causal generalisations and the causal stories in which they are embedded, according to the subjective judgment of each individual rather than a uniform criterion, that individual forms a minimal belief in the truth of the first story. The greater the difference in degree of coherence, the stronger the degree of belief. This is the method of argument and proof employed by lawyers and judges in actual litigation.[59]

In the mesothelioma cases, the competing causal generalisations to which Wright refers would be 'Mesothelioma caused by exposure X', 'Mesothelioma caused by exposure Y' and so on. The only particularistic evidence available on such facts is that relating to each claimant's various exposures to asbestos and, according to Wright, this is insufficient for coming to a decision about which possible causal

---

[57] S Gold 'Causation in Toxic Torts: Burdens of Proof, Standards of Persuasion, and Statistical Evidence' (1986) 96 *Yale Law Journal* 376, 395.

[58] *Sienkiewicz v Greif* [2011] UKSC 10, [2011] 2 AC 229 at [222] 303 (per Lord Dyson). This is by no means a new controversy, with theoretical and definitional disagreement reaching back to its conception in the decade around 1660: see I Hacking, *The Emergence of Probability* (Cambridge, CUP, 1975) ch 2. Even now, there are 'personalists ... who have said that propensities and statistical frequencies are some sort of mysterious "pseudo-property", that can be made sense of only through personal probability. There are frequentists who contend that frequency concepts are the only ones that are viable' (at 14).

[59] R Wright, 'Proving Causation: Probability v Belief' in R Goldberg (ed), *Perspectives on Causation* (Oxford, Hart Publishing, 2011) 209.

story has been completely instantiated in any given case.[60] A conclusion devoid of a decision, however, is not one to which a court is entitled to come, and so it must identify something other than the relationship between particularistic evidence and causal generalisations, as a basis on which it can make a decision.

Susan Haack uses the concept of 'warrant' to describe the degree of credence which a rational individual would give to any evidence, a degree which rises with any increase in that evidence (a concept which, as Haack points out, echoes that identified as a 'degree of credibility' by Bertrand Russell in 1948). It is crucial, she maintains, to distinguish between mathematical probabilities and degrees of credibility:[61]

> [T]he concept of warrant is too subtle, and too worldly, to allow a viable epistemological theory of a purely formal kind. Moreover, degrees of warrant don't satisfy the axioms of the standard probability calculus. The probability of *p* and the probability of not-*p* must add up to 1; so when *p* has a low degree of probability, not-*p* must have a high degree. But when there is insufficient evidence either way, *neither p nor* not-*p* may be warranted to any appreciable degree.[62]

As has been established in previous chapters, it is a phenomenon akin to this 'warrant' or 'belief probability', which provides the basis on which courts can, and do, reach conclusions in the absence of ideal evidence. Sometimes the evidence will be good, and sometimes it will be less good, but this means only that a decision is easier or harder to reach, not that it is more or less likely to happen. Mathematical probabilities are, in a forensic context, a means and not an end.[63] In mesothelioma cases, therefore, the fact that it is currently scientifically impossible to link a particular asbestos exposure to a particular claimant's illness does not mean that it is impossible to formulate a *rational* belief[64] that a preponderance of 'breach' exposures makes it more likely than not that the claimant's illness was caused by a defendant (or indeed that a preponderance of non-breach exposures makes it more likely than not that the claimant's illness was part of her normal course of events). If we assume indifference between the causal valence of each exposure (because duration and intensity have not been acknowledged to make any such causative difference), and treat each as a unitary and uniform causal factor, we have a means of splitting the difference in what is really a very difficult (as opposed to hard) case. Thus, we also have a means of avoiding the spiralling liability consequences which continue, in the absence of such a means, to trouble our highest courts.

---

[60] Email correspondence, 18 November 2013, for which I am very grateful.
[61] S Haack, 'The Embedded Epistemologist: Dispatches from the Legal Front' (2012) 25 *Ratio Juris* 206, 217.
[62] ibid, 217. Porat and Stein make the same point about what they describe as the 'inductivist' framework, to be distinguished from their conception of an 'aleatory' framework, which is based on pure frequentist probabilities. See Porat and Stein, *Tort Liability*, above n 33 at 46–47.
[63] See Ch 2 and Ch 8.
[64] Haack makes it very clear that 'legal degrees of proof are not degrees of credence; they are degrees of *rational credibility* or *warrant*': Haack, 'The Embedded Epistemologist', above n 61 at 219.

This approach might appear either to be an overly technical, or indeed naive, analysis of these particular cases (or both). It manages, however, to address the bigger-picture concern that started to emerge once the 'exceptional' *Fairchild* principle was tested on facts which, although related to those in the landmark case itself, were different in a normatively significant way. In *Sienkiewicz*, for example, there was no intractable problem of multiple duty-breaching employers, each of whom, by their very existence, appeared to make themselves and their fellow defendants judgment-proof.[65] Rather, it was more a case of a grave disease with a tricky aetiology, which might have been caused by industrial exposure, but might not have been. Lord Brown recognises this:

> The point I make is that it is hardly to be thought that, had the House, on the occasion of the Fairchild hearing, been considering not the facts of those three appeals but instead the facts of the present appeals, the claimants would have succeeded and the law have developed as it has.[66]

Judicial concerns about how the *Fairchild* principle[67] would play out across different fact patterns were first discernible in *Barker*, and were obvious by the time that *Sienkiewicz* reached the Supreme Court. There was a sense in both of these cases that the decision in *Fairchild* had started a process as much as it had stated a principle, and that, outside of the particular facts of the case in which it was conceived, the principle would have very different and not necessarily favourable implications.[68] In fact, in *Sienkiewicz*, Lord Phillips went so far as to describe its effect, combined with section 3 Compensation Act 2006, as 'draconian' as far as defendants were concerned.[69] The statutory provision to which he refers was the means by which Parliament reversed the aliquot liability aspect of the decision in *Barker* only a few months after it had been handed down:[70]

3 Mesothelioma: damages
  (1) This section applies where
    (a) a person ('the responsible person') has negligently or in breach of statutory duty caused or permitted another person ('the victim') to be exposed to asbestos,
    (b) the victim has contracted mesothelioma as a result of exposure to asbestos,
    (c) because of the nature of mesothelioma and the state of medical science, it is not possible to determine with certainty whether it was the exposure mentioned

---

[65] The same problem encountered in cases of overdetermination – see Ch 4.
[66] See *Sienkiewicz v Greif* [2011] UKSC 10, [2011] 2 AC 229 at [179] 292–93 (per Lord Brown).
[67] This has been referred to elsewhere, most notably by J Stapleton, 'Lords a'Leaping' (above n 5), as the *McGhee/Fairchild* principle. Since, however, it is my contention that *McGhee* was incorrectly decided, it seems more appropriate to refer to it here as the *Fairchild* principle.
[68] See, eg, *Barker v Corus* [2006] UKHL 20, [2006] 2 AC 572 at [43] 592 (per Lord Hoffmann), [61] 598–99 (per Lord Scott), [117] 614 (per Lord Walker) and [128] 617 (per Baroness Hale).
[69] *Sienkiewicz v Greif* [2011] UKSC 10, [2011] 2 AC 229 at [58] 257 (per Lord Phillips).
[70] The alacrity with which this was achieved can be explained by the fact that the Bill which ultimately became the Compensation Act 2006 was already being considered when the House of Lords decided *Barker*. Section 3 was added after several Members of Parliament expressed outrage at the decision. For a detailed discussion, see J Lee, 'Inconsiderate Alterations in Our Laws: Legislative Reversal of Supreme Court Decisions' in J Lee (ed), *From House of Lords to Supreme Court: Judges, Jurists and the Process of Judging* (Oxford, Hart Publishing, 2011) 79–86.

in paragraph (a) or another exposure which caused the victim to become ill, and
(d) the responsible person is liable in tort, by virtue of the exposure mentioned in paragraph (a), in connection with damage caused to the victim by the disease (whether by reason of having materially increased a risk or for any other reason).
(2) The responsible person shall be liable
 (a) in respect of the whole of the damage caused to the victim by the disease (irrespective of whether the victim was also exposed to asbestos
  (i) other than by the responsible person, whether or not in circumstances in which another person has liability in tort, or
  (ii) by the responsible person in circumstances in which he has no liability in tort), and
 (b) jointly and severally with any other responsible person.

The effect of this section is to make joint and several liability the only possible consequence of an application of the *Fairchild* principle *where the damage is mesothelioma and the harmful agent is asbestos*, regardless of whether there has been any exposure whatsoever[71] other than that resulting from the defendant's breach of duty. Any analogous, but not identical, facts which attract the application of that principle (such as, for instance, dermatitis and brick dust) remain subject to the common law rule of apportionment derived from *Barker*. Lord Phillips' choice of adjective, therefore, was prompted by the consequence of this; that any employer who is deemed by a court to have materially increased the risk of a claimant's developing mesothelioma[72] must be held liable *in solidum*, despite the fact that its contribution to that risk could well have been minimal[73] (and, lest it be forgotten, whose contribution to a case of mesothelioma might in fact have been non-existent).

It is, however, Lord Brown's speech in *Sienkiewicz* which most clearly articulates the troublesome implications of extending *Fairchild* beyond the problem it was originally intended to address, and applying it in a way which divests it entirely of even aggregate But For content.[74] Of *Barker*, he says

> [O]ne finds the House having to face up to some of the problems it had left open with *Fairchild* and, as it seems to me, beginning to have second thoughts both as to the juristic basis for this special rule of causation which Fairchild held to apply in certain toxic tort cases and as to where the abandonment of the 'but for' principle was taking the law ... It is to my mind quite clear that the preparedness of the majority of the court in *Barker* to extend the reach of the *Fairchild* principle this far was specifically dependent upon there being aliquot liability only.[75]

---

[71] This is so whether or not the other exposures resulted from (anyone's) breach(es) of duty – s 3(2)(a)(i) and (ii).

[72] Meaning the contribution of anything not insignificant in comparison with other factors – see *Barker v Corus* [2006] UKHL 20, [2006] 2 AC 572 at [111] 312–13.

[73] And the claimant's other exposures to asbestos could dwarf the risk created by the defendant's breach.

[74] Which, as we have seen, was what was lacking in both *McGhee* and *Sienkiewicz*.

[75] *Sienkiewicz v Greif* [2011] UKSC 10, [2011] 2 AC 229 at [181]–[182] 293. The legislative reversal of only one limb of that decision (that liability should be apportioned rather than joint and several) leaves open the question of whether the House would have reached the same conclusion on the other

The Necessary Breach Analysis and Evidentiary Gaps 141

His Lordship's broader concern for the integrity of the law is both apparent and well founded:

> [M]esothelioma cases are in a category all their own. Whether, however, this special treatment is justified may be doubted . . . The unfortunate fact is, however, that the courts are faced with comparable rocks of uncertainty in a wide variety of other situations too and that to circumvent these rocks on a routine basis – let alone if to do so would open the way, as here, to compensation on a full liability basis – would turn our law upside down and dramatically increase the scope for what hitherto have been rejected as purely speculative compensation claims. Although, therefore, mesothelioma claims must now be considered from the defendant's standpoint a lost cause, there is to my mind a lesson to be learned from losing it: the law tampers with the 'but for' test of causation at its peril.[76]

Finally, the wording of Lord Kerr's conclusion in *Sienkiewicz*[77] 'that there is no basis on which the *Fairchild* exception should not be applied in these cases' is telling. It suggests, as does the tenor of much of the decision, that the *Fairchild* principle has somehow snowballed into an irresistible legal force which courts now struggle not to apply, regardless of their misgivings.[78] Writing extra-judicially, Lord Hoffmann seems to have taken the view that this has occurred because the exception was not sufficiently justified by principle: 'In retrospect, I think we failed this test quite badly'.[79] Had Viktor Frankenstein been in a position to advise the House of Lords, he would doubtless have warned them how easy it is to lose control of such daring creations. Indeed, the creature himself would most probably have agreed: 'Yet you, my creator, detest and spurn me, thy creature, to whom thou art bound by ties only dissoluble by the annihilation of one of us'.[80] That is not to suggest that the House of Lords, when formulating the *Fairchild* principle, was unaware of the inevitable pressure that would be exerted upon it by the march of the common law.[81] Lord Bingham was explicit in his recognition of this in what is now one of the most well-known sentences in the entire judgment:

> It would be unrealistic to suppose that the principle here affirmed will not over time be the subject of incremental and analogical development. Cases seeking to develop the principle must be decided when and as they arise.[82]

---

limb (the applicability of the Fairchild principle to such a situation) had joint and several liability been the result. See J Morgan, 'The English – and Scottish – Asbestos Saga' in Goldberg (ed), *Perspectives on Causation*, above n 59 at 79. See also Hoffmann, '*Fairchild* and After', above n 3 at 67.

[76] *Sienkiewicz v Greif* [2011] UKSC 10, [2011] 2 AC 229 at [186] 294.
[77] *Sienkiewicz v Greif* [2011] UKSC 10, [2011] 2 AC 229 at [203] 298.
[78] See *Sienkiewicz v Greif* [2011] UKSC 10, [2011] 2 AC 229 at [167] 288–89 (per Baroness Hale), [189] 295 (per Lord Mance) and [58] 257 (per Lord Phillips).
[79] Hoffmann, '*Fairchild* and After', above n 3.
[80] MW Shelley, *Frankenstein* (1818).
[81] J Morgan, 'Asbestos Saga', above n 75 at 64–65, where he refers to *Fairchild* as 'an essentially legislative attempt to carve out an exceptional category, doomed to failure by the common law's ineluctable method of reasoning by analogy'.
[82] [2002] UKHL 22, [2003] 1 AC 32 at [34] 68.

142 *Material Increase in Risk*

It is unfortunate, however, that Lord Bingham's own formulation of the principle was not the one to inform such incremental and analogical development. Had it been, the *Fairchild* principle would most likely have remained exceptional, defensible and more closely aligned with the objectives of the tort of negligence. It was Lord Bingham, and only he, who insisted that the exception should apply only where the following criteria are met:

> (1) C was employed at different times and for differing periods by both A and B, and (2) A and B were both subject to a duty to take reasonable care or to take all practicable measures to prevent C inhaling asbestos dust because of the known risk that asbestos dust (if inhaled) might cause a mesothelioma, and (3) both A and B were in breach of that duty in relation to C during the periods of C's employment by each of them with the result that during both periods C inhaled excessive quantities of asbestos dust, and (4) C is found to be suffering from a mesothelioma, and (5) any cause of C's mesothelioma other than the inhalation of asbestos dust at work can be effectively discounted, but (6) C cannot (because of the current limits of human science) prove, on the balance of probabilities, that his mesothelioma was the result of his inhaling asbestos dust during his employment by A or during his employment by B or during his employment by A and B taken together . . .[83]

This version of the *Fairchild* principle does minimal violence to But For causation. Whilst it must of necessity dispense with *individualised* But For causation, because no single exposure can be linked to any one claimant's case, Lord Bingham's formulation nonetheless requires all sources of asbestos to involve breaches of duty to the claimant. In doing so, it retains *aggregate* But For causation, and means that claimants will only be compensated where it is more likely than not that at least one breach of duty has altered their normal course of events for the worse. This element of necessity is crucial if the law in this area is to retain any semblance of consistency and coherence. Insisting at the very least on aggregate But For causation would, for instance, have avoided the situation lamented by Lords Brown and Rodger in *Sienkiewicz*,[84] in which mesothelioma cases resulting from asbestos exposure are arbitrarily excepted not only from the requirement to prove individualised But For causation, but also from the principle that disallows recovery for purely speculative claims. The law's distaste for such claims has, quite correctly, been reaffirmed twice at the highest level, on facts which show no obvious substantive distinction from those which currently receive extraordinary treatment from the *Fairchild* principle in combination with section 3 of the Compensation Act 2006.[85]

Aggregate necessity, the minimum requirement of the NBA, is *the least* that a claimant should be able to prove in order to have a successful claim in negligence. Without at least one breach of a duty of care being necessary for a claimant's damage to have occurred, the claimant would have been no better off in the absence of

---

[83] [2002] UKHL 22, [2003] 1 AC 32 at [2] 40.
[84] Above, text to n 19, [2011] UKSC 10, [2011] 2 AC 229 at [85] 264 and [186] 294.
[85] This distinction without a difference will be discussed in detail in Ch 7.

any breach. Since the tort of negligence is concerned with compensating those who have been injured as a result of another's breach of duty,[86] and not with unrelated misfortunes, this particular point of discrimination is crucial. There is little doubt that duties of care are breached regularly and frequently, but it is only when these breaches can be sufficiently linked to another's damage that the tort of negligence is engaged.[87] Basic though this point may seem in a general and theoretical sense, it becomes increasingly difficult to bear in mind as defendants' behaviour becomes less and less palatable, and claimants' injuries more and more unpleasant. The mesothelioma cases, therefore, are an ideal testing ground for it.

Under an NBA, an affirmative answer to the question of whether at least one breach was necessary for the claimant's damage to have occurred leads to a further discriminatory stage; that which establishes a link between the operation of a specific breach and the claimant's damage (or part of that damage). Liability imposed on the satisfaction of both stages is the ideal position, at least from a corrective justice standpoint.[88] In the exceptional circumstances, however, in which an affirmative answer to the first question has been reached, but no further specific linkage is possible, the court is faced with two choices: either hold all of those in breach of duty liable on the basis that the claimant's normal course of events has been altered for the worse by at least one breach, or hold none of them liable, and leave the claimant (whose normal course of events has been altered for the worse by at least one breach), to bear the loss. Essentially, this amounts to a straightforward decision between giving the benefit of the doubt to an injured claimant, or to a duty-breaching defendant. The extraordinary characteristics of the *Fairchild* scenario (in the form of multiple defendants and an insurmountable evidentiary gap) demand that the causal inquiry drills down as far as this question. Despite the fact that the tort of negligence, by definition, is not motivated by moral culpability concerns, when the choice is as stark as it was in *Fairchild*, it is not hard to see why the House of Lords came to the decision it did. The damage, on the facts of *Fairchild* as presented to the Court, would not have happened absent any negligent behaviour: this is (axiomatically) a state of affairs which engages the tort of negligence.

At this level, the concern becomes, crudely, one of claimants as a class versus defendants as a class. A no-liability result in *Fairchild* could have been said to be unfair to claimants as a class,[89] whilst the principle, *as it was applied subsequently to Sienkiewicz*, could be said to be unfair to defendants as a class. As between claimants and defendants as classes of individuals, however, requiring the necessity of at least one breach to be proven on the balance of probabilities splits the evidential

---

[86] See, inter alia, T Honoré, 'The Morality of Tort Law: Questions and Answers' and D Owen, 'Philosophical Foundations of Tort Law' in D Owen (ed), *Philosophical Foundations of Tort Law* (Oxford, Clarendon Press, 1995).
[87] See, for instance, *JD v East Berkshire Community Health NHS Trust* [2005] UKHL 23, [2005] 2 AC 373 at [100] 410 (per Lord Rodger): 'The world is full of harm for which the law furnishes no remedy'.
[88] E Weinrib, *Corrective Justice* (Oxford, OUP, 2012) 87–91.
[89] Particularly by certain MPs in the House of Commons, as the debates preceding the addition of s 3 Compensation Act suggest. See above n 70.

difference between the two in terms of limited knowledge and risk of error.[90] Ideally, a court would associate individual breaches of duty with specific instances of damage,[91] but, where this is not possible, it is the next best thing to ask which class should bear the loss *when what we can establish is that it is more likely than not that a member of one has been adversely affected by at least one member of the other*.[92] This is what the result in *Fairchild* did.[93] In this way, it 'may be seen as a compromise between our interests as agents in freedom of action, and our interests as victims in security of person and property and the repair of adverse outcomes'.[94] Given the fact that we are all potential defendants and potential claimants, the splitting of such differences is not so much the prerogative as the obligation of the common law: the aggregation of those who damage as well as those who are damaged 'distribute[s] the costs of epistemological uncertainty'[95] across both classes.

Where, however, it has not been established, even on this arithmetical, non-qualitative[96] But For model, that it is more likely than not that at least one breach of duty made a difference to the claimant's normal course of events, the situation and its attendant considerations are entirely different. On facts such as those in *Sienkiewicz*, for example, in which one source of asbestos was a breach of duty, and one was not the result of any breach, it cannot be established on the balance of probabilities that *any* defendant has affected the claimant, let alone which one. This set of facts does not, therefore, present the court with the same stark choice it had in *Fairchild*, because there is no comparable reason to choose to decide in favour of defendants, as opposed to claimants. The same is true of *McGhee*. This is slightly less apparent than it is in the case of *Sienkiewicz*, simply because, in *McGhee*, both sources of brick dust were produced by the defendant. The point remains, however, that only one resulted from a breach of a duty of care, meaning that only in relation to one source was the Coal Board a defendant, and only in

---

[90] See M Geistfeld, 'The Doctrinal Unity of Alternative Liability and Market-Share Liability' (2006) 155 *University of Pennsylvania Law Review* 447 at 461–62.

[91] See E Weinrib, *The Idea of Private Law* (Cambridge, MA, Harvard University Press, 1995) 153.

[92] See A Ripstein and B Zipursky, 'Corrective Justice in an Age of Mass Torts' in G Postema (ed), *Philosophy and the Law of Torts* (Cambridge, CUP, 2001) 235 and S Levmore, 'Probabilistic Recoveries, Restitution and Recurring Wrongs' (1990) 19 *Journal of Legal Studies* 691.

[93] See *Barker v Corus* [2006] UKHL 20, [2006] 2 AC 572 at [128] 617 (per Baroness Hale).

[94] Cane, *Responsibility in Law and Morality*, above n 54 at 123.

[95] ibid, 125. See also the arguments put forward by Porat and Stein, *Tort Liability*, above n 33 at ch V: essentially that 'Corrective justice would be adequately attained as long as the wrongdoers pay their dues, the victims are compensated for their losses, and the payments made by the wrongdoers are used for compensating the victims'. Their suggested concept of 'collective liability', however, which they justify on these grounds, is more of an active principle than the exceptional concession to aggregate analysis made by the NBA, and it has further-reaching damages implications (see ibid, 138). It also needs to be considered alongside their conception of 'evidential damage' (see ibid, chs VI and VII), which is the doctrine they would apply to asbestos-mesothelioma cases.

[96] Such an approach to be contrasted with the idea, referred to above, of attempting to apportion responsibility amongst those in breach of duty, according to their relative contributions to the adverse outcome. As pointed out, such a qualitative assessment is exactly what cannot be carried on in genuine evidentiary gaps cases. See MD Green, 'The Future of Proportional Responsibility' in S Madden (ed), *Exploring Tort Law* (New York, NY, CUP, 2005) 353, 358.

relation to one was the claimant entitled to any recovery *in negligence*.[97] In *Fairchild* and *Barker* the claimants could establish at least aggregate But For causation. This is an exceptional and minimal burden, but one which is defensible in the light of the evidentiary gap/multiple defendant matrix of these cases. In *McGhee* and *Sienkiewicz*, the claimant's evidence fell short of this, meaning that liability was imposed on the basis of nothing more than a purely speculative claim that the illnesses therein should fall to the account of defendants, as opposed to claimants. The significance of this was recognised by the Supreme Court of Canada in *Clements v Clements*,[98] a case in which it refused to apply a material contribution to risk analysis because only one of the potential causal factors involved was a breach of duty:[99]

> The special conditions that permit resort to a material contribution approach were not present in this case. This is not a case where we know that the loss would not have occurred 'but for' the negligence of two or more possible tortfeasors, but the plaintiff cannot establish on a balance of probabilities which negligent actor or actors caused the injury. This is a simple single-defendant case: the only issue was whether 'but for' the defendant's negligent conduct, the injury would have been sustained.[100]

Quite apart from the theoretical implications of extending *Fairchild* beyond even aggregate But For causation, and thereby dispensing with the need for breaches of duty to have made a difference to claimants' lives, the result in *Sienkiewicz* has serious practical implications. It is a sad fact that mesothelioma cases are a significant phenomenon of modern times, as a result of the long latency period of the disease, and the era in which asbestos was commonly used. Given that many of those likely to develop the disease were exposed to asbestos as a result of their employment, the implications for the insurance industry of the legal treatment of these cases are huge. It is not uncommon for this point to be dismissed as

---

[97] Which says nothing, of course, about his entitlement to compensation under any other scheme such as insurance or welfare, but behaviour which does not amount to a breach of duty remains irrelevant for the purposes of the tort of negligence, even where it is exhibited by those who have breached a duty by performing another action. Negligence only compensates those who have suffered as a result of another's breach to them, not as a result of any behaviour towards them by someone who happens to be a tortfeasor in another context. See Weinrib, *Corrective Justice*, above n 88 at 87–91 and R Stevens, *Torts and Rights*, above n 34 at 129 but cf Lord Hoffmann in *Barker v Corus* [2006] UKHL 20, [2006] 2 AC 572 at [16] 584.

[98] [2012] SCC 32, [2012] SCR 181 at [30] (per McLachlan CJ).

[99] Since the facts of that case do not fit the narrow criteria outlined above for the material contribution to risk principle, it was not an appropriate case for its application in any event – there was no 'rock of uncertainty': the experts were unable to say *in this case* what the contributions of the relevant factors were, but it is not impossible as *a matter of principle* to ascertain the causal contributions of such factors to such events. Nevertheless, the recognition of the implications of a single breach of duty for But For causation is to be welcomed (as was the approach of the same court in *Resurfice v Hanke* [2007] SCC 7, [2007] 1 SCR 333 in recognising that the material contribution to risk approach is exceptional, and not therefore generally applicable). See also RW Morasiewicz, 'The Supreme Court of Canada Narrows the "Material Contribution" Causation Test' (2013) 71 *The Advocate* 213. The result in *Clements* also concurs with the outcome that would be reached on an NBA, since the claimant could not establish that it was more likely than not that any damage occurred which would not have done so when it did but for at least one breach (the first stage, therefore, would not be satisfied).

[100] [2012] SCC 32, [2012] SCR 181 at [50] (per McLachlan CJ).

unimportant, or at least unimportant if one views it as a case of pitching the interests of insurers as against those of the victims of this inevitably fatal disease. Who, when faced with that choice, would feel sympathy for an insurance company? But this is not an authentic comparison, and is both naive and dangerous in its simplicity. Insurers and victims are not independent interest groups. Insurers only exist because claimants have interests for them to protect. It is not, therefore, simply a case of favouring the interests of one over the other, since these are interdependent. Every time an insurer dissolves, so does the prospect of recovery for any number of individuals who have been wrongfully exposed to asbestos by an insured, but either are not yet suffering from the disease, or do not yet realise that they are. Failing to pay heed to this fact has the simple, but invidious, effect of subordinating the claims of those who happen to develop the disease later to the claims of those who developed it earlier.[101]

Whilst, therefore, *Sienkiewicz*, and s 3 Compensation Act 2006 have the appearance of being claimant friendly, this is only true if a discrete snapshot of claimants within a particular time frame is being considered. Over the long term, non-apportioned liability imposed in the absence of even an aggregate But For causal link merely front-loads assistance to claimants. Parliament's legislative reversal of the apportionment approach taken in *Barker* was carried out principally because MPs were horrified at the prospect of mesothelioma sufferers being denied 'full compensation'. However, because of the haste with which this was done, the lack of any detailed consideration of the broader issues and, quite frankly, the absence of much understanding of the legal issues involved,[102] all that the Act managed to achieve was a postponement of the problem. The House of Lords' approach in *Barker* would have ensured that more claimants received at least some compensation. Parliament's intervention means that some claimants will now receive full compensation, but many more will receive nothing at all.

Despite the 'draconian' effect of the *Fairchild* principle combined with s 3 Compensation Act, it is clear from *Sienkiewicz* that all is not lost, and that the tide is capable of being stemmed, if not reversed:

> The Court of Appeal [2010] QB 370 treated section 3(1) as enacting that, in cases of mesothelioma, causation can be proved by demonstrating that the defendant wrongfully 'materially increased the risk' of a victim contracting mesothelioma. This was a misreading of the subsection. Section 3(1) does not state that the responsible person *will be* liable in tort if he has materially increased the risk of a victim of mesothelioma. It states that the section applies *where* the responsible person is liable in tort for materially increasing that risk. Whether and in what circumstances liability in tort attaches to one who has materially increased the risk of a victim contracting mesothelioma remains a question of common law.[103]

It would be desirable for the law, therefore, to employ an aggregate But For test, akin to the first stage of the NBA, in order to limit *Fairchild* liability, and therefore

---

[101] See S Green, 'Winner Takes All (2004) 120 *LQR* 566.
[102] Morgan, 'Asbestos Saga', above n 75 at 77–82.
[103] *Sienkiewicz v Greif* [2011] UKSC 10, [2011] 2 AC 229 at [70] 260–61 (per Lord Phillips).

the application of s 3 Compensation Act, to only those situations in which there are more 'breach' sources of asbestos exposure than there are 'non-breach' sources.[104] This would have the effect of making it harder than it is currently for a mesothelioma claimant to establish causation, but would mean that successful claimants, both now and in the future, are far more likely to recover full damages. It also means that some claimants, who *might* have been injured by one or more defendants, will recover nothing, whilst others recover their loss in full. Whilst this is not a utopian situation, it does, however, have the redeeming characteristic of being consistent with the position of the law in relation to any other type of indivisible injury that might be caused by another's negligence:[105]

> There is a rough justice about the law of personal injury liability as a whole. To compensate a claimant in full for a lost finger because there was a 60:40 chance that he would have worn protective gloves had they been made available to him may be regarded as rough justice for defendants. But it is balanced by the denial of compensation to a claimant who cannot establish that he would probably have worn the gloves – or whose finger the judge concludes was probably already doomed because of frostbite. Save only for mesothelioma cases, claimants should henceforth expect little flexibility from the courts in their approach to causation ... I have ... difficulty ... in accepting that the courts should now, whether on this or any other basis, be thinking of creating any further special rules regarding the principles governing compensation for personal injury. The same logic which requires that the claims of these claimants succeed to my mind requires also that the courts should in future be wary indeed before adding yet further anomalies in an area of law which benefits perhaps above all from clarity, consistency and certainty in its application.[106]

Mesothelioma is an awful disease, but it is still difficult to see why it is regarded as uniquely so. As Lord Brown in *Sienkiewicz* articulated, the 'rock of uncertainty' is not limited to such cases and, when considered against the full spectrum of horrendous injuries and difficulties of proof facing claimants in tort, the need for such dramatically different treatment for one specific injury is not obvious. Requiring aggregate But For liability as a condition of imposing joint and several liability would mean that the difference between the *Fairchild* principle and orthodox causation would be as small as it needs to be in order to avoid injustice, and would prevent the exception from being able to swallow up the rule.[107] Since, however, the formulation of such an approach is not precluded by *Fairchild*, or by *Barker*, but would be inconsistent with the decision in *Sienkiewicz*,[108] it could only now be

---

[104] Thereby ameliorating the current situation without dispensing with the *Fairchild* principle, which Lord Brown feared would be the only way of avoiding the 'draconian consequences' of that principle combined with s 3 Compensation Act 2006. See *Sienkiewicz v Greif* [2011] UKSC 10, [2011] 2 AC 229 at [184] 294. Since, as explained above, this principle serves a defensible purpose on the facts of the original case, this is not necessarily a desideratum.
[105] For further discussion, see Ch 5.
[106] *Sienkiewicz v Greif* [2011] UKSC 10, [2011] 2 AC 229 at [187] 294–95 (per Lord Brown).
[107] Stapleton would regard such an approach, which factors in exposures from sources other than the defendant's when evaluating a claim, as a 'danger' and as being based on 'flawed reasoning' – see J Stapleton, 'The Fairchild Doctrine: Arguments on Breach and Materiality' (2012) 71 *CLJ* 32, 34.
[108] See *Williams v University of Birmingham* [2011] EWCA Civ 1242, [2011] All ER (D) 25 (Nov) at [72] (per Aikens LJ).

achieved by legislation or by an audacious Supreme Court. Whatever the means, they would be justified in this case by the ends.

It is not as simple, however, as a choice between [aggregate But For causation + joint and several liability] or [more-than-minimal-contribution-to-risk + joint and several liability].[109] It remains the case that in any situation featuring a single agent,[110] multiple defendants and an evidentiary gap, but which does *not* involve mesothelioma as a result of asbestos exposure, claimants can succeed by establishing more than minimal risk contribution, but the liability of defendants will then be apportioned.[111] This is arguably a compromise between the two positions already discussed, since it makes claimants' tasks easier in terms of standard of proof, but also goes some way to protecting defendants from excessive and disproportionate liability.[112] Whilst it is preferable to the current and extreme position of [more-than-minimal-contribution-to-risk + joint and several liability], it remains inferior to [aggregate But For causation + joint and several liability] because it dispenses with any connection to necessity in the form of But For causation. This means that not only does it allow recovery in cases which should not be the concern of the tort of negligence (that is, those in which it is more likely than not that at least one breach of duty has made a difference to the claimant's normal course of events), but, in doing so, it amounts to a direct contradiction of other important negligence decisions made at the highest level.[113]

Single Agent

Since *Fairchild* and *Barker* there has been much academic focus on a supposedly critical distinction between so-called 'single agent' and 'multiple agent' cases, the suggestion being that the former more readily lend themselves to special rules of causation than the

---

[109] Often, when these cases are discussed, reference is made to the decision of the California Supreme Court in *Sindell v Abbott Laboratories* 607 P 2d 924 (Cal 1980) in which liability was apportioned amongst the defendant drug-producers according to their share of the market. That case, however, whilst undoubtedly problematic, exhibits a different problem to the one presented by *Fairchild*: given that not all the defendants were before the court in *Sindell*, it was not even possible for the claimant to establish that any one of them exposed her to risk, (see also *Fairchild v Glenhaven Funeral Services* [2002] UKHL 22, [2003] 1 AC 32 at [130] 100 (per Lord Hoffmann)). *Sindell* cannot, therefore, be classified as a material contribution to risk case when this was not the basis for the liability there imposed. It was, however, possible to assert that each defendant's breach would have been severally sufficient, in the absence of the others (in contrast to the facts of *Fairchild*). The facts of that case make it more analogous to an overdetermination problem, and so it is discussed in Ch 4.

[110] See next section.

[111] *Barker v Corus* [2006] UKHL 20, [2006] 2 AC 572. This part of the judgment was reversed by s 3 Compensation Act 2006 explicitly and only in relation to mesothelioma and exposure to asbestos.

[112] I have argued in favour of this in the past as a means of minimising the effects of *Fairchild* where not all potential sources of asbestos are tortious – see Green, 'Winner Takes All', above n 101.

[113] Most obviously the House of Lords' decision in *Gregg v Scott* [2005] UKHL 2, [2005] 2 AC 176. See Ch 7 for a detailed discussion.

latter. For my part I have difficulty even in recognising the distinction between these categories, at any rate in some cases.[114]

With respect, this distinction is indeed critical. For a start, 'the claimant must prove that his injury was caused by the eventuation of the kind of risk created by the defendant's wrongdoing'.[115] In NBA terms, this is necessary in order for the effects of a defendant's breach of duty to be deemed 'operative' in terms of the claimant's damage.[116] The relevance of this distinction was addressed explicitly by Lord Hoffmann in *Barker*: ultimately, despite not having recognised its importance in *Fairchild*,[117] His Lordship conceded that, for the exceptional principle to apply, 'the mechanism by which it caused the damage, whatever it was, must have been the same',[118] and that he was 'wrong' to think otherwise.[119] Lord Rodger, by contrast, had already identified the significance of single, as opposed to multiple, sources of risk at the time the *Fairchild* principle was formulated:

> [T]he principle does not apply where the claimant has merely proved that his injury could have been caused by a number of different events, only one of which is the eventuation of the risk created by the defendant's wrongful act or omission. *Wilsher* is an example.[120]

Wilsher remains good law, and long may this be true. The issue in that case, in which the defendant's breach was one of five different potential causes[121] of the claimant's injury, was that the *agent* of cause was indeterminate. This is not true of the mesothelioma cases, in which it is accepted that the causal agent is asbestos, and so it is the *source* of the causal agent which is indeterminate. In the mesothelioma cases, therefore, we know more. We know, for instance, that the defendant's behaviour contributed to *the* specific risk which came to fruition in the form of the claimant's ultimate injury. In *Wilsher*, this was not known; all that was established therein was that the defendant had created *a* risk which might have resulted in the claimant's injury. For a distinction that calls for such a minor textual alteration, it has major implications. Deciding in favour of liability in *Wilsher*-type situations

---

[114] *Sienkiewicz v Greif* [2011] UKSC 10, [2011] 2 AC 229 at [187] 294–95 (per Lord Brown).
[115] *Fairchild v Glenhaven Funeral Services* [2002] UKHL 22, [2003] 1 AC 32 at [170] 118–19 (per Lord Rodger).
[116] This is the function of the second stage. See also Stapleton, 'Factual Causation', above n 46 at 225.
[117] *Fairchild v Glenhaven Funeral Services* [2002] UKHL 22, [2003] 1 AC 32 at [73] 77.
[118] *Barker v Corus* [2006] UKHL 20, [2006] 2 AC 572 at [24] 587. Although see the wording used by Lord Mance in *Durham v BAI (Run-off Ltd)* [2012] UKSC 14, [2012] 3 All ER 1161 at [65] 1194: 'In the present state of scientific knowledge and understanding, there is nothing that enables one to know or suggest that the risk to which the defendant exposed the victim actually materialised. What materialised was at most a risk of the same kind to which someone, who may or may not have been the defendant, or something or some event had exposed the victim'. With respect, this wording is a little misleading because all sources of asbestos create the same risk; that of causing an individual to develop mesothelioma. What scientific knowledge cannot yet establish is whether a particular defendant's *contribution* to that single risk was necessary for a particular claimant to develop mesothelioma.
[119] *Barker v Corus* [2006] UKHL 20, [2006] 2 AC 572 at [23] 587.
[120] *Fairchild v Glenhaven Funeral Services* [2002] UKHL 22, [2003] 1 AC 32 at [170] 118–19 (per Lord Rodger).
[121] The other four of which were not breaches of duty. For a Canadian perspective, see *Smith v Moscovich* (1989) 40 BCLR (2d) 49 (BC SC).

would effectively mean imposing liability for risk creation, regardless of whether or not that risk actually resulted in injury. Not only would this rail against the law of torts' (correct) refusal to award damages for risk creation *simpliciter*,[122] but, in practical terms, it would mean potentially crushing liability for defendants, and particularly those likely by dint of their nature to create such risks on a regular basis, such as the NHS.[123] Under the second stage of the NBA, as has already been made clear, only breaches of duty whose effects are operative on the claimant when the injury occurs are deemed to be causes of that injury. This, the means by which pre-emptive factors can be distinguished from pre-empting factors, ensures that defendants will not be held liable when it cannot be established that the claimant's ultimate injury fell within *the* risk created by their negligent behaviour. The application of this stage of the NBA also means that liability on the basis of aggregate But For causation will only occur where all defendants have contributed to *the* single risk which came to fruition in the claimant's case. This limitation, in making the exception more exceptional, also makes it more acceptable: first, only rarely will it apply, and, second, it is far easier to defend the aggregation of defendants where they have all contributed to the creation of the same risk, which has then eventuated in harm to the claimant.[124]

Crucially, liability under the *Fairchild* principle is not liability for the creation of risk *simpliciter*. Although, in *Barker*, Lord Hoffmann presented the basis of the principle as one which recognises risk creation, it is clear that this was a means by which the imposition of proportionate liability for an indivisible injury could be both carried out and justified, and would only occur *where that risk has eventuated in damage to the claimant.*[125] In *Durham v BAI (Run-off) Ltd*,[126] Lord Mance confirmed this interpretation:

> In reality, it is impossible, or at least inaccurate, to speak of the cause of action recognised in *Fairchild's* case and *Barker's* case as being simply 'for the risk created by exposing' someone to asbestos. If it were simply for that risk, then the risk would be the injury; damages would be recoverable for every exposure, without proof by the claimant of any (other) injury at all. That is emphatically not the law . . . The cause of action exists because the defendant has previously exposed the victim to asbestos, because that exposure *may* have led to the mesothelioma, not because it did, and because mesothelioma has been suffered by the victim . . . The actual development of mesothelioma is an essential element of the cause of action. In ordinary language, the cause of action is 'for' or 'in respect of' the mesothelioma, and in ordinary language a defendant who

---

[122] See *Rothwell v Chemical & Insulating Co Ltd* [2007] UKHL 39, [2008] AC 281 and S Green, 'Risk Exposure and Negligence' (2006) 122 *LQR* 386 (a note on the Court of Appeal decision, but referred to by the House of Lords at [55]).

[123] See J Stapleton, 'Unnecessary Causes' (2013) 129 *LQR* 39, 58.

[124] A solution which, in these extraordinary circumstances, will be of use to the lawyer, *pace* Stevens, *Torts and Rights*, above n 34 at 131.

[125] *Barker v Corus* [2006] UKHL 20, [2006] 2 AC 572 at [35] 589–90, [36] 590 and [48] 593–94. This is not the same as liability for risk creation *simpliciter*, where, by definition, no harm has yet befallen the claimant, and may well never do so. See also Hoffmann, '*Fairchild* and After', above n 3 at 67.

[126] *Durham v BAI (Run-off) Ltd* [2012] UKSC 14, [2012] 3 All ER 1161. See also [68] 1195, [72] 1197, [73] 1197, [77] 1202–03, [85] 1204–05, [87] 1205 and [90] 1205.

exposes a victim of mesothelioma to asbestos is, under the rule in *Fairchild's* case and *Barker's* case, held responsible 'for' and 'in respect of' both that exposure and the mesothelioma.[127]

This accords with the House of Lords' decision in *Rothwell v Chemical & Insulating Co Ltd*,[128] in which the existence of symptomless pleural plaques, the risk of developing an asbestos-related disease, and consequent anxiety were deemed not to qualify as the gist of a claim in negligence in the absence of compensable physical damage.[129]

The reasons for not allowing recovery for such 'pure' risk exposure are several. The first is conceptual. By definition, a risk cannot be an indicium of an existing injury; in auguring the possibility of future harm, it necessarily indicates the current absence of that harm. The second, a corollary of the first, is practical: allowing claimants to recover for the possibility that they will be injured in the future means that, inevitably, those individuals who do not develop an illness will receive a windfall, whilst those who do will remain inadequately compensated for their loss. The third reason, based on policy, is the need to avoid a situation that would allow for speculative claims, and to ensure that, in each case, 'what is at stake justifies the use of the process' (at [21]).[130] Such reasons alone are sufficient to justify a finding of no liability in a case where the risk to which the defendant has exposed the claimant has not eventuated in actual harm.[131]

The *Fairchild* principle, howsoever applied, constitutes a more or less exceptional approach to the causal inquiry. It does not, however, have any unorthodox effect on the gist of the action in tort for negligence, and does not allow claimants to recover for having been exposed to risk *simpliciter*.[132] Nor, as this chapter has made clear, is it synonymous with a material contribution to injury analysis.

---

[127] [2012] UKSC 14, [2012] 3 All ER 1161 at [65] 1194.
[128] *Rothwell v Chemical & Insulating Co Ltd* [2007] UKHL 39, [2008] 1 AC 281. In particular, see [1] 287–88, [2] 288, [12] 291, [17] 292, [49] 300, [50] 300, [60] 303, [68] 305–06, [73] 307, [74] 308, [88]–[90] 311–12 and [103] 314–15.
[129] See also *Gregg v Scott* [2005] UKHL 2, [2005] 2 AC 176 and *Hicks v Chief Constable of the South Yorkshire Police* [1992] 2 All ER 65 (HL).
[130] *Rothwell v Chemical & Insulating Co Ltd* [2006] EWCA Civ 27, [2006] 4 All ER 1161 at [21] 1169.
[131] Green, 'Risk Exposure and Negligence', above n 122 at 387. See, further, the contrasting views on this point in E Weinrib, '*Causation and Wrongdoing*' (1987) 63 *Chicago-Kent Law Review* 407 and CH Schroeder, '*Corrective Justice and Liability for Increasing Risk*' (1990) 37 *UCLA Law Review* 439.
[132] As confirmed in *Durham v BAI (Run off) Ltd (Trigger Litigation)* [2012] UKSC 14, [2012] 3 All ER 1161. For an explanation of why this interpretation of the principle is so crucial in practice, see Merkin and Steele, *Insurance and the Law of Obligations*, above n 22 at 373–81: essentially, if liability is based on the reformulated gist of increased risk, it cannot therefore be associated with a particular policy year, because there is no 'single indivisible loss' in each year of cover (which is the accepted artifice by which claims in asbestos-mesothelioma cases are made post-*Trigger*), and, consequently, coverage can be hard to establish. This is a dire outcome for claimants, and amounts to the same 'remarkable' outcome which the Supreme Court in *Trigger* aimed to avoid (see [89] 1205). Conversely, on the interpretation approved in *Trigger*, which retains mesothelioma as the gist of the damage caused, albeit by modified causal connections, each and every insurer on risk is liable. See also *Phillips v Syndicate 992 Gunner* [2003] EWHC 1084 (QB), [2003] All ER (D) 168 (May). It has also now been established that the application of the *Fairchild* principle does not modify the formulation of the duty of care: see *Williams v University of Birmingham* [2011] EWCA Civ 1242, [2011] All ER (D) 25 (Nov).

7

Lost Chances

Illustrative Cases: (Genuine) *Allied Maples, Chester*; (Illusory) *Hotson, Gregg*

Factual Basis

This analysis should be applied where

- a chance exists independently of the breach of duty, so that the defendant's breach affects the claimant's ability to avail herself of that chance, but does not affect the substance of the chance itself.

This analysis has been applied erroneously where

- the chance and the breach of duty are interdependent because the breach affects the existence and content of the chance itself.

This chapter is concerned with the thorny question[1] of recovery in negligence for a lost chance. This issue is commonly divided into two supposedly distinct categories: the lost chance of a better physical outcome, and the lost chance of a better financial outcome.[2] Since, as we shall see, it *appears* to be the case that courts are willing to allow claimants to recover for the latter but not for the former, this means of classification has led unsurprisingly to some moral disquiet.[3] This is unfortunate, not least because it is unnecessary. There is indeed a distinction to be made between two 'types' of factual situation which have been presented to the courts as appropriate vehicles for loss of chance claims. The true distinction, however, does not lie in whether the claim is for a lost chance of financial gain or for a 'lost chance'[4] of a better physical outcome.[5] Rather, such claims divide more

---

[1] The fact that both Andrew Burrows and Harvey McGregor regard it as difficult suggests that this is indeed the case. See A Burrows, 'Uncertainty about Uncertainty: Damages for Loss of a Chance' [2008] *Journal of Personal Injury Law* 31 and H McGregor, 'Loss of Chance: Where Has It Come From and Where Is It Going?' [2008] *Professional Negligence* 1.

[2] Which includes both the making of a gain and the avoidance of a loss.

[3] See, for instance, *Gregg v Scott* [2005] UKHL 2, [2005] 2 AC 176 at [25] 185 (per Lord Nicholls), and S Steel, 'Rationalising Loss of a Chance in Tort' in S Pitel, J Neyers and E Chamberlain (eds), *Challenging Orthodoxy in Tort Law* (Oxford, Hart Publishing, 2013) 258.

[4] A concept which is not without definitional difficulty: see below.

[5] Nor whether the chance depends on the actions of the claimant herself or a third party.

appropriately into those in which the chance is independent of, and therefore unaffected by, the relationship between the claimant and the defendant, and those in which the chance is dependent on, and therefore determined by, that relationship. Crucially, what is common to both types of case analysed here is that the 'chance', howsoever it is understood, is the focus of the claim itself. We are not here concerned, for instance, with the myriad examples of cases in which claimants have lost chances *consequent upon* damage of an orthodox actionable type, such as the lost opportunity to earn the salary of a professional footballer as a result of a personal injury.[6] These are conventional quantification questions, and as such, have no bearing on the causal inquiry.[7]

In the cases presented as involving the lost chance of a better physical outcome, the 'lost chance' is not in legal terms a chance at all, but merely a repackaging of the forensic margin of error.[8] In cases concerned with lost financial chances, however, (in which there is an independent chance to be lost), that chance is a measure of the degree to which the claimant is worse off as a result of the defendant's breach. In the latter, the element of unpredictability relates only to the extent of the claimant's loss, whereas in the former, the 'chance' in question is a *causally relevant* uncertainty, thus having a bearing on whether the claimant has lost anything at all in legal terms. Clearly, this looks a lot like a distinction between causation and quantification, which is precisely what it is.[9] This points, therefore, to the real difficulty that applies to negligence cases in which there is an independent chance capable of being lost: if the chance itself represents the measure of damages to be awarded, the question remains as to what exactly is the actionable damage which the defendant's breach has caused. As the following analysis of the relevant cases shows, this damage can only sensibly be identified as the infringement of individual autonomy, since this best describes what has happened to a claimant

---

[6] *Collett v Smith* [2009] EWHC Civ 583, 153 Sol Jo (No 24) 34. See also, eg, *Hughes v McKeown* [1985] 3 All ER 284 (QB); *Langford v Hebran* [2001] EWCA Civ 361; and *Doyle v Wallace* [1998] 30 LS Gaz R 25, [1998] PIQR Q 146 (CA).
[7] See H McGregor, M Spencer and J Picton, *McGregor on Damages*, 18th edn (London, Sweet & Maxwell, 2009) 8.II.5(4).
[8] Cf J King Jr, 'Causation, Valuation and Chance in Personal Injury Torts involving Pre-Existing Conditions and Future Consequences' (1981) 90 *Yale Law Journal* 1353.
[9] See A Beever, *Rediscovering the Law of Negligence* (Oxford, Hart Publishing, 2009) 504 for the point that *Gregg v Scott* [2005] UKHL 2, [2005] 2 AC 176 is not really a loss of chance case at all. Beever's concern, however, where lost chances are claimed, is, similar to both Steel, 'Rationalising Loss', above n 3, and H Reece, 'Losses of Chances in the Law' (1996) 59 *MLR* 188, that the 'chance' claimed by the claimant in that case is purely epistemic (as opposed to objective) and is, therefore, no chance at all. In legal terms, this does not matter, since the *nature* of the uncertainty has no bearing on the fact that the forensic process needs to take a view on whether one outcome is more likely than another. In any event, as Perry has pointed out in S Perry, 'Risk, Harm and Responsibility' in D Owen (ed), *Philosophical Foundations of Tort Law* (Oxford, OUP, 1994) 324 and 333, this is a distinction which turns on 'predictability in principle, not predictability in practice' and that '[o]bjective risk cannot be directly observed. We must always rely on our best estimate of the objective risk, using whatever evidence is to hand. We necessarily operate, in other words, with the notion of epistemic risk'. Since the forensic process has to presume all such uncertainties are being viewed epistemically, objective possibilities are practically irrelevant in their own right. See also *Sellars v Adelaide Petroleum NL* [1994] HCA 4, (1994) 179 CLR 332 and G Masel, 'Damages in Tort for Loss of a Chance' (1995) 3 *Torts Law Journal* 43, 49.

who has been unable to take a particular chance. Without this, there is no 'damage' on which the lost chance, itself independent of the parties' relationship, can stand.[10] It would seem, therefore, that there really is no such thing as a 'lost chance' case in negligence, because even in those cases in which there is a genuine chance to be lost, what the claimant is really claiming for is a loss of autonomy.

These are all issues which have been obscured by the way in which recent cases have been argued and categorised. The following criteria suggest a different manner of classification – one which will lead to a clearer and more coherent means of dealing with the relevant issues. They suggest the division of factual situations into two types:

Type 1: those in which the chance exists independently of the breach of duty,[11] so that the breach affects a claimant's ability to avail herself of that chance, but not the substance of the chance itself. These are cases in which there is a genuine chance to be lost.[12] Liability in such cases is possible on an NBA.

Type 2: those in which the chance and the breach of duty are interdependent because the breach affects the existence and content of the chance itself. These are not in fact loss of chance cases at all, because to claim that a chance has been lost here is to beg the question. Liability in such cases is not possible on an NBA.[13]

A closer look at the differences between these two types of case reveals how significant the distinction is.

Type 1 Cases Explained

In cases of this type, the uncertainty with which the law is concerned is *whether* the claimant has been denied access to an opportunity which exists independently of the relationship between the parties. Given that this is the *legally relevant* uncertainty on these facts, it is one which must be established on the balance of probabilities.[14] Unless the claimant can do this, she has not proved that she is worse off

---

[10] As we shall see below, this is not a problem in the law of contract, since breaches of contract are actionable per se.

[11] See *Chester v Afshar* [2004] UKHL 41, [2005] 1 AC 134 at [81] 161 (per Lord Hope): 'the risk of which she should have been warned was not created by the failure to warn. It was already there, as an inevitable risk of the operative procedure itself however skilfully and carefully it was carried out'.

[12] Although, as we shall see below, they are more accurately described as 'loss of autonomy' cases.

[13] At least, where the balance of probabilities has not been met. Where it has, however, cases of this kind are unlikely to have been formulated as 'lost chance' cases, since the claimants therein could have recovered on orthodox grounds.

[14] A question which necessarily involves the claimant establishing that she would, on the balance of probabilities, have taken the chance had she been able to: *McWilliams v Sir William Arroll Co Ltd* [1962] 1 All ER 623 (HL) and McGregor, Spencer and Picton, *McGregor on Damages*, above n 7 at 8-049–8-05. This is what the claimant failed to establish in *Sykes v Midland Bank Executor & Trustee Co* [1971] 1 QB 113 (CA) and justifies the finding therein in favour of the defendant. Cf Steel, 'Rationalising Loss' above n 3 particularly at 247–53, in which he seems to suggest that the uncertainty inherent *in the chance itself* is a problem for proving causation; something that analysing in this way would avoid.

as a result of the breach, and such facts would not satisfy the first stage of the NBA. If, however, a claimant can establish on the balance of probabilities that she has been denied access to an opportunity, the consequent damages will be calculated according to the magnitude of the chance which has been lost, regardless of whether it exceeds 50 per cent.[15] This is because the uncertainty inherent in the chance itself is not affected by anything the defendant has done, and it is not, therefore, necessary for the law to resolve it one way or another: in legal terms, such a chance is relevant only in terms of quantification. This quantification is nonetheless pivotal, since, on one view, it is only by proving that the opportunity denied is a valuable one, and different from any available in spite of the breach, that a claimant should be able to recover.[16]

A Right to Autonomy?

As outlined above, a potential problem with emphasising the distinction between causation and quantification in this way arises when classifying the nature of the loss which, on the balance of probabilities, the defendant is supposed to have caused. If the magnitude of the independent chance constitutes the quantification, the question of what constitutes the harm itself remains open.[17] An answer can be found, however, in both *Chester v Afshar*[18] and *Allied Maples v Simmons & Simmons*.[19] In substantive terms, these are Type 1 cases, and decisions which amount in effect to common law recognition of an individual's right to autonomy. A proper analysis of those cases in which there is an independent chance to be lost (Type 1), leads to the inescapable conclusion that the damage in such cases is really the claimant's ability to make a free choice. Any quantification subsequent to this recognition should then compare the chance (if any) available to the claimant as a result of the breach with the chance which would have been available but for the breach,[20] thereby establishing whether the claimant is any worse off in consequential terms for having been denied a chance. Lord Hoffmann, in alluding to this in his dissenting judgment in *Chester*,[21] was the only panel member involved in that

---

[15] As long as it is more than speculative: see *Allied Maples v Simmons & Simmons* [1995] 4 All ER 907 (CA) at 919 (per Stuart-Smith LJ).
[16] Currently, the House of Lords' decision in *Chester v Afshar* [2004] UKHL 41, [2005] 1 AC 134 suggests that this is not the case. In effect, substantial damages were granted therein for interference with the claimant's autonomy. Without more. See below. See also R Stevens, *Torts and Rights* (Oxford, OUP, 2007) 77–79.
[17] This is a function of the tort of negligence, of which damage is the gist. It does not affect, for instance, the law of contract, which is actionable per se: A Burrows, *Remedies for Torts and Breach of Contract*, 3rd edn (Oxford, OUP, 2004) 59 and H Luntz, 'Loss of Chance' in A Freckleton and D Mendelson (eds), *Causation in Law and Medicine* (Aldershot, Ashgate, 2002) 153.
[18] *Chester v Afshar* [2004] UKHL 41, [2005] 1 AC 134.
[19] *Allied Maples v Simmons & Simmons* [1995] 4 All ER 907 (CA) discussed below. See also *Sellars v Adelaide Petroleum NL* [1994] HCA 4, (1994) 179 CLR 332.
[20] This, as we shall see, is where the judgment in *Chester v Afshar* [2004] UKHL 41, [2005] 1 AC 134, went wrong.
[21] Which ultimately, of course, he did not support.

decision who recognised (at least explicitly) the implications of a decision in favour of liability.

In *Chester v Afshar* the defendant performed elective surgery upon the claimant in order to alleviate her severe back pain. Although he did so without negligence, she suffered significant nerve damage and was consequently left partially paralysed. The defendant breached his duty of care by failing to warn his patient of the 1–2 per cent risk of such paralysis occurring as a result of the operation. The causal problem arose in this case because the claimant did not argue that, had she been warned of the risk, she would *never* have had the operation, or even that, duly warned, she would have sought out another surgeon to perform the operation.[22] Her argument was simply that, had she been properly warned of the risks inherent in the procedure, she would not have consented to having the surgery within three days of her appointment, and would have sought further advice on alternatives. The House of Lords (Lords Bingham and Hoffmann dissenting) held Mr Afshar liable on the basis that, since the ultimate injury suffered by the claimant was a product of the very risk of which she should have been warned, it could therefore *be regarded* as having been caused by that failure to warn. In the course of his dissent, Lord Hoffmann made the following point:

> Even though the failure to warn did not cause the patient any damage, it was an affront to her personality and leaves her feeling aggrieved. I can see that there might be a case for a modest solatium in such cases.[23]

The very fact that the majority decision on the facts in *Chester* was one in favour of liability suggests that what the claimant was being compensated for was the denial of her right to make a free choice, since that denial led in her case to no consequential loss.[24] It is not, however, a straightforward exercise to justify *Chester* in this way. First, although in *Rees v Darlington Memorial NHS Trust*[25] there is an explicit recognition at the highest level that an infringement of autonomy can amount to actionable damage in negligence, none of the judgments in *Chester* made reference to that decision.[26] Second, the House of Lords in *Chester* departed substantively from the *Rees* approach in any event by awarding substantial damages for the infringement, far in excess of the £15,000 conventional award granted in the earlier case. Finally, these contextual issues aside, the outcome in *Chester v Afshar* is simply not presented as one based on the idea of autonomy as a freestanding right; rather, it is presented as a conclusion reached on causal grounds.

---

[22] cf *Chappel v Hart* [1998] HCA 55, (1998) 195 CLR 232, in which Kirby and Gaudron JJ found that the claimant's hypothetical actions in that case, in seeking out a more experienced surgeon, would have decreased her risk of injury as a result of the procedure. There was no agreement on this point.

[23] [2004] UKHL 41, [2005] 1 AC 134 at [33] 147.

[24] Because the chance available to her as a result of the breach was identical to the one that would have been available to her but for the breach. See Stevens, *Torts and Rights*, above n 16 at 76–78 and *Chappel v Hart* [1998] HCA 55, (1998) 195 CLR 232 at [40]–[43] 249–50 (per McHugh J).

[25] *Rees v Darlington Memorial NHS Trust* [2003] UKHL 52, [2004] 1 AC 309.

[26] Despite the fact that both counsel in *Chester* referred directly to *Rees* in their respective submissions. For an assessment of the implications of *Rees*, see D Nolan, 'New Forms of Damage in Negligence' (2007) 70 *MLR* 59, 70–80.

Similarly, as we shall see, *Allied Maples*, although decided before *Rees*, is deficient in explicit references to the concept of autonomy. Notwithstanding these objections, Type 1 situations such as these can *only* be satisfactorily explained by reference to a claimant's right of autonomy. Such situations should be described therefore as 'lost autonomy' cases, after the nature of the loss itself. To use the 'lost chance' label in such a context is somewhat misleading, since it refers only to the (potential) loss consequent upon the wrong.

Unrealised Chances

An essential feature of Type 1 cases is that the independent chance in question must be one which exists only as a 'pure' chance. In other words, the opportunity in question must not be a possibility which has been realised, but only one whose outcome can never, as a result of the breach, be known.[27] The following situation, for example, does not illustrate a true loss of autonomy case:

> A and B are both driving cars in opposite directions along a narrow road. B is distracted by C, who is acting negligently. B's car crosses over the middle of the road and collides with A. A sues C on the basis that had C not distracted B, then there was a chance of 1 in 5 that B would have avoided colliding with A. A sues C on the basis that had C not distracted B, then there was a chance of 1 in 5 that B would have avoided colliding with A.[28]

Here, the claimant's damage is not a lost chance at all, because the relevant chance was taken, and it is the *outcome* of that chance (the personal injury) which forms the damage to the claimant on these facts. The fact that a chance must remain unrealised, and therefore hypothetical, if it is to be recognised as such, fits with the well-known distinction made by Lord Diplock in *Mallet v McMonagle* between those uncertainties on which the law needs to take a certain view, and those which it can accept as uncertain:

> The role of the court in making an assessment of damages which depends upon its view as to what will be and what would have been is to be contrasted with its ordinary function in civil actions of determining what was. In determining what did happen in the past a court decides on the balance of probabilities. Anything that is more probable than not it treats as certain. But in assessing damages which depend upon its view as to what will happen in the future or would have happened in the future if something had not happened in the past, the court must make an estimate as to what are the chances that a particular thing will or would have happened and reflect those chances, whether they are more or less than even, in the amount of damages which it awards.[29]

---

[27] See also Steel, 'Rationalising Loss', above n 3 at 236.

[28] M Cannon QC, 'Allied Maples and the Clever Sheep', kindly sent by personal correspondence, which suggests that this is a loss of a chance case, and goes on to lament how it 'involves a radical change to the burden of proof and as to what is recognised as actionable damage'. Were such facts to represent a case in which damages were recoverable against C for the loss of a chance (as opposed to the ultimate injury as a joint tortfeasor), this argument would hold water. As it is, however, it does not.

[29] *Mallet v McMonagle* [1970] AC 166 (HL) at 176. See also *Malec v JC Hutton Pty Ltd* [1990] HCA 20, (1990) 169 CLR 638 and Masel, 'Damages in Tort', above n 9.

Similarly, Lord Reid said in *Davies v Taylor*:

> You can prove that a past event happened, but you cannot prove that a future event will happen and I do not think that the law is so foolish as to suppose that you can. All that you can do is to evaluate the chance. Sometimes it is virtually 100 per cent.: sometimes virtually nil. But often it is somewhere in between.[30]

This point also demonstrates why the material contribution to risk cases, discussed in Chapter 6, cannot be conceptualised as lost chance cases, despite the fact that it might not be obvious on the face of it how a material increase in risk differs from a lost chance.[31] As was made clear in the earlier discussion, however, liability premised on a material increase in risk is *not* liability for the creation of risk, and is parasitic upon the occurrence of the resultant harm. As Sandy Steel succinctly puts it, damages awarded in the material contribution to risk context 'are for the chance of causation, not causation of a lost chance'.[32]

Type 2 Cases Explained

The relevant issue in these situations is not whether the defendant has infringed the claimant's right to take advantage of an independently established opportunity, but whether the claimant has had any right infringed at all. Proof on the balance of probabilities is necessary in this context because the very uncertainty which the law has to resolve here is whether the defendant's breach of duty has made the claimant worse off in a *legally recognised* way. Such legal recognition can only be achieved by proving that it is more likely than not that the defendant's breach *has* affected the claimant's right: making the outcome 'more likely' without making it 'more likely than not' would not satisfy the first stage of the NBA, meaning a causal link would not be established. Type 2 cases present, therefore, the orthodox causal question. It is only as a consequence of claimants' attempted reformulation of this question, in response to a no-liability result, that such factual situations have been represented as involving lost chances.[33] On the basis of such a reformulation, however, almost every negligence scenario could be recast as one in which a so-called chance has been lost, because the forensic process does not deal in certainties.[34] Consequently, any conclusion established on the balance of

---

[30] *Davies v Taylor* [1974] AC 207 (HL) at 213.

[31] There are in fact several reasons for this, another being the existence of the 'rock of uncertainty', which is exclusive to material contribution to risk situations, and means that such cases are not synonymous with those involving a 'loss of a chance', howsoever it is understood, but see Burrows, 'Uncertainty about Uncertainty', above n 1 and E Peel, 'Lost Chances and Proportionate Recovery' (2006) *LMCLQ* 289. See also Ch 6, text to n 13.

[32] Steel, 'Rationalising Loss', above n 3 at 242.

[33] As in *Hotson v East Berkshire AHA* [1987] AC 750 (HL) and *Gregg v Scott* [2005] UKHL 2, [2005] 2 AC 176.

[34] As recognised in *Gregg v Scott* [2005] UKHL 2, [2005] 2 AC 176 at [224] 233 (per Baroness Hale). See also L Hoffmann, 'Causation' in R Goldberg (ed), *Perspectives on Causation* (Oxford, Hart Publishing, 2011) 8.

probabilities will necessarily admit of there being a 'chance' that it is not an accurate representation of actual events.[35]

Once this Type 1/Type 2 distinction is recognised, the apparent incoherence of the common law's current position disappears: claims in which there is an independent chance capable of being lost (Type 1 cases) can be successful where it can be shown that the defendant's breach of duty affected the claimant's *access* to an independent opportunity. On the other hand, those which are simply orthodox claims otherwise bound to fail (Type 2 cases in which the defendant's breach has not been established on the balance of probabilities to have changed the claimant's course of events for the worse) are not saved by relabelling them as claims for lost chances. Since, as we shall see, most cases falling within the Type 2 category are medical negligence claims, it appears as if the law is skewed towards favouring material interests over issues of physical integrity. A correct analysis, however, reveals this not to be the case: *Chester v Afshar*, for instance, is both a medical negligence case and one which should properly be understood as one in which there was an independent chance to be lost as a result of the claimant's autonomy being infringed. Although the argument here is that the result in that case should have been one of no liability, this has nothing to do with the interest at stake, and everything to do with the particular physical sequence of events that transpired, in which, ultimately, no chance was actually denied to the claimant.

Allied Maples v Simmons and Simmons[36] is one of the clearest examples of a Type 1 case. The defendants therein were solicitors, whose negligence in drafting an acquisition contract had deprived the claimants of the opportunity to negotiate for more advantageous terms. The claimants, relying on the negligently drafted agreement, believed they were protected from liabilities arising from the acquisition when in fact they were not. Had they known of their vulnerability, they would have attempted to acquire such protection from the vendor before the contract was concluded. The Court of Appeal decided in favour of the claimants, and determined that damages should be assessed by reference to the chance of any negotiations (had they been possible) being successful. It concluded that, as long as there was a real, as opposed to speculative, chance of success, there was no need for a positive outcome to be more likely than not. These facts fall squarely within the Type 1 classification because the solicitors' breach of duty had no effect whatsoever on the substance or content of the chance itself; its existence was extraneous to, and independent of, the solicitors' actions. The phenomenon of the chance *eo ipso*, therefore, is of no relevance to the question of causation, but only to the subsequent question of quantification. The breach of duty did, however, affect the claimant's ability to avail itself of that chance, and it is this specific question of the defendant's effect on the claimant's autonomy which represents the true *causal* question on such facts. Consequently, it is the answer to this question which must be (and was) established by the claimant on the balance of probabilities:

---

[35] See Ch 2.
[36] [1995] 4 All ER 907 (CA).

On the evidence before him the judge was justified in concluding that the defendants' breach of duty did have a causative impact upon the bargain which the plaintiffs and the vendors struck. He was entitled to find that, if the plaintiffs had negotiated further, they had a measurable chance of negotiating better terms which would have given them at least some protection against the liability on assigned leases which they were to assume on the draft agreement as it then stood, and as ultimately signed.[37]

In this type of case,[38] what the claimant has to prove is that the defendant's breach of duty has prevented her from taking advantage of an opportunity which exists independently of the claimant–defendant relationship and the interaction therein. This, therefore, becomes a straightforward, but specific, causal question: is it more likely than not that, but for the defendant's breach, the claimant would have *had access to* an opportunity from which she is currently excluded? If such access has been denied by the defendant's breach, the effect has been to interfere with the claimant's autonomy, and it is this interference which constitutes the relevant damage for the purposes of the causal inquiry.

Hotson and Gregg: Why They Are Type 2, and Not Loss of Autonomy Cases

The difference between loss of autonomy (Type 1) cases and those properly classified as Type 2 situations can be illustrated by reference to the two English decisions currently (but incorrectly) associated with claims for lost chances: *Hotson v East Berkshire Area Health Authority*,[39] and *Gregg v Scott*.[40] In the first case, the claimant, a 13-year-old boy, fell from a tree and injured his left hip.[41] The defendant's hospital, from which he sought treatment, negligently failed to diagnose or treat him correctly for five days. Ultimately, the claimant suffered avascular necrosis of the epiphysis, involving disability of the hip joint with the virtual certainty that osteoarthritis would later develop. At trial, Simon Brown J found that, even had the injury been properly diagnosed and treated in a timely manner, there remained a 75 per cent risk that avascular necrosis would have developed, but he awarded the claimant damages corresponding to the 25 per cent chance of which the defendant's negligence had supposedly deprived him. Whilst the Court of Appeal concurred, the House of Lords decided in favour of the defendant and held that the trial judge's finding that, at the time of the fall there had already been a 75 per cent chance of avascular necrosis developing, amounted to a finding on the balance of probabilities that the fall was the sole cause of the injury. The Court did not, however, expressly exclude the possibility that 'loss of a chance', as

---

[37] [1995] 4 All ER 907 (CA) at 925 (per Hobhouse LJ).
[38] Other examples include *Kitchen v RAF Association* [1958] 2 All ER 241 (CA) and *Yardley v Coombes* (1963) 107 Sol Jo 575 (QB).
[39] *Hotson v East Berkshire Area Health Authority* [1987] AC 750 (HL).
[40] *Gregg v Scott* [2005] UKHL 2, [2005] 2 AC 176.
[41] More specifically, his left femoral epiphysis.

it was presented therein, could ever form the basis of a successful claim in negligence.[42]

In *Gregg v Scott*, the claimant visited the defendant GP, complaining of a lump under his left arm, which the defendant diagnosed as a benign lipoma. In failing to refer the claimant to a specialist at that point, the defendant was held to have been in breach of his duty of care. It was not until a biopsy was carried out by a specialist, following a referral by another GP nine months later, that the claimant discovered that he had cancer in the form of non-Hodgkin's lymphoma. The trial judge found that the claimant's chance of being 'cured' (defined in this context as a period of 10 years' remission) was 42 per cent when he made his visit to the defendant, but that the nine-month delay, consequent upon the defendant's negligent failure to diagnose his illness correctly, reduced his chance of being cured to 25 per cent. As the claimant had only a 42 per cent chance of a cure in the first place, however, he was unable to prove on the balance of probabilities that the defendant's negligence caused him to be in a worse state than he would have been in, had his treatment not been delayed by nine months. In the light of this fact, the claimant argued that he had suffered the loss of a chance of being cured as a result of the defendant's negligence. In so doing, he invited the court to address a similar question to the one first considered by the House in *Hotson* as to whether or not such a loss should be recoverable.

By a majority of three to two (Lord Hope and Lord Nicholls dissenting), the House of Lords dismissed the claimant's appeal and held that it was (still) not prepared to extend loss of a chance claims to such cases. Whilst, on the facts as found by the trial judge, it had been established that the defendant's breach of duty had reduced the epidemiological likelihood of survival by 17 per cent, the House of Lords correctly declined to recognise this as actionable damage. As Lord Hoffmann put it: 'A wholesale adoption of possible rather than probable causation as the criterion of liability would be so radical a change in our law as to amount to a legislative act'.[43] In refusing to depart from the orthodox approach to causation, the Court recognised that it is our epistemic limitations which pose the most consistent problem for the causal inquiry and, to use Lord Hoffmann's words once more: 'What we lack is knowledge and the law deals with lack of knowledge by the concept of the burden of proof'.[44]

In other words, although the law cannot expect to deal in certainties, the least it can do is expect outcomes to be more likely than not. Common to both *Hotson* and *Gregg* is the fact that the claimants could not have established their claims on the basis of the orthodox approach to causation, since in neither case was it more likely than not that the claimant was any worse off as a result of the defendant's breach of duty. Neither would have satisfied the first stage of the NBA. The

---

[42] [1987] AC 750 (HL) at 786.
[43] [2005] UKHL 2, [2005] 2 AC 176 at [90] 198. See also *Tabet v Gett* [2010] HCA 12, (2010) 240 CLR 537; *Naxakis v Western General Hospital* [1999] HCA 22, (1999) 197 CLR 269; and *Laferrière v Lawson* [1991] 1 SCR 541 (SCC).
[44] [2005] UKHL 2, [2005] 2 AC 176 at [79] 196.

formulation of both claims, misleadingly couched in terms of 'lost chances', was an attempt to sidestep the standard of proof on the basis that the claimants had lost something of value to them in having their 'already likely to suffer an adverse outcome' position made, by the defendant's breach, into 'even more likely to suffer an adverse outcome'. That this argument was made is, at least from a human interest point of view, easily understandable, since most individuals would class even the tiniest percentage chance of avoiding an adverse physical outcome as being something of significant value to them.[45] In legal terms, however, such a 'chance' is less a prediction of what would have happened to a particular claimant than it is an approximation of the forensic margin of error:

> If it is proved statistically that 25 per cent. of the population have a chance of recovery from a certain injury and 75 per cent. do not, it does not mean that someone who suffers that injury and who does not recover from it has lost a 25 per cent. chance. He may have lost nothing at all. What he has to do is prove that he was one of the 25 per cent. and that his loss was caused by the defendant's negligence. To be a figure in a statistic does not by itself give him a cause of action. *If the plaintiff succeeds in proving that he was one of the 25 per cent and the defendant took away that chance*, the logical result would be to award him 100 per cent of his damages.[46]

As the emphasis shows, what is uncertain in these cases is whether the claimant ever had a greater-than-evens chance of recovery, and whether the defendant's breach affected the substance of that possibility, making it into a less-than-evens chance.[47] The balance of probabilities approach is necessary here because this is an uncertainty which affects the causal issue to be resolved. In Type 2 factual situations such as these, a claimant is deemed to have had a chance of avoiding a detriment only where she *starts off* in a position in which it is more likely than not that she will avoid the adverse outcome. If a breach of duty reduces this chance to a level at which it is still greater than evens, the defendant in question is not liable in negligence, despite having affected the likelihood of that outcome occurring. If a breach of duty brings the claimant's chances below the evens threshold, however, that defendant will be liable in negligence for having made the claimant worse off in the eyes of the law. Since the assessment of the claimant's chances in Type 2 cases is intrinsic to the question of *whether* she deserves compensation from the defendant, the degree to which her chances have been reduced is not as important as the comparative effect of the defendant's breach. In *McGhee v National Coal Board*,[48] Lord Salmon gave the following well-known illustration:

> Suppose ... it could be proved that men engaged in a particular industrial process would be exposed to a 52 per cent risk of contracting dermatitis even when proper washing facilities were provided. Suppose it could also be proved that that risk would be increased

---

[45] See Steel, 'Rationalising Loss', above n 3 at 263–68.
[46] *Hotson v East Berkshire Health Authority* [1987] AC 750 (CA) at 769 (per Croom-Johnson LJ), emphasis added.
[47] Another reason why the common law's balance of probabilities standard renders the award of damages proportional to the 'chance' lost inappropriate. See Ch 4 text to n 86 and Ch 8.
[48] *McGhee v National Coal Board* [1972] 3 All ER 1008 (HL).

to, say, 90 per cent when such facilities were not provided. It would follow that . . . an employer who negligently failed to provide the proper facilities would escape from any liability to an employee who contracted dermatitis notwithstanding that the employers had increased the risk from 52 per cent to 90 per cent. The negligence would not be a cause of the dermatitis because even with proper washing facilities, ie without the negligence, it would still have been more likely than not that the employee would have contracted the disease – the risk of injury then being 52 per cent. If, however, you substitute 48 per cent for 52 per cent the employer could not escape liability, not even if he had increased the risk to, say, only 60per cent Clearly such results would not make sense; nor would they, in my view, accord with the common law.[49]

Fortunately, in *Sienkiewicz v Greif*,[50] Lord Phillips made direct reference to this argument, and said of it:

I can understand why Lord Salmon considered that to base a finding of causation on such evidence would be capricious, but not why he considered that to do so would be contrary to common law. The balance of probabilities test is one that is inherently capable of producing capricious results.[51]

Although this comment has the obvious merit of aligning the result in Lord Salmon's example with the orthodox common law position, it also has the unfortunate effect of fortifying the view that such a position is capricious. It is not. Whilst the balance of probabilities approach might sometimes produce results which seem harsh either to a particular claimant or defendant considered discretely, those results will at least be consistent across the spectrum of *causal relationships between parties*. That is, a defendant found liable in negligence for having caused a 4 per cent reduction in a claimant's chances (say, from 52 per cent to 48 per cent) might look hard done by, as compared to another who was found not liable, despite having caused a 42 per cent reduction (say, from 48 per cent to 6 per cent) in another claimant's chances. Comparing the positions of defendants alone, however,[52] is not an authentic means of evaluating a mechanism intended to allocate the risk of error as between defendants and claimants as distinct, but related, classes. If the positions of claimants are considered alongside that of the defendants with whom they are correlative,[53] results will be consistent, and this is the comparison which really matters. For every losing defendant whose breach reduces a claimant's chances from 51 per cent to 49 per cent, for instance, there will be a losing claimant whose chances have been reduced by, say, 45 per cent, but who only ever had a chance amounting to 46 per cent. In other words, the potential for harsh results cut both ways. But it *always* cut both ways, so it cannot accurately be described as 'capricious'. Indeed, since we are all potentially claimants in negligence as much as we are potential defendants, splitting the risk of error in this way is the *least* capricious way of dealing with the inherent imperfections of the

---

[49] [1972] 3 All ER 1008 (HL) at 1018.
[50] *Sienkiewicz v Greif* [2011] UKSC 10, [2011] 2 AC 229.
[51] [2011] UKSC 10, [2011] 2 AC 229 at [26] 245–46.
[52] Or indeed claimants, to whom the argument applies with equal force.
[53] See E Weinrib, *Corrective Justice* (Oxford, OUP, 2012) 20.

forensic process.⁵⁴ Consequently, it is not open to claimants to re-characterise a claim which does not reach this evidentiary standard as being a claim for a lesser *degree* of loss,⁵⁵ because the legal result of falling short of this standard is that no loss has been suffered. Claimants in Type 2 cases, therefore, are not those who have lost a less-than-evens chance, but those for whom there is a less-than-evens chance that they have lost anything at all.

Hotson and *Gregg* could be analytically distinguished from each other on the basis that, in the former, the adverse outcome had already befallen the claimant whereas, in the latter, Mr Gregg was still alive at the time of his appeal to the House of Lords, 10 years after receiving the negligent diagnosis. When viewed (erroneously) as cases in which the claimant might have 'lost a chance', this is a distinction which could matter, because it differentiates between realised and unrealised chances. Since, however, the 'chance' in both cases has no existence independent of the relationship between the parties to the negligence action, to ask whether it has been realised is to ask the wrong question. Since the 'chance' in question is really just a measure of the probability that the defendant's breach altered the claimant's course of events for the worse, the only question of legal relevance is whether this can be established on the balance of probabilities. In other words, was it more likely than not that, but for the breach, the claimant would have avoided the adverse outcome *and*, as a result of the breach, is it no longer more likely than not that such an outcome will be avoided? Any evidence which goes to the question of whether the outcome actually will occur, or has occurred, is then relevant only to quantification in the orthodox way,⁵⁶ where and only where the defendant's breach has turned a 'more likely to avoid than not' scenario into a 'less likely to avoid than not' scenario.

As *Allied Maples* demonstrates, the claim that a breach of duty has caused the loss of a chance is not one that is alien to the law of tort. It is also definitively clear in the law of contract that claimants can recover damages which are calculated according to the percentage chance of making a particular gain, where the defendant's breach has denied them the opportunity of making that gain. In *Chaplin v Hicks*,⁵⁷ for example, the claimant was granted damages corresponding to the one in four chance that, had she been granted the appointment with the defendant to which her contract entitled her, she would have been chosen⁵⁸ as one of the 12

---

⁵⁴ See *Sienkiewicz v Greif* [2011] UKSC 10, [2011] 2 AC 229 at [187] 294–95 (per Lord Brown).

⁵⁵ As the claimants in both *Hotson* and *Gregg* did; they claimed not for the full extent of their final injury, but for a proportion of it, calculated according to the 'chance of avoiding it' they claimed to have lost.

⁵⁶ In which courts assess in the standard way the likelihood of the future effects of a defendant's tort upon the claimant. See McGregor, Spencer and Picton, *McGregor on Damages*, above n 7 at ch 35, II.1 and 2.

⁵⁷ *Chaplin v Hicks* [1911] 2 KB 786 (CA). This is the clearest statement of the point, but see also *Hotson v East Berkshire Area Health Authority* [1987] AC 750 (CA) at 785 (per Lord Mackay) and H Reece, 'Losses of Chances in the Law' (1996) 59 *MLR* 188, 197.

⁵⁸ Although, see Burrows, *Remedies*, above n 17 at 54, n 4 for the point that, had the Court of Appeal recognised that the contingency turned on the decision of the defendant himself, as opposed to an independent panel, no damages at all should have been awarded, on the basis of the principle that a

most attractive finalists in his beauty/talent competition. Whilst recovery for loss of a chance in contract is understandably a lot less contentious than it is in tort, owing to the fact that breaches of contract are actionable per se,[59] the factual basis of both types of claim is the same: we can seek answers in relation to what the opportunity was, what it consisted of, and how it might be assessed. All of these inquiries are conducted without reference to the defendant's breach, or to the relationship between claimant and defendant. Their answers, therefore, will tell us nothing about the interaction or causal link between the defendant's breach and the claimant's damage. Since, in these cases, there exists a chance independent of the relationship between the parties, from which the actions of the defendant could exclude the claimant, it is legitimate to ask whether, on the balance of probabilities, the actions of the defendant caused the claimant to 'lose a chance'. In other words, did the defendant prevent the claimant from availing herself of the discretely identifiable opportunity? In *Allied Maples*, for example, the claimants' relationship with their solicitors had no influence whatsoever on the content of the chance of their being able to elicit concessions from their vendor in the hypothetical world in which they could attempt this.[60] What was important to establish there was whether, on the balance of probabilities, the defendants had in fact denied the claimants the opportunity to take advantage of such a chance.

This sort of conclusion is not one, however, which can be made on the basis of Type 2 factual scenarios because, in these situations, there is no such thing as an independently quantifiable opportunity which can be divorced from the question of whether the defendant's breach made the claimant worse off. In the Type 2 cases of *Gregg* and *Hotson*, for instance, the claimants' chances of avoiding an adverse physical outcome were inextricably bound up with the effects of the defendant's negligent diagnosis. The relevant 'chance' therefore is not assessable independently of the breach, since it is defined and determined by it. Given that there is nothing extraneous from which the claimant can be excluded, the question is not whether, on the balance of probabilities, the claimant has been denied access to an opportunity, but whether the claimant ever had such an opportunity in the first place, and whether the defendant's breach deprived her of it. This is the crux of the question in Type 2, but not in Type 1 cases.

---

court is entitled to assume that a defendant will make the decision most favourable to herself. See also Burrows, 'Uncertainty about Uncertainty', above n 1 at n 44.

[59] There is, therefore, no need in contract for a claimant to establish any injury to her autonomy (although, in substance, this will describe what has happened). See Burrows, *Remedies*, above n 17 at 59.

[60] The fact that they could not attempt it in the real world explains why they had lost their chance.

Chester v Afshar: Why It Only Makes Sense As a Lost Autonomy Case

The majority in *Chester*[61] decided that, although Miss Chester could not recover on the basis of conventional principles of causation,[62] her claim should nevertheless be successful. This decision was supported by two principal arguments; that her injury lay within the scope of the surgeon's duty of care and that, were she to be denied recovery, such a duty to warn would be drained of meaningful content. With respect, neither of these claims adequately supports the radical departure from established principles of causation demanded by that conclusion. First, the fact that the injury fell within the scope of the surgeon's duty of care is not a *substitute* for causal involvement; rather, it is a *limiting* device which applies once causally relevant factors have been identified. So, a causally relevant factor can be deemed legally irrelevant because it causes a result which falls outside of a defendant's duty of care. It does not follow, however, that a factor which has no causal relevance to an outcome can be made legally relevant because that outcome (which was not caused by the factor in question) just happens to be the mischief against which the defendant's duty of care was intended to guard. This is a clear *non sequitur*, and is not, unsurprisingly, an argument that has been repeated elsewhere in the tort of negligence.[63] In *Chester*, the defendant's breach had not been established, on the balance of probabilities, to have played any historical role in the claimant's injury because, but for the failure to warn, she would have run exactly the same risk (the 1–2 per cent risk of cauda equina syndrome inherent in the procedure itself, however carefully performed) on a different day. The fact, therefore, that the failure to warn did not make Miss Chester any worse off renders the *scope* of the defendant's duty irrelevant: in negligence, individuals do not have a duty to compensate for damage that they do not cause.

Second, a finding of no liability which follows a failure to establish a factual causal link between a breach of a duty of care and a claimant's damage has no effect whatsoever on the content of that duty of care. The form of the negligence inquiry is such that a breach of a duty of care is a necessary but not sufficient element of a successful negligence claim. This inevitably means that duties will be breached with impunity from negligence liability, so long as no damage has thereby been caused (or at least so long as no damage can be established on the balance of probabilities to have thereby been caused).[64] This does not detract from the point or the

---

[61] Lords Hope, Steyn and Walker.

[62] Because she did not argue that, but for the failure to warn, she would not have had the procedure at any point. Had she done so successfully, she would have established causation on orthodox grounds: see *McWilliams v Sir William Arroll Co Ltd* [1962] 1 All ER 623 (HL).

[63] Although Lords Walker and Steyn attempted to make an analogy with the causal exception in *Fairchild v Glenhaven Funeral Services* [2002] HCA 22, [2003] 1 AC 32, the two situations are, as demonstrated below, far from analogous.

[64] Stevens, *Torts and Rights*, above n 16 at 44.

worth of the duty of care concerned. A patient's dignity and right to decide is protected by the law of tort's recognition that a medical professional has a duty to warn, not by a readiness to override causal considerations in the claimant's favour. If a breach of that duty to warn causes the claimant no loss, then a finding of no liability does not violate that right. It merely serves as an acknowledgement that the patient's inability to exercise that right did not, on this occasion, cause any loss. In his dissenting judgment in *Chester*, Lord Bingham is very clear about the importance of this point:

> I do not for my part think that the law should seek to reinforce that right by providing for the payment of potentially very large damages by a defendant whose violation of that right is not shown to have worsened the physical condition of the claimant.[65]

It is of course trite that damage is the gist of a claim in negligence and, whilst Mr Afshar was negligent, and Miss Chester damaged, the two were not connected in the way this axiom anticipates. The fact that both the negligence and the damage occurred within the same factual matrix was no more than coincidental. In this context, an outcome is described as coincidental if the breach of duty is not one which increases the general risk of that outcome materialising. An oft-cited example is that of the claimant who, having had his leg broken by the defendant, is being taken to hospital when the ambulance in which he is travelling is struck by lightning. The fact that the claimant is killed by the lightning strike is coincidental in terms of the defendant's actions, since breaking someone's leg does not generally increase the risk of their being killed by lightning. In *Chester v Afshar*, the risk which eventuated in the injury to Miss Chester (paralysis brought about by cauda equine syndrome) was integral to the surgical procedure she underwent, and was not a risk which was, or could be, increased by a surgeon's failure to warn a patient of its existence.

Stapleton disagrees with this, and argues that an increase in failures to warn patients of such risks will lead to a greater number of surgical procedures being undertaken, which will then in turn lead to a greater number of cases of paralysis occurring.[66] Since the net result will be a greater *incidence* of such injuries, the relationship between the failure to warn and cauda equine syndrome is not coincidental. In so doing, Stapleton makes it very clear that it is the incidence of such injuries, and not the degree of risk of the injury occurring on any given occasion which will thereby be increased. Unfortunately, this is precisely the point which undermines her argument. If the tort of negligence were concerned chiefly with the incidence of injuries, and had as one of its avowed aims the optimisation of risks, this would be a persuasive argument. It is what Law and Economics scholars would argue for, but it is not the premise on which the English law of negligence rests; a tort which, in its current form, serves corrective ends at the expense of

---

[65] [2004] UKHL 41, [2005] 1 AC 134 at [9] 142.
[66] See J Stapleton, 'Occam's Razor Reveals an Orthodox Basis for *Chester v Afshar*' (2006) 122 *LQR* 426, 441.

distributive values.⁶⁷ The question, therefore, of whether *this* defendant increased the risk of injury to *this* claimant trumps any macro-level considerations about damage across a population. In these terms, the eventuation of the risk of cauda equine syndrome in Miss Chester's unfortunate case was coincidental upon Mr Afshar's failure to warn her. This was a fact acknowledged by all of those in the majority,⁶⁸ who nevertheless decided that there existed sufficient 'policy' concerns to impose liability in spite of it.⁶⁹

The fact that the occurrence of the claimant's damage in *Chester v Afshar* was coincidental upon the defendant's breach says something important about the correct way to analyse the case. One of the many remarkable features of this decision is that it is not generally thought to fit within any of the recognised analytical categories into which negligence cases divide. As such, it has been characterised as a 'failure to warn' case and is often analysed as if it were *sui generis*.⁷⁰ In actual fact, however, *Chester v Afshar* is a loss of autonomy case. Classifying *Chester* in this way not only achieves the most consistency in terms of the broader tort of negligence, but it also facilitates the clearest analysis of the issues outlined above.

The 'chance' in *Chester v Afshar* was the 1–2 per cent risk of developing cauda equine syndrome, and this risk was *inherent in the surgical procedure*.⁷¹ The same risk could of course be represented as a 98–99 per cent chance of the procedure *not* having this adverse outcome. As Stuart-Smith LJ makes clear in *Allied Maples*, it matters not how such an uncertainty is perceived: 'I can see no difference in principle between the chance of gaining a benefit and the chance of avoiding a liability'.⁷² It was this 98–99 per cent likelihood of avoiding injury which forms the independent chance element in *Chester*, analogous to the claimants' opportunity of negotiating a more favourable settlement in *Allied Maples*, or the young actress's prospects of gaining a lucrative contract in *Chaplin v Hicks*.⁷³ Just as in these cases, the content and magnitude of this chance was not affected by anything that occurred between the parties to the dispute in *Chester*: rather, their relationship

---

⁶⁷ This is an enormous question, and one well beyond the scope of the current discussion. Fortunately, the task has been tackled by those far better suited to the task than I: see J Gardner, 'What is Tort Law For? Part 1: The Place of Corrective Justice' (2011) 30 *Law and Philosophy* 1; E Weinrib, *The Idea of Private Law*, revised edn (Oxford, OUP, 2012); J Coleman, *The Practice of Principle* (Oxford, OUP, 2011); and A Beever, *Rediscovering the Law of Negligence* (Oxford, Hart Publishing, 2009).

⁶⁸ [2004] UKHL 41, [2005] 1 AC 134 at [22] 145, [81] 162 and [101] 166. See also Stevens, *Torts and Rights*, above n 16 at 165.

⁶⁹ [2004] UKHL 41, [2005] 1 AC 134 at [22] 145 (per Lord Steyn), at [87] 162–63 (per Lord Hope) and at [101] 166 (per Lord Walker).

⁷⁰ It is categorised as a 'Particular Causation Problem' in J Steele, *Tort Law: Text, Cases and Materials*, 2nd edn (Oxford, OUP, 2009); as 'Coincidental Loss' in Stevens, *Torts and Rights*, above n 16; under the heading 'What Would Have Happened' in WVH Rogers, *Winfield and Jolowicz on Tort*, 18th edn (London, Sweet & Maxwell, 2010); and in WHB Lindsell, AM Dugdale and MA Jones, *Clerk & Lindsell on Torts*, 20th edn (London, Sweet & Maxwell, 2010) 2-14, the case is said to stand in a 'third category' of its own (the first one being made up of situations in which properly advised claimants would have followed the same path regardless, and the second covering those who would have acted differently).

⁷¹ And not, therefore, affected in its magnitude by the way in which the surgery was conducted.

⁷² [1995] 4 All ER 907 (CA) at 916.

⁷³ On the version of facts as ultimately accepted by the Court, but see A Burrows, *Remedies*, above n 17 at 54, n 4.

determined only whether the claimant could avail herself of the chance in question. Mr Afshar's failure to warn his patient meant that he performed her operation on Monday 21 November 1994 as opposed to a date sometime later. Consequently, Miss Chester ran the 1–2 per cent risk of developing cauda equine syndrome. (She also, and simultaneously, availed herself of the 98–99 per cent chance of avoiding that eventuality.) But for Mr Afshar's negligence, Miss Chester would probably, according to her own evidence, have run an identical risk, and taken an identical chance, on a different day. The breach of duty in this case, therefore, did not affect her ability to take advantage of an independent chance in the way that the breaches in *Allied Maples* and *Chaplin v Hicks* were found to have done. It remains a Type 1 case, however, because the breach *could* have had such an effect and *would* have done if, for instance, she could have shown that she would never have undergone the surgical procedure, thereby giving herself a 100 per cent chance of avoiding cauda equine syndrome.

Of course, it is true that, on 21 November 1994, Miss Chester was unfortunate enough to succumb to the relatively small risk of injury, and that the consequences for her were dire. This is, however, irrelevant in negligence terms. The tort is not one which seeks to compensate those who suffer loss as a result of misfortune: ideological considerations aside, it is staggeringly ill-equipped to do so.[74] The causal element of the negligence inquiry is what binds it to corrective, as opposed to distributive, ends and for courts to choose to override this on an ad hoc basis is to do the common law a disservice:

> To be acceptable the law must be coherent. It must be principled. The basis on which one case, or one type of case, is distinguished from another should be transparent and capable of identification. When a decision departs from principles normally applied, the basis for doing so must be rational and justifiable if the decision is to avoid the reproach that hard cases make bad law.[75]

With the greatest of respect to those in the majority in *Chester*, it is neither 'rational' nor 'justifiable' in a Type 1 case to take account of what *actually happened* as a result of the chance taken by the claimant. The key question on such facts is whether the claimant had her access to that chance, and therefore her autonomy, denied by the defendant's actions. A defendant in such a case, in which the content of the chance in question is completely independent of his actions, should not be held liable for the way that chance turned out. The very fact that we know how things turned out suggests that no chance has been lost because, by definition, it must have been taken. To compare, as judges and commentators have done,[76] the

---

[74] See *Chester v Afshar* [2004] UKHL 41, [2005] 1 AC 134 at [34] 147 (per Lord Hoffmann). See also Gardner, 'What Is Tort Law For?', above n 67 at 26: '"corrective justice" tells us ... what it is that the law of torts is supposed to be efficient *at*. It is supposed to be efficient at securing that people conform to a certain (partly legally constituted) moral norm of corrective justice. If it is not efficient at this job then, from the point of view of corrective justice itself, the law of torts should be abolished forthwith'.

[75] *Fairchild v Glenhaven Funeral Services Ltd* [2002] HCA 22, [2003] 1 AC 32 at [36] 68 (per Lord Nicholls).

[76] See, for example, J Stapleton, 'Occam's Razor', above n 66, and *Chester v Afshar* [2004] UKHL 41, [2005] 1 AC 134 at [21] 139 (per Lord Steyn).

1–2 per cent a priori risk of injury with the 100 per cent *ex post* knowledge[77] that the injury occurred is an inauthentic exercise. To state the truism that, had the operation been performed some days later, the injury would probably not have occurred (since it was 98–99 per cent likely not to have done so) but then to conclude from this that the defendant's breach thereby caused the injury because it exposed the claimant to an identical risk *which is now known to have eventuated*, is not to compare like with like. It would have been unlikely that injury would have resulted on 28 November, but *no more* unlikely than it was on 21 November. The only difference between the two events is that we (now) know what happened as a result of one of them, but it is an eventuality which is both independent of, and coincidental to, the defendant's actions.[78] Put simply, the defendant made no difference to the claimant's normal course of events, and should therefore not be subject to negligence liability.

How Far Does Hypothetical Third Party Action Take Us?

The approach to lost chances in *Allied Maples* made specific reference to the fact that the hypothetical outcome therein was partly dependent on how an independent third party might have behaved.[79] Although this has come to be regarded as a defining characteristic of the decision,[80] it is not obvious why it should be so.[81] What matters in the context of Type 1 cases is not whether the independent chance to which the claimant wanted access is determined by the actions of third parties, but only whether it exists independently of the interaction between the claimant and the defendant. That this is true is demonstrated by *Chester v Afshar* itself,[82] which, although a Type 1 case, is one in which the relevant chance was in no way determined by any human behaviour.

The Court of Appeal decision in *Dixon v Clement Jones*[83] further corroborates the point and suggests that what was important about the third party distinction made in *Allied Maples* was really that the action of the claimant herself should

---

[77] Or 'uniquely instantiating' evidence – see Ch 2, text to n 23.
[78] This is to go an inferential step further than that anticipated by S Fischoff in 'Heuristics and Biases in Hindsight' in D Kahneman, P Slovic and A Tversky (eds), *Judgment Under Uncertainty: Heuristics and Biases* (Cambridge, CUP, 1982), in which he discusses the phenomenon of situations being regarded as inherently deterministic once their outcome is known.
[79] [1995] 4 All ER 907 at 927 (per Stuart-Smith LJ).
[80] Lindsell, Dugdale and Jones, *Clerk & Lindsell*, above n 70, 2-23.
[81] See also Stevens, *Torts and Rights*, above n 16 at 48. See also Lord Hoffmann in *Gregg v Scott* [2005] UKHL 2, [2005] 2 AC 176 at [83] 197; McGregor, Spencer and Picton, *McGregor on Damages*, above n 7 at 8-056; *Benton v Miller* [2005] 1 NZLR 66 (NZCA) and *Accident Compensation Corporation v Ambros* [2007] NZCA 304, [2008] 1 NZLR 340.
[82] [2004] UKHL 41, [2005] 1 AC 134.
[83] *Dixon v Clement Jones* [2004] EWCA Civ 1005, (2004) Times, 2 August.

always be subject to the balance of probabilities standard. After all, without such proof that the claimant would have acted differently absent a breach, it is not possible to argue in a legal sense that her autonomy has been adversely affected. In *Dixon*, the claimant was suing her solicitors for their negligent failure to serve a statement of claim, thereby causing her negligence action against her accountants to be struck out. The trial judge had concluded from the evidence before him that it was only 30 per cent likely that she would have decided against embarking upon the business venture on which she had sought advice, even if that advice had been carefully given. On orthodox causal principles, therefore, following *McWilliams v Sir William Arroll*,[84] her prospective negligence claim *against her accountants*, would fail on such evidence. Notwithstanding this, the Court of Appeal allowed her claim against her solicitors for the loss of the 30 per cent chance of succeeding in her litigation against her accountants. Since the 30 per cent assessment clearly does not meet the balance of probabilities standard, it might seem at first glance as if *Dixon* is wrong. In actual fact, however, the decision is consonant with the principles identified herein: the immediate effect of her solicitors' negligence was to deny her the possibility of succeeding against her accountants. The 30 per cent likelihood was *not an assessment of whether she would have sued her accountants*, which would have needed to be established on the balance of probabilities in order to establish that the breach of duty had adversely affected her autonomy. Rather, it was an assessment of whether that action would have been successful, which was not something over which she had any effective control at that point. As a possibility, however, that was completely independent of the relationship between the claimant and the defendant solicitors, it was not necessary for the court to massage it into a certainty one way or another (which it would do using the balance of probabilities). As an independent possibility, consequent upon the legal certainty which the court *did* find (that the solicitors deprived her of ever realising the 30 per cent chance and therefore infringed her autonomy), that 30 per cent chance should remain an unrealised possibility, and thereby quantify the damages accordingly. The Court of Appeal recognised in *Dixon* that, were it to look behind the 30 per cent chance, it would effectively be conducting a 'trial within a trial'.[85] This might be overstating the point somewhat, but it is certainly true that the independent 30 per cent chance of success in the hypothetical litigation constituted a question that was not theirs to answer.

It would seem, therefore, that cases presented, or subsequently classified, as those involving 'lost chances' are more constructively described in other ways. By distinguishing properly between those cases in which there is an independent chance to be lost, and those in which otherwise unsuccessful claimants are attempting to reformulate their claim in a legally inauthentic way, we can begin to see what amounts to a lost chance in the tort of negligence. By taking the necessary next step and identifying independent chance cases for what they really are in terms of the

---

[84] *McWilliams v Sir William Arroll* [1962] 1 All ER 623 (HL).
[85] [2004] EWCA Civ 1005, (2004) Times, 2 August at [27], [29] and [36].

claimant's damage, it becomes apparent how the law of negligence can accommodate genuine lost chances. Once such lost chances are recognised as being the consequences of compromised autonomy, it seems clear that the claimant/third party distinction is irrelevant so long as the chance itself is independent of the relationship between the parties to the negligence action.

8
Concluding Thoughts

There is little more to be said. The aim of this book has been to simplify a concept which has long occupied the minds of scholars and devoured the resources of the common law. That is not to claim that causation is, or ever could be, substantively easy to understand or to establish in every factual situation with which the tort of negligence is faced. Rather, it is to make the point that a bespoke and systematic approach to the causal inquiry will mean that its results are at least clear, consistent and coherent.

In considering (a fraction of) the colossal body of literature on this subject, the question of why causation has proved particularly thorny raised its head on numerous occasions. Some reasons for this, such as the technicality of the evidence often required, the presence of multiple potential factors for each outcome, and the challenges presented by counterfactual analysis, have been the subject of much of this book. Another reason, which has also been considered here, is the fact of our fundamental epistemic limitations. Within the tort of negligence, it is unsurprising that the question of cause, so intrinsically dependent at some level upon factual discovery, should fall hardest at this hurdle. The balance of probabilities standard of proof, adopted to deal with these limitations, does not resolve the problem, but ensures instead that both sides of the adversarial divide are as likely to be winners as losers. Formally, the equity of the balance of probabilities standard is hard to deny; a balance being precisely what it achieves in terms of the risk of error in its conclusions. The balance of its method is, however, absent from its result, which is binary in nature: either X caused Y or X did not cause Y:

> The law operates a binary system in which the only values are 0 and 1. If the evidence that something happened satisfies the burden of proof . . . then it is assigned a value of 1 and treated as definitely having happened. If the evidence does not discharge the burden of proof, the event is assigned a value of 0 and treated as definitely not having happened. There is no forensic space for the conclusion that something which has to be proved may have happened.[1]

Consequently, the analysis in this book has emphasised that, under a balance of probabilities standard, the difference between 'more likely than not' and 'definitely not' cannot be repackaged as a 'lost chance'. For similar reasons, it has also

---

[1] L Hoffmann, 'Causation' in R Goldberg (ed), *Perspectives on Causation* (Oxford, Hart Publishing, 2011) 8.

made clear that proportional liability (understood as awards being quantified according to the probability that the defendant caused the damage in question), is not something which can be accommodated without doing some violence to the conceptual structure of the common law. In focusing on other common law jurisdictions that employ the same standard, however, the comparative aspects of this book have not for the most part isolated the effects that such a standard itself has, when compared with alternative proof requirements. A glance in the direction of Europe provides a more informative contrast:

> [A] powerful reason why proportional liability has so far remained exceptional in English law is the common law's traditional standard of proof: the balance of probabilities. A causal link that is more likely than not is treated as a certainty; a causal link that does not satisfy this requirement is a nullity. In English law, proportional liability must infringe at least one (and arguably both) of these fundamental principles. The same difficulty does not arise in systems where the standard of proof is a certainty or a substantial certainty, as there exists in such systems a wide area of uncertainty between (substantially) certain causation and (substantially) certain non-causation, and it is here that proportional liability can play a role without undermining the established approach to proof. To render proportional liability fit for the common law, as something more than an ad hoc response to particular cases, may therefore involve a fundamental review of the law of evidence and proof, and not just the substantive law of causation.[2]

The (academic) impetus behind a move towards proportional liability appears to be growing.[3] This suggests an increasing dissatisfaction with the polarity of outcomes generated by the balance of probabilities standard, and this is so particularly where our knowledge and understanding of scientific processes is incomplete (which means not just that a particular claimant might have difficulty in proving causation, but that all similarly placed claimants will). It is perhaps the forensic fiction inherent in a system which purports to give certain answers on the basis of uncertain evidence which generates such disquiet, and makes proportional liability appear on its face to be more appealing. (Currently, a defendant can part-cause an injury, or cause part of an injury, but what he cannot do is maybe cause an injury.)

> Evidence is never perfect; uncertainty always exists. Why not recognise the uncertainty entailed in any attempt to reconstruct history, particularly with the difficulties of the necessarily counterfactual inquiry required by causation? Shouldn't law frankly acknowledge the probabilistic nature of factual assessments such as causation and adjust the extent of liability accordingly?[4]

---

[2] K Oliphant, 'Causal Uncertainty and Proportional Liability in England and Wales' in I Gilead, M Green and B Koch (eds), *Proportional Liability: Analytical and Comparative Perspectives* (Vienna, De Gruyter, 2013) 139.

[3] It is apparently strongest in the US. See Ch 5, n 110 and J Makdisi, 'Proportional Liability: A Comprehensive Rule to Apportion Tort Damages Based on Probability' (1989) 67 *NCL Review* 1063, n 3. But, see also I Gilead, M Green and B Koch (eds), *Proportional Liability: Analytical and Comparative Perspectives* (Vienna, De Gruyter, 2013).

[4] M Green, 'The Future of Proportional Responsibility' in M Madden (ed), *Exploring Tort Law* (Cambridge, CUP, 2005) 354.

As has been made clear in the course of this work, however, and as Oliphant points out in the extract above, even if proportional liability were a fairer or more effective approach to take,[5] it is simply not compatible with a system which has the balance of probabilities standard at its core. What is more, this book's detailed examination of causation in negligence has led definitively to the conclusion that the 'fundamental review of the law of evidence and proof' which would be required to accommodate proportional liability should not be pursued. An adversarial system is better served by the balance of probabilities standard, and the production of all-or-nothing results. That way, not only do the winners know they are winners and the losers know they are losers, but, more often than not, both will have their deserts. By contrast, a system of proportional liability is highly unlikely ever fully to compensate or to correct.[6] In the Introduction to this book, the point was made that the law cannot indulge in the luxury of indecision. Whilst it might be argued that granting an award on the basis of proportional liability is a decision because it effects some form of transfer, that decision is one defined by its own uncertainty, and amounts therefore to a form of forensic capitulation. Were any form of proportional approach to be the norm, this argument would lack force, but whilst the default standard remains that of all-or-nothing, the making of an exception based on proportionality undermines its whole conceptual basis. To equate the probability of causation with the quantity of damages awarded to the claimant is to misunderstand both the nature of the causal inquiry in negligence, and the role of probability within it:

> Probability theory is intended as a mathematical description of the world. Its goal is to bring the uncertainty in our world view as closely as possible into congruence with the uncertainty in the world. Our mathematicians have done a great job of constructing and elaborating mathematical systems and theorems towards that end. Our scientists have made considerable progress in discovering which physical systems obey which models. But no adequate philosophical explication of probability theory exists as yet, nor can one do so until we learn a good deal more logic and a good deal more physics than we presently know. We may never be sure we have it right, perhaps, until we possess a general theory of rationality and know for certain whether or not God plays dice with the Universe.[7]

Probability is a tool and not a conclusion. Eliding the two is the same as attributing greatness to Ashkenazy's piano, Tendulkar's bat or Shakespeare's quill: unfortunately, a tool itself generates no outcome. Probability should not, therefore, be used to quantify what it has not been able to justify.

---

[5] And this is by no means a universally accepted truth: ibid, 385.
[6] J Gardner, 'What is Tort Law For? Part 1: The Place of Corrective Justice' (2011) 30 *Law and Philosophy* 1, 21 and E Weinrib, *Corrective Justice* (Oxford, OUP, 2012) 87–96.
[7] R Weatherford, *Philosophical Foundations of Probability Theory* (London, Routledge and Kegan Paul, 1982) 252.

INDEX

acts of claimants 50
acts of God 27
agents
   causal agents 33–4, 79, 89, 119, 135, 140, 159
   moral luck 92–3
   multiple agent cases 27, 148–9
   single agent aspects 6, 123–4, 148–51
aggregation
   But For causation 10, 13–14, 17, 20–1, 26, 98, 142, 146–7, 150
   disaggregation 17
   initial aggregation 10
   material increase in risk 142–4, 146–7, 150
   necessity 142–4
   overdetermination 13, 71–2, 75
Allied Maples case 155–7, 159, 164–5, 168–70
apportionment
   damages 72, 85, 101–3, 128–9, 131–2, 134, 140, 142–8
   discriminatory apportionment 132, 143
   material contribution to injury 97–8, 101–3
   material increase in risk 128–9, 131–2, 134, 140, 142–3, 146–8
asbestos 115–16, 120–2, 125–51
at least one breach of duty *see* one breach of duty, at least
Australia 85
autonomy, loss of 7, 153–4, 155–7, 160–70, 171

Bailey v Ministry of Defence 94, 107–9
Baker v Willoughby case 39–41, 43, 46, 51
balance of probabilities
   basic principles 32, 33–4, 36–8
   beyond a reasonable doubt 11
   But For causation 10–12
   chance, mathematical calculus of 11
   civil proceedings 10–11
   difference to normal course of events, making 32, 33–4, 36–8
   duplicative causation 111
   indifference, principle of 69, 70–1
   liability only for difference breach makes to course of events 32, 36–8
   lost chances 155, 158–9, 161–7, 173–4
   material increase in risk 127–8, 131–3, 136–7, 143–4
   medical negligence 33–4, 36, 111
   overdetermination 66, 69, 70–1
   proportional liability 174–5
   stage 1 (more likely than not test) 10–12, 66, 69, 109, 173–4

Barker v Corus 35, 128, 131–4, 139–40, 145–51
basic principles 32–57
   balance of probabilities 32, 33–4, 36–8
   difference to normal course of events, making 32, 33–4, 36–8
   liability only for difference breach makes to course of events 32, 36–8
   material contribution to injury analysis 32–5
   material contribution to risk analysis 32–5
   significance of risk eventuation 47–57
   stage 2 (operative effect of individual's breach) 43–57
   steamlined and less complex causal inquiry 5
   victim as found, taking 32, 38–43
Beever, Allan 15, 27–31, 66, 81–2
belief probabilities 12, 114, 119–20, 131, 136–8
Bernouilli, James 69
beyond a reasonable doubt 11
Bonnington Castings case 34, 63, 94–7, 115
breach of contract 164–5, 168–9
Broadbent, Alex 116–18
burden of proof, reversal of 73
But For causation 5, 8–31
   aggregation 10, 13–14, 20–1, 26, 98, 142, 146–7, 150
   balance of probabilities 10–12
   Beever 27–31
   breach of duty 9, 20–2
   corrective justice 8, 17, 29–30
   counterfactuals 10, 17–19, 60
   current perspectives of But For causation 19–31
   damage 26–8
   damages 14, 20–1, 24–5, 27, 30–1, 107
   de minimis rule 109
   duplicative causation 22–4, 58–9
   generic physical cause 28
   importance of But For causation, reasons for 8–10
   indivisible injuries 104, 107–8
   latitudinal and explanatory, as 23
   legal causal relevance 20, 23–4, 26, 28–30
   material contribution to injury 97–8, 103–5, 107–8
   material increase in risk 127, 142, 145–8, 150
   multiple causes 9, 17, 20–1
   necessity 9, 13, 19–20, 23, 27–30
   NESS test 9–10, 22–5, 28–9, 81
   non-tortious factors and tortious factors, combination of 68

But For causation *cont*
  one breach of duty, at least 9, 13, 17–18, 26–7, 135, 142
  overdetermination 9–10, 13, 17, 9–10, 19–20, 28–9, 58, 68
  pre-emption 17–18, 58, 78–81
  principled approach 29–31
  remoteness 43–4, 56
  specific causation 10, 15–17, 22–4, 28
  stage 1 (more likely than not test) 10–12, 15–19, 109
  stage 2 (operative effect of individual's breach) 10, 14, 17–21, 43–4
  Stapleton 19–22, 24
  Stevens 24–7, 30, 54
  substitutive causal factors 58
  sufficiency 19, 27–30, 62, 63
  Wright 22–4, 28–9

Canada 72–3, 145
cancer 38–9, 44, 113–15, 120, 161
Cane, Peter 88–90, 93, 136
causal agents 33–4, 79, 89, 119, 135, 140, 159
causal relevance *see* legal causal relevance
causation, definition of *see* definition of causation
chain of causation *see* intervening events
chance, loss of a *see* lost chances
Chester v Afshar 7, 155–6, 159, 166–70
chronological approach 15–17, 78
Civil Liability (Contribution) Act 1978 100–1
clinical negligence *see* medical negligence
clubs, votes to expel members from 19–20, 61, 129
coincident intervening events 46–57
collisions at sea 49
combination of tortious and non-tortious factors 66–77
common law 16, 30, 56, 66, 87, 159, 163, 174
Compensation Act 2006 139–42, 146–7
concurrent factors 95–7, 105
consecutive factors 96–8
contract, breach of 164–5, 168–9
contribution *see* contributory negligence; material contribution to injury
contributory negligence 47–9, 131–2
Cook v Lewis 72–3
corrective justice
  autonomy, loss of 167–9
  But For causation 8, 17, 29–30
  counterfactual to factual, move from 17
  lost chances 167–9
  manage own risk, effect on claimant's ability to 53–4
  material increase in risk 127, 143
  moral luck 92
  Necessary Breach Analysis (NBA) 3

non-tortious factors and tortious factors, combination of 76
overdetermination 61, 76
victim as found, taking 42–3
counterfactual analysis 10, 17–19, 48, 60, 173
criminal proceedings, standard of proof in 11
crumbling skull doctrine 105–6

damages
  aggregation 14
  apportionment 72, 85, 101–3, 128–9, 131–2, 134, 140, 142–8
  But For causation 14, 20–1, 24–5, 27, 30–1, 107
  Compensation Act 2006 139–42, 146–7
  consequential loss 27
  contributory negligence 49
  corrective justice 8
  counterfactual to factual, move from 18
  difference breach has made, liability to extent 36–7
  false imprisonment 35
  future or hypothetical events 107
  heads of damages 51–2
  indifference, principle of 71
  indivisible injuries 100–7
  intervening acts 52
  legal causal relevance 112
  lost chances 153, 155–64, 167, 171
  material contribution to injury 6, 97–108, 112
  material increase in risk 126–9, 131–2, 134, 139–41, 142–3, 146–51
  medical negligence 112
  no better off principle 20
  nominal damages 30
  overcompensation 30
  proportional liability 175
  quantification 30, 51–2, 107, 153–64, 171, 175
  stage 1 (more likely than not test) 18
  substitutive damages 25, 27, 55
  vicissitudes of life principle 41–2
  victim as found, taking 38–42, 106
de minimis rule 109
deafness 37
death 64–5, 103–4
defective products *see* product liability
definition of causation
  But For causation 21
  causal law, definition of 11
  cause, definition of 16
  Necessary Breach Analysis (NBA) 4
  non-legal definition 21
dermatitis 34, 123–5, 129–30, 134–5, 140, 162–3
desert traveller scenario 25–6, 64–5
Dickins v O2 94, 101
difference to normal course of events, where breach has made a
  basic principles 32, 33–4, 36–8

Index

liability only for difference made 32, 36–8
divisible injuries 63, 94–8, 102, 108–9, 115–16
double omissions 65–6, 83, 85
doubling of risk (DRT) test 6, 113–22
duplicative causation 58–93 *see also*
  overdetermination (real duplicative causation);
  pre-emption (potential duplicative causation)
  aggregation 14
  balance of probabilities 111
  But For causation 22–4, 58–9
  legal causal relevance 14
  multiple causal factors 58–93
  Necessary Breach Analysis (NBA) 5–6, 58–93
  NESS test 22–4
  significance of risk eventuation 47–8
  sufficiency 111

egg shell skull rule *see* victim as found, taking
Eggleston, Richard 70
employers' liability *see* industrial diseases/injuries
end, bringing infringement to an 46, 51–2
Ennis, Robert H 90
epidemiology 6, 104, 132–3, 161
epistemic limitations 126, 133, 144, 161, 173
evidence *see also* balance of probabilities; standard of proof
  epidemiological evidence 6
  evidentiary gap 124, 126–48
  material contribution to injury 96
  material increase in risk 6, 123–4, 126–48
  unavailable but causation theoretically possible to prove 6, 123

factual analysis
  counterfactual analysis 10, 17–19, 48
  duplicative causation 58–9
  identification of all factual contributing causes 22–3
  lost chances 152–4
  material contribution to injury 94–109
  material increase in risk 123–30
  overdetermination 129
  probabilities 12
  significance of risk eventuation 48
  stage 1 (more likely than not test) 68–9
  types of causal problems 1, 5
factual causation *see* But For causation
false imprisonment, damages for 21
Fairchild case
  aggregate necessity 142–4
  class of claimants, unfairness to 143–4
  Compensation Act 2006 139–42, 146–7
  joint and several liability 140, 147–8
  material contribution to injury 35
  material increase in risk 6, 35–6, 123–32, 134, 139–51

single agent aspects 6, 148–51
financial outcomes, lost chances and 7, 152–3
fire cases 50–1, 67, 69–70
Firestein, Stuart 133
Fischer, David A 83, 90
foreseeability 35–6, 38, 47–8, 51, 53–7, 125
Fumerton, Richard 34, 83
fungibles 61–2, 75–6, 104

Gardner, John 3, 87, 90
Gatecrasher paradox 118
Geistfeld, Mark 73–4
general causation 15, 23, 28, 117–18, 137–8
Gödel, Kurt 133
Gold, Steve 11–12, 136
Greenland, Sander 119, 121
Gregg v Scott 160–5

Haack, Susan 138
Hart, HLA 50, 76–7, 81
Heisenberg, Werner 133
Himalayan porters scenario 116–17
Hoffmann, Leonard 16, 45–6
Hohfeld, Wesley Newcom 24
Honoré, Tony 8, 50, 76–7, 81, 87–9, 91–3
Hotson case 160–5
hunting/shooting scenarios 9, 13, 48, 62, 72–4, 129
hypothetical scenarios
  counterfactuals to factual 18
  damages, quantification of 107
  lost chances 157, 165, 170–2
  NESS test 23
  no-damage scenarios 13
  overcompensation 29–30
  pre-emption 78–81, 84–6
  Wagon Mound test 55

ignorance 69, 133–4
improvement of claimant's position 46, 51–2
increase in risk *see* material increase in risk
independent causal factors 58, 62, 63, 78, 84, 86
indeterminate causes 12, 33–4, 92, 149
indifference, principle of 69–71, 131
indivisible injuries 58, 62–5, 94–109, 111–12, 114, 129, 147
industrial diseases/injuries
  contributory negligence 49
  deafness 37
  dermatitis 34, 123–5, 129–30, 134–5, 140, 162–3
  doubling of risk test 113–14, 119–22
  material contribution to risk 34, 95–102, 105–6, 113–14, 119–22
  material increase in risk 123–51
  pneumoconiosis 34, 95–8
  predispositions to injury 38–40
  psychiatric injury 99–102, 105–6, 112–13

industrial diseases/injuries *cont*
  smoking 113–14, 120–2
  suicide 49
  victim as found, taking 38–42, 105–6
insufficient but necessary part of an unnecessary but sufficient condition test (INUS test) 4
insurance 76, 97, 145–6
intervening events
  acts of claimant 50
  coincident intervening events 46–57
  contributory negligence 48–9
  end, bringing infringement to an 46, 51–2
  foreseeability 47–8
  improvement of claimant's position 46, 51–2
  manage own risk, effect on claimant's ability to 47, 52–4
  medical negligence 110
  significance of risk eventuation 48–51
  timeframe, foreseeable damage occurring within reasonable 47–8, 53–7
  victim as found, taking 40
INUS test 4

Jobling case 39–41
joint and several liability 47–8, 76, 140, 147–8
joint enterprises 73–5

Keynes, John Maynard 69–70
Kress, Ken 24, 83

Law and Economics movement 167–8
legal causal relevance
  basic principles 32
  But For causation 20, 23–4, 26, 28–30
  damages 112
  duplicative breaches 14
  lost chances 153
  material contribution to injury 103–4, 112
  non-tortious factors and tortious factors, combination of 73
  overdetermination 61
  pre-emption 80
  specific concept of cause 15, 28
liability only for difference breach makes to course of events 32, 36–8
lost chances 152–72
  access to opportunities 159–60, 165, 170
  autonomy, loss of 7, 153–4, 155–7, 160–70, 171
  balance of probabilities 155, 158–9, 161–7, 173–4
  breach of contract 164–5, 168–9
  Chester v Afshar 7, 155–6, 166–70
  corrective justice 167–9
  damages 153, 155–64, 167, 171
  factual basis 152–4
  financial outcomes 7, 152–3

  forensic margin of error 153, 162
  genuine chance to be lost 154
  Gregg case 160–5
  Hotson case 160–5
  hypothetical scenarios 157, 165, 170–2
  independent of breach of duty but does not affect substance of chance, chance exists (type 1 situations) 152–9, 165–72
  interdependence of chance and breach of duty where breach affects existence and content of chance (type 2 situations) 152, 154, 158–70
  legal recognition of being worse off 158
  legally relevant uncertainty 154–5
  material contribution to risk cases 158
  medical negligence 156, 159–70
  Necessary Breach Analysis (NBA) 154–5, 158, 161–2, 164
  no-liability result 158
  physical outcomes 7, 152–3
  pure chance 157
  quantification of damages 153, 155–9
  stage 1 (more likely than not test) 161–2, 164
  type 1 situations 152–9, 165–72
  type 2 situations 152, 154, 158–70
  unrealised chances 157–8
luck *see* **moral luck**

manage own risk, effect on claimant's ability to 47, 52–4
market share liability 76–7
material contribution to injury 94–122
  acceptable evidentiary basis 96
  actual effect on claimant's position at trial, tortious factors have 94
  aggregation 98
  apportionment 97–8, 101–3
  basic principles 32–5
  But For causation 97–8, 107–9
  concurrent factors 95–7, 105
  consecutive factors 96–8
  crumbling skull doctrine 105–6
  damages 6, 97–108, 112
  death 103–4
  difference to normal course of events, making 33–5
  divisible in principle but not possible to attribute constituent parts to particular factors 94–8
  doubling of the risk (DRT) test 6, 113–22
  duplicative causation 111
  epidemiology 6, 104
  factual basis 94–109
  indivisible injuries 63, 94–112
  industrial diseases/injuries 34, 123–5, 129–30, 134–5, 140, 162–3
  intervening acts 110
  legal causal relevance 103–4, 112

lost chances 158
medical negligence 33–4, 107–13
multiple causes 94–122
Necessary Breach Analysis (NBA) 99–100, 105, 109, 111–12
non-breach factors 101–2, 109–10
orthodox principles, proper application of 6
overdetermination 94, 104, 108
part-cause of divisible injury 98
part-cause of indivisible injury 98, 108–9, 111–12, 114
pre-emption 85–6, 94, 104–5, 108, 112
pre-existing conditions 105–7
psychiatric injury 99–102, 105–6, 112–13
stage 1 (more likely than not test) 94–5, 98–109, 111
stage 2 (operative effect of individual's breach) 111–13
temporal dimension of causal inquiry 103–5
victim as found, taking 105–6
material contribution to injury 32–5
material increase in risk 123–51
aggregation 142–4, 146–7, 150
apportionment 128–9, 131–2, 134, 140, 142–3, 146–8
balance of probabilities 127–8, 131–3, 136–7, 143–4
belief probabilities 131, 136, 138, 148
But For causation 127, 142, 145–8, 150
class of claimants, unfairness to 143–4
Compensation Act 2006 139–42, 146–7
contributory negligence 131–2
corrective justice 127, 143
damages 126–9, 131–2, 134, 139–41, 142–3, 146–51
evidence unavailable but causation theoretically possible to prove 6, 123
evidentiary gap 124, 126–48
exceptional causal principle, as 6, 123–4, 128, 141, 149–50
factual basis 123–30
Fairchild case 6, 35–6, 123–32, 134, 139–51
ignorance 133–4
indifference, principle of 131
indivisible injury 129, 147
industrial diseases/injuries 123–51
joint and several liability 140, 147–8
mathematical probabilities 132, 135–8, 144
medical negligence 149–50
McGhee v NCB case 6, 123–4, 126, 128–9, 131, 134, 144–5
multiple causes 123–51
Necessary Breach Analysis (NBA) 131–50
overdetermination 129
precautions 124–5
pre-emption 129, 150
probabilities 131–2, 135–7, 144
qualitative evaluation 133

risk creation 150–1
rock of uncertainty 6, 123, 126–7, 147
simultaneous effects 129
single agent aspects 6, 123, 148–51
stage 1 (more likely than not test) 136–7
stage 2 (operative effect of individual's breach) 149–50
statistical probabilities 132, 135–7, 144
sufficiency 129
mathematical probabilities 11, 72, 132, 135–8, 144, 175
McGhee case 6, 34, 123–4, 126, 128–9, 131, 134, 144–5, 162
medical negligence 109–13
aggregation 14
autonomy, loss of 156, 160–70
balance of probabilities 33–4, 36, 111
damages 112
difference to normal course of events, making 33
duplicative causation 111
indivisible injuries 108
intervening acts 110
legal causal relevance 112
lost chances 156, 159–70
material contribution to injury 33–4, 107–13
material increase in risk 149–50
multiple factors 112
Necessary Breach Analysis (NBA) 112
non-breach factors 109–10
precautions 34
pre-emption 85–6, 112
psychiatric injury 112–13
stage 1 (more likely than not test) 14
stage 2 (operative effect of individual's breach) 111–13
significance of risk eventuation 53
warn, failure to 85–6, 156, 166–70
Moore, Michael 44, 87
medication 14, 74–7, 115–16
moral culpability 143
moral luck
dispositional luck 90–1, 93
functional analysis 88, 89, 91–3
outcome responsibility 87–9, 91–2
personhood, understanding 88, 91
pre-emption 6, 86, 87–93
reciprocity 92–3
more likely than not test *see* **stage 1 (more likely than not test)**
multiple agent cases 27, 148–9
multiple causes *see also* **overdetermination (real duplicative causation)**
agents 148–9
aggregation 13
But For causation 9, 17, 20–1
double omissions 65–6
duplicative causation 58–93

multiple causes *cont*
  indivisible injuries 99, 108
  material contribution to injury 94–122
  material increase in risk 123–51
  medical negligence 112
  overdetermination 58, 59–77
  pre-emption 58, 77–8
  victim as found, taking 39–43
multiple defendants 17–19, 72, 139, 143, 145, 148

Nagel, Thomas 79, 87, 90, 93
natural events 41, 66–7, 78, 128
Necessary Breach Analysis (NBA) 1–6 *see also* basic principles; stage 1 (more likely than not test); stage 2 (operative effect of individual's breach)
  breach of duty 2–4
  But For causation 5, 8–31
  combination of tortious and non-tortious factors 71–7
  difference to normal course of events, making 36
  duplicative and potentially duplicative causation 5–6, 58–93
  evidentiary gap 131–48
  forensic necessity 45–6
  indifference, principle of 71
  indivisible injuries 99–100, 105, 109
  intervening acts 49–50
  INUS test 4
  lost chances 155, 158, 161–2, 164
  material contribution to injury 99–100, 105, 109, 111–12
  material increase in risk 131–50
  NESS test 4, 80–1
  non-tortious factors and tortious factors, combination of 66–77
  overdetermination 58, 59–77
  philosophical analysis 1–2
  pragmatic reasoning 1–2
  pre-emption 58, 77–93
  significance of risk eventuation 47–50
  simultaneous effects 64
  specificity of causal question 2–3
  stage 1 (more likely than not test) 4
  stage 2 (operative effect of individual's breach) 4, 5 chapter 3
  sufficiency 62, 81
  timeframe, foreseeable damage occurring within reasonable 56
  vicissitudes of life principle 41–2
  victim as found, taking 39–42
necessary element in a sufficient set test *see* NESS test
necessity *see also* Necessary Breach Analysis (NBA); NESS test
  aggregation 13, 142–4

  But For causation 9, 13, 19–20, 23, 27–30
  duplicate necessity 19–20
  forensic necessity 45–6
  necessary condition test 8
  one breach, at least 13
  overcompensation 30
  pre-emption 58, 77–8
  sine qua non approach 23
NESS test
  But For causation 9–10, 22–5, 28–9, 81
  causal law, definition of 22
  competing or duplicate factors 22–4
  consistency of principle 22
  factual causation 24
  generalist, as 23
  identification of all factual contributing causes 22–3
  longitudinal and algorithmic, as 23, 82–3
  minimal sufficiency 22
  Necessary Breach Analysis (NBA) 4, 80–1
  overdetermination 9–10, 28
  pre-emption 80–4
  sufficiency 22, 81–3
  vertical sets 23
  weak necessity concept 81–3
no better off principle 20
nominal damages 30
non-breach events 66–77, 101–2, 109–10, 134–5
novus actus interveniens see intervening events
nuclear tests group action by ex-military 35–6
nuisance 26–7, 70

Oliphant, Ken 175
omissions 65–6, 83, 85, 128, 149
one breach of duty, at least
  balance of probabilities 131, 134, 143–4
  But For causation 9, 13, 17–18, 26–7, 135, 142
  factual basis 105
  joint and several liability 76
  material increase in risk 142–4, 148
  operative, definition of 46
  overdetermination 59–61
  significance of risk eventuation 47
  victim as found, taking 40–2
operative, definition of
  coincidental intervening events 46–7, 51–7
  remoteness 43–6
  significance of risk eventuation 47–57
  stage 2 (operative effect of individual's breach) 5, 45, 46–57
operative effect of individual's breach *see* stage 2 (operative effect of individual's breach)
outcome responsibility 87–9, 91–2
overdetermination (real duplicative causation) 5–6, 59–77
  aggregation 13, 71–2, 75
  balance of probabilities 66, 69, 70–1
  But For causation 9–10, 13, 17, 19–20, 28–9, 58, 59–62, 68

Index   183

combination of tortious and non-tortious factors 66–77
corrective justice 61, 76
death 64–5
double omissions 65–6
duplicate necessity 19–20
factual situations 68–9, 129
independent causal factors 58, 62, 63
indifference, principle of 69–71
indivisible injuries 58, 62, 63, 64–5, 104, 108
joint enterprises 73–5
legal causal relevance 61, 73
market share liability 76–7
material contribution to injury 63, 94, 104, 108
material increase in risk 129
multiple causal factors 58, 59–77
Necessary Breach Analysis (NBA) 58, 59–77
NESS test 9–10, 28
omissions 65–6
pre-emption 6, 59, 64–6, 77–8, 84–5
psychological injuries and physical injuries 63
same type of damage, causal factors leading to 58, 62, 64
severally sufficient causal factors 58, 61–5, 68–9
simultaneous effect aspects 58, 62, 64–5, 69–70
sine qua non test 23, 28, 30, 59
stage 1 (more likely than not test) 66–73
stage 2 (operative effect of individual's breach) 17
subset of factors 62
substitutive causal factors 58
sufficiency 28–9, 58, 61–5, 68–9
threshold conditions 70
under-inclusiveness 20
what constitutes an overdetermined event 61–5

part-causes 98, 104–5, 108–9, 111–12, 114
Performance Cars case 36, 39, 94, 97
Perry, Stephen 88
pneumoconiosis 34, 95–8
pollution 26–7, 70
potential duplicative causation *see* pre-emption (potential duplicative causation)
precautions 34, 124–5
predispositions to injury 38, 41–2 *see also* victim as found, taking
pre-emption (potential duplicative causation) 77–93
But For causation 17–18, 58, 78–81
chronological approach 78
death 64
double omissions 66
independent and separate causal factors 58, 78, 84, 86

legal causal relevance 80
material contribution to injury 85–6, 94, 104–5, 108, 112
material increase in risk 129, 150
medical negligence 85–6, 112
moral luck 6, 86, 87–93
multiple causal factors 58, 77–8
Necessary Breach Analysis (NBA) 5–6, 58, 77–93
NESS test 80–4
non-duplicative effect on claimant, causal factors having 58, 84–6
operative effect, factors never having 77–8, 80
overdetermination 6, 59, 64–6, 77, 84–6
pre-empting and pre-empted factors 78, 82
rights model 79–80
risk of injury, exposure to 78
simultaneous effects 64–5
stage 2 (operative effect of individual's breach) 17–18
substitutive causal factors 58
sufficiency concept 81–3
what constitutes pre-emption 84–6
principles *see* basic principles
probabilities *see also* balance of probabilities
belief probabilities 12, 114, 119–20, 131, 136–8
doubling of the risk test 115, 120–1
equal probabilities 69
fact probabilities 12
intervening events 48, 51
lost chances 157, 161, 170
material increase in risk 125–6, 130, 122, 136
non-tortious and tortious factors, combination of 67–8, 76
material increase in risk 131–2, 135–7, 144
mathematical probabilities 11, 72, 132, 135–8, 144, 175
statistical probabilities 11–12, 132, 135–7, 144
product liability
doubling of risk test 115–16
material increase in risk 135
medication 74–7, 115–16
overdetermination 74–7
warn, failure to 66
proportional liability 27, 148, 150, 174–5
psychiatric injury 63, 99–102, 105–6, 112–13

Rahman case 94, 99–100, 102, 110
real duplicative causation *see* overdetermination (real duplicative causation)
relevance *see* legal causal relevance
remoteness 43–6, 56
Restatement (Third) of Torts (United States) 68, 71, 72, 77, 113
rights model 26, 30, 55, 79–80
Ripstein, Arthur 92–3
risk allocation devices 38–40

road accidents
  brakes, faulty 47–8, 65–6, 78–87
  double omissions 65–6
  extent of difference breach had made, defendant liable only to 36–7
  intervening acts 49–50
  material contribution to injury 106–7
  moral luck 87
  overdetermination 84–5
  pre-emption 17–18, 78–87
  significance of risk eventuation 47–52
  victim as found, taking 39–43
rock of uncertainty 6, 123, 126–7, 147
Russell, Bertrand 138

same type of damage, causal factors leading to 58, 62, 64
Saunders Systems case 47–8, 65–6, 78–86
shooting/hunting scenarios 9, 13, 48, 62, 72–4, 129
Sienkiewicz case 35–6, 115, 121, 126–7, 130–6, 139–47, 163
significance of risk eventuation 47–57
simultaneous effects 58, 62, 64–5, 69–70, 129
single agent aspects 6, 123–4, 148–51
smoking 113–14, 120–2
solicitors, negligence by 159–60, 171
specific causation
  But For causation 10, 15–17, 22–4, 28
  chronological approach 15–17
  death 103–4
  doubling of the risk 118–19
  indeterminate causes 33–4
  intervening events 50
  lost chances 159–60
  material contribution to injury 118
  material increase in risk 143–4, 147, 149
  Necessary Breach Analysis (NBA) 2–3
  non-tortious and tortious factors, combination of 68–9
  temporal specificity 104
stage 1 (more likely than not test)
  aggregation 13–14, 17
  balance of probabilities 10–12, 66, 69, 109, 173–4
  But For causation 10–12, 15–19, 109
  factual causes 68–9
  lost chances 161–2, 164
  material contribution to injury 94–5, 98–109, 111
  material increase in risk 136–7
  more likely than not, definition of 10–12
  non-tortious factors and tortious factors, combination of 66–71
  overdetermination 66–71
stage 2 (operative effect of individual's breach)
  basic principles 43–57
  factual causation 43–4

material contribution to injury 111–13
material increase in risk 149–50
multiple defendants 18–19
Necessary Breach Analysis (NBA) 4, 5
non-tortious factors and tortious factors, combination of 72–3
operative, definition of 5, 45, 46–57
overdetermination 17, 72–3
pre-emption 17–18
significance of risk eventuation 48
standard of proof 11 *see also* balance of probabilities
Stapleton, Jane 2, 15, 19–22, 24, 43–4, 56, 61, 83, 108, 133, 167
statistical probabilities 11–12, 132, 135–7, 144
Steel, Sandy 158
Stevens, Robert 15, 24–7, 30, 54–5, 67–8, 79
stress at work 101–2, 105–6
substitutive causal factors 58, 86
sufficiency
  But For causation 19, 27–30, 62, 63
  dependence 63
  difference to normal course of events, making 33
  duplicative causation 111
  independent causal factors 58, 62, 63
  indivisible injuries 58, 62, 63, 64–5
  material increase in risk 129
  medical negligence 111
  minimal sufficiency 22
  Necessary Breach Analysis (NBA) 62, 81
  NESS test 22, 81–3
  non-necessary but sufficient factor 30
  non-tortious factors and tortious factors, combination of 68–9
  overdetermination 28–9, 58, 61–5, 68–9
  pre-emption 81–3
  same type of damage, causal factors leading to 58, 62, 64
  severally sufficient causal factors 58, 61–5, 68–9, 111, 129
  simultaneous causes 58, 62, 64–5
  threshold conditions 61
suicide 49, 57
Summers case 72–3, 85

taxi cab problem 118
Teff, Harvey 76
temporal dimension of causal inquiry 103–5
thin skull rule *see* victim as found, taking
timeframe, foreseeable damage occurring within reasonable 47–8, 54–7
transparency 22, 169

United States
  aggregate causation 113
  alternative liability 73
  burden of proof, reversal of 73

joint enterprises 73–5
market share liability 76–8
medical negligence 112–13
Multiple Sufficient Causes 68
non-tortious factors and tortious factors, combination of 67–77
preponderance of the evidence 11–12
Restatement (Third) of Torts 68, 71, 72, 77, 113

valuations of property 52–4
vicissitudes of life principle 41–2
victim as found, taking
  corrective justice 42–3
  crumbling skull doctrine 105–6
  damages 38–42, 106
  industrial diseases/injuries 38–42, 105–6
  intervening acts 40
  material contribution to injury 105–6
  multiple torts 39–43
  Necessary Breach Analysis (NBA) 32, 38–43
  risk allocation device, as 38
  two separate injuries 39–43
  vicissitudes of life principle 41–2
voluntariness 51

Wagon Mound test 47, 54–6
warn, failure to 66, 85–6, 156, 166–70
Weatherford, Roy 69
Weinrib, Ernest J 17, 37, 42
Weir, Tony 100, 103
Wilsher case 33–4, 36, 149–50
Wright, Richard 9–10, 11–12, 13, 15, 22–4, 28–9, 72, 81–3, 117–19, 137–8